ECLECTIC EDUCATIONAL COURSE.

RAY'S HIGHER ARITHMETIC.

THE PRINCIPLES OF ARITHMETIC,

ANALYZED AND PRACTICALLY APPLIED.

FOR ADVANCED STUDENTS

BY JOSEPH RAY, M.D.

LATE PROFESSOR OF MATHEMATICS IN WOODWARD COLLEGE.

EDITED BY CHAS. E. MATTHEWS, M.A.

REVISED STEREOTYPE EDITION.

CINCINNATI:

SARGENT, WILSON & HINKLE.

NEW YORK: CLARK & MAYNARD.

THE BEST AND CHEAPEST

MATHEMATICAL WORKS.

BY PROFESSOR JOSEPH RAY.

Each BOOK *of the Arithmetical Course, as well as the Algebraic, is Complete in itself, and is sold separately.*

PRIMARY ARITHMETIC.—RAY'S ARITHMETIC, FIRST BOOK: simple mental Lessons and tables for little learners.

INTELLECTUAL ARITHMETIC.—RAY'S ARITHMETIC, SECOND BOOK: the most interesting Intellectual Arithmetic extant.

PRACTICAL ARITHMETIC.—RAY'S ARITHMETIC, THIRD BOOK: for schools and academies; a full and complete treatise, on the inductive and analytical methods of instruction.

KEY TO RAY'S ARITHMETIC, containing solutions to the questions; also of test Examples for the Slate and Blackboard.

RAY'S HIGHER ARITHMETIC.—The Principles of Arithmetic, analyzed and practically applied. For advanced classes.

ELEMENTARY ALGEBRA.—RAY'S ALGEBRA, FIRST BOOK; for common schools and academies; a simple, progressive, and thorough elementary treatise.

HIGHER ALGEBRA.—RAY'S ALGEBRA, SECOND BOOK; for advanced students in academies, and for colleges; a progressive, lucid and comprehensive work.

KEY TO RAY'S ALGEBRA, FIRST AND SECOND BOOKS. In one vol.

Preparing for Publication,

I.—**THE ELEMENTS OF GEOMETRY**, embracing plain and solid geometry, with numerous practical exercises. II.—**TRIGONOMETRY AND MENSURATION**, containing logarithmic computations, plane and spherical trigonometry, with their applications, mensuration of planes and solids, with logarithmic and other tables. III.—**SURVEYING AND NAVIGATION**; surveying and leveling, navigation, barometric hights, &c.

To be followed by others, forming a complete Mathematical Course for schools and colleges.

CHAS. E. MATTHEWS, M. A. for many years a pupil of, and subsequently an associate Instructor with the late DR. RAY in the Mathematical department of Woodward College, will edit and superintend the publication of the unfinished parts of the series.

Stereotyped by C. F. O'Driscoll & Co.

PREFACE.

As my name appears in connection with that of the late Dr. Joseph Ray, on the title-page of this work, it is proper to account for the circumstance by referring briefly to matters personal in their nature. As an author, Dr. Ray is known to the public at large, through the medium of his excellent mathematical publications. As an able and faithful teacher, his merits are deeply impressed on those who have been so fortunate as to come within reach of his immediate instructions and example. In every line of duty he was conspicuous for unremitting industry, and in all the relations of life, his first desire was to be of service to others.

To me, his memory is endeared by many acts of friendship and confidence, extending over a period of twenty years, sixteen of which were spent with him as pupil and colleague, in Woodward College and High School.

In his last illness, Dr. Ray expressed a wish that I should prepare for publication, his Higher Arithmetic, then unfinished, and directed his manuscripts and materials to be handed to me for that purpose.

In performing the part thus assigned to me, I have endeavored to preserve the spirit and pursue the course indicated in the other Arithmetics of this series. With this view, the language of the author has been retained without material change as far as the manuscript was complete and ready for publication. Whenever a change has been made, it has been done with great caution and with an anxious desire to carry out the general plan of the work. New matter has been occasionally introduced when it seemed necessary to render a demonstration more clear and conclusive, or the treatment of a subject more full and satisfactory.

This has been especially the case in the Contractions in Multiplication, Contractions in Division, Greatest Common Divisor and Least Common Multiple both of whole numbers and fractions.

Pure Arithmetic, comprising Simple Numbers, Common and Decimal Fractions has been discussed before any of its applications. This arrangement is philosophical, and not open to objection in a work of this character.

In questions of proportion, and generally throughout the book, the analytic method of solution has been preferred to mere formal and irrational directions; for no true development of the intellectual powers or satisfactory knowledge of any science can be attained, unless the spirit of every operation is clearly seen through its form.

Where the importance of the subject seemed to demand it, as in Insurance, Simple Interest, Compound Interest and their applications, brief methods of operation have been given, and practical rules deduced for the use of accountants and men of business.

The difficulties and obscurities attending some of the applications of Simple and Compound Interest are inseparable from the subjects themselves, and it is better to meet and overcome these obstacles in the school-room than in the counting-room. No apology is therefore offered for introducing these subjects and discussing them at large.

The exercises are numerous, most of them new and interesting, and have been prepared with a view to practical utility. While they afford a full and thorough exercise in the principles of Arithmetic, at the same time, they enable the pupil to make use of his own knowledge to the best advantage.

Circulating Decimals and other matters rather curious than extensively useful, may be omitted until a review, if the want of time or the character of the pupils make it necessary; and any examples considered too difficult at first may also be postponed in the same way.

THE EDITOR.

STEREOTYPE EDITION.

The marked approbation extended to this volume, and its wide introduction into the best Schools of the country, having led to the sale of several editions in a few months, the work has been thoroughly revised, and is now presented in a permanent stereotype form. In the revision, by a careful abridgment of language, the ideas are presented with greater precision and perspicuity, the book condensed, and a large number of new and interesting examples added.

CONTENTS.

ARITHMETIC.

I. INTRODUCTION.

ARTICLE 1. QUANTITY is any thing which can be increased or diminished. Thus, numbers, lines, space, motion, time, and weight, are quantities.

ART. 2. MATHEMATICS is the science of quantity.

ART. 3. The fundamental branches of Mathematics are Arithmetic, Algebra, and Geometry.

ART. 4. ARITHMETIC is the science of numbers.

ART. 5. A PROBLEM is a question proposed for solution.

ART. 6. A THEOREM is a truth to be proved.

ART. 7. Problems and Theorems are both called Propositions.

ART. 8. A COROLLARY is a truth deduced from a preceding proposition.

ART. 9. A DEMONSTRATION is a process of reasoning by which a proposition is shown to be true.

ART. 10. A DIRECT DEMONSTRATION is one which commences with known truths, and by a chain of reasoning establishes the proposition to be proved.

REVIEW.—1. What is quantity? Give examples. 2. What is Mathematics? 3. What are its fundamental branches? 4. Define Arithmetic. 5. What is a problem? 6. A theorem? 7. What common name is applied to both? 8. What is a corollary? 9. A demonstration? 10. A direct demonstration?

ART. **11**. An INDIRECT DEMONSTRATION is one which assumes the proposition to be *false*, and then proves that some absurdity will necessarily follow. This is sometimes called *Reductio ad absurdum.*

ART. **12**. An AXIOM is a self-evident truth; that is, a proposition so evident that it can not be made plainer by any demonstration. The following are among the most important axioms.

1. If the same or equal quantities be added to equals, the sums will be equal.
2. If the same or equal quantities be subtracted from equals, the remainders will be equal.
3. If the same or equal quantities be multiplied by the same number, the products will be equal.
4. If the same or equal quantities be divided by the same number, the quotients will be equal.
5. Generally, if the same identical operation be performed on two equal quantities, the results will be equal.

MATHEMATICAL SIGNS.

ART. **13**. For brevity, characters, called *signs*, are used in Mathematics. Those most used in Arithmetic, are

$$+, \quad -, \quad \times, \quad \div, \quad =, \quad (\ )\ \text{or}\ \overline{\quad}.$$

ART. **14**. The sign $+$, read *plus*, is the sign of *Addition*; it shows that the numbers between which it is placed are to be added together. Thus, $3 + 5$ equals 8.

ART. **15**. The sign $-$, read *minus*, is the sign of *Subtraction;* the number which follows it is to be subtracted from that which precedes it. Thus, $7 - 4$ equals 3.

Plus and *minus* are Latin words, signifying *more* and *less*.

ART. **16**. The sign $\times$, read *times*, is the sign of *Multiplication.* The numbers between which it is placed are to be multiplied together. Thus, 4×5 equals 20.

ART. **17**. The sign $\div$, read *divided by*, is the sign of *Division.* The number which precedes it is to be divided by that which follows it. Thus, $20 \div 4$ equals 5.

REVIEW.—11. An indirect demonstration? 12. What is an axiom? Name the most important axioms. 13–19. Describe the signs most frequently used in Arithmetic, and give examples of their use.

ART. 18. The sign =, read *equals*, or *is equal to*, is the sign of *Equality;* the quantities between which it is placed are equal to each other. Thus, $5 + 3 = 9 - 1$.

ART. 19. A *parenthesis* ( ), or *vinculum* ——, shows that two or more numbers are to be considered as one. Thus, $(7+4) \times 3 = 33$; or $\overline{8-5} \times 4 = \overline{19+5} \div 2$.

ARITHMETICAL DEFINITIONS.

ART. 20. A *unit* is a single thing; a cent, a hat, &c.

ART. 21. A *number* is a unit, or a collection of units classed under the same name, and answers to the question, *How many?*

The *unit* of a number is *one* of the things it expresses: thus, in *five cents*, *one cent* is the *unit;* in *ten* apples, *one apple* is the *unit.*

Units are sometimes only *relative* in their *character;* thus, *one foot* is a unit in regard to *feet*, but it is only a part of a unit in regard to yards. One dime is a unit, that is, *one* in regard to *dimes*, but it is ten in regard to *cents.*

ART. 22. Numbers are either *abstract* or *concrete.* An *abstract number* is one in which the kind of unit is not designated, as *one*, *two*, *three*, &c. A *concrete number* is one in which the kind of unit is designated, as *one cent*, *two apples*, *ten bushels*, &c. Concrete numbers are frequently called *denominate numbers.*

ART. 23. Arithmetic is founded on NOTATION, and its operations are carried on by means of ADDITION, SUBTRACTION, MULTIPLICATION, and DIVISION. These are termed the fundamental rules of Arithmetic.

II. NUMERATION AND NOTATION.

ART. 24. NUMERATION is the art of *naming* numbers.

NOTATION is the art of representing numbers by characters called *figures* or *digits.*

REVIEW.—20. What is a unit? 21. What a number? Show the relative character of some units. 22. What are abstract numbers? Concrete? Give examples. 23. On what is Arithmetic founded? What are its fundamental rules? 24. What is numeration? What is notation?

ART. **25**. The first nine numbers are each represented by a single figure, thus:

1	2	3	4	5	6	7	8	9
one.	two.	three.	four.	five.	six.	seven.	eight.	nine.

All other numbers are represented by combinations of these and another figure, 0, called *zero*, *naught*, or *cipher*.

REMARK.—The cipher, 0, is used to indicate *no value*. The other figures are called *significant* figures, because they indicate *some value*.

ART. **26**. The number next higher than 9 is named TEN, and is written with two figures thus, 10: in which the cipher, 0, merely serves to show that the unit, 1, on its left is different from the unit, 1, standing alone, which represents a *single thing*, while this represents a *single group of ten things*.

The numbers succeeding Ten are written and named as follows:

11	12	13	14	15	16
eleven.	twelve.	thirteen.	fourteen.	fifteen.	sixteen.

17	18	19
seventeen.	eighteen.	nineteen.

In each of which, the 1 on the left, represents a *group of ten things*, while the figure on the right, expresses the units or *single things* additional, required to make up the number.

REMARK.—The words *eleven* and *twelve* are supposed to be derived from the Saxon, meaning *one left after ten*, and *two left after ten*. The words *thirteen*, *fourteen*, &c., are contractions of *three and ten*, *four and ten*, &c.

The next number to nineteen, (*nine* and *ten*), is *ten* and *ten*, or two *groups of ten*, written 20, and called *twenty*.

The next are *twenty-one*, 21; *twenty-two*, 22; &c., up to *three tens*, or *thirty*, 30; *forty*, 40; *fifty*, 50; *sixty*, 60; *seventy*, 70; *eighty*, 80; *ninety*, 90.

REVIEW.—25. How are the first nine numbers represented? How are numbers above nine represented? *Rem.* For what is the cipher used? What are the other figures called? 26. How is Ten written? What does the cipher show? What does the 1 represent? Write the next nine numbers. Explain their names. Show what the figures denote in each.

The highest number that can be written with two figures is 99, called *ninety-nine;* that is, nine tens and nine units.

The next higher number is 9 tens and ten, or *ten tens*, which is called a *hundred*, and written with *three* figures, 100; in which the two ciphers merely show that the unit on their left is neither a single thing, 1, nor a group of ten things, 10, but a *group of ten tens*, being a unit of a higher grade than either of those already known.

In like manner, 200, 300, &c., express *two hundreds*, *three hundreds*, and so on, up to *ten hundreds*, called a *thousand*, and written with *four* figures, 1000, being a unit of a still higher order.

ART. 27. From what has been said, it is clear that a figure in the 1st place, with no others to the right of it, expresses *units* or single things; but standing on the left of another figure, that is, in the 2d place, expresses *groups of tens;* and standing at the left of two figures, or in the 3d place, expresses *tens* of *tens*, or hundreds; and in the 4th place, expresses *tens* of *hundreds* or *thousands*. Hence, counting from the right hand,

The order of *Units*	is in the 1st place .	1
The order of *Tens*	is in the 2d place .	10
The order of *Hundreds*	is in the 3d place .	100
The order of *Thousands*	is in the 4th place .	1000

By this arrangement, *the same figure has different values according to the place* in which it stands. Thus, 3 in the first place is 3 *units;* in the second place 3 *tens*, or thirty; in the third place 3 *hundreds;* and so on.

ART. 28. The word UNITS may be used in naming all the orders as follows:

Simple units are called . . .	*Units of the* 1*st order.*
Tens	*Units of the* 2*d order.*
Hundreds	*Units of the* 3*d order.*
Thousands	*Units of the* 4*th order.*
&c.	&c.

REVIEW.—26. What are 20, 30, 40? What is the highest number of two figures? What is the next called? How is it written? Explain its figures. What is a Thousand? How written? 27. What does a figure in the 1st place express? a figure in the 2d place? in the 3d place? in the 4th place? How does the value of the same figure vary?

NOTE.—When units are named without reference to any particular order, units of the first order are always intended.

ART. 29. The preceding articles show the method of expressing numbers less than one thousand. For example, in the number *four hundred and twenty-five*, there are 4 hundreds, 2 tens or twenty, and 5 units; the number is therefore written, 4 2 5.

In the number *three hundred and nine*, there are 3 hundreds, 0 tens, and 9 units; or 3 units of the third, and none of the second, and 9 of the first order; hence, the number is represented thus, 309.

ART. 30. SUMMARY OF PRINCIPLES.

1. *All numbers are represented by the nine digits and zero.*

2. *Zero has no value; its use is to fill vacant places.*

3. *The base of our system of notation is ten; ten units of any order making one unit of the order next higher.*

4. *The same figure has different values according to the place it occupies.*

ART. 31. TABLE OF ORDERS.

9th.	8th.	7th.	6th.	5th.	4th.	3d.	2d.	1st.
Hundreds of Millions	Tens of Millions	Millions	Hundreds of Thousands	Tens of Thousands	Thousands	Hundreds	Tens	Units

ART. 32. For convenience in reading and writing numbers, periods of three orders each, are used. The first 3 orders, UNITS, TENS, and HUNDREDS, constitute the first or UNIT PERIOD. The second 3 orders form the second or THOUSAND PERIOD; the third 3 orders, the third or MILLION PERIOD.

REVIEW.—29. Write the number four hundred and twenty-five. Write the number three hundred and nine.

ART. 33. List of the Periods, according to the common or French method of Numeration.

First Period . Units.	Sixth Period . Quadrillions.
Second . . . Thousands.	Seventh . . . Quintillions.
Third . . . Millions.	Eighth . . . Sextillions.
Fourth . . . Billions.	Ninth . . . Septillions.
Fifth . . . Trillions.	Tenth . . . Octillions.

The next twelve periods are, Nonillions, Decillions, Undecillions, Duodecillions, Tredecillions, Quatuordecillions, Quindecillions, Sexdecillions, Septendecillions, Octodecillions, Novendecillions, Vigintillions.

ART. 34. Division of the orders into periods.

of Trillions.			of Billions.			of Millions.			of Thousands.			of Units.		
Hundreds	Tens	Units	Hundreds	Tens	Units	Hundreds	Tens	Units	Hundreds	Tens	Units	Hundreds	Tens	Units
6	5	4	3	2	1	9	8	7	6	5	4	3	2	1
5th Period.			4th Period.			3d Period.			2d Period.			1st Period.		

ART. 35. RULE FOR NUMERATION.

Begin at the right, and point the number into periods of three figures each. Commence at the left, and read in succession each period with its name.

REMARK.—Numbers may also be read by merely *naming each figure with the name of the place in which it stands.* This method, however, is rarely used except in teaching beginners. Thus, the numbers expressed by the figures 205, may be read *two hundred and five*, or *two* hundreds *no* tens and *five* units.

Express in words, the number which is represented by 608921045. The number, as divided into periods, is 608'921'045; and is read six hundred and eight millions nine hundred and twenty-one thousand and forty-five.

REVIEW.—30. What are the principles of Notation and Numeration? 31. Repeat the Table of Orders. 32. What are periods and their use? 33. Name the first ten. 34. Give an example of the use of periods in a particular case. 35. What is the rule for Numeration? What other method may be used?

EXAMPLES IN NUMERATION.

7	12345	1375482	29347283
40	68380	6030564	37053495
85	94025	7004025	45004024
503	70500	8025607	50340726
278	165247	9030040	60025709
1345	350304	6002007	343827544
2450	204026	4300201	904207080
3708	500050	8603004	700200408
4053	808080	2030405	502003070
7009	730003	6005010	830070320

832045682327825000000321
80070060050040030020010000000
600300200900800700500600030070
504030209102800703240703250207

ART. **36**. RULE FOR NOTATION.

Write, first the number of the highest period, then, of the other periods in their proper succession, filling vacant places with ciphers.

Express in figures the number four millions twenty thousand three hundred and seven. *Ans.* 4020307.

Write 4 in *millions* period; place a dot after it to separate it from the next period: then, write 20 in *thousands* period; place another dot: then write 307 in *units* period. This gives 4˙ 20˙307. As there are but *two* places in the thousands period, a cipher must be put before 20 to complete its orders, and the number correctly written, is 4020307.

NOTE.—Every period, except the highest, must have its *three* figures; and if any period is not mentioned in the given number, supply its place by three ciphers.

PROOF.—Apply to the number, as written, the rule for numeration, and see if it agrees with the number given.

EXAMPLES IN NOTATION.

1. Seventy-five.
2. Ninety.
3. One hundred and thirty-four.
4. Two hundred and forty-four.
5. Two hundred and forty.
6. Two hundred and four.

7. Three hundred and seventy.
8. Six hundred and ten.
9. Eight hundred and one.
10. One thousand two hundred and thirty-four.
11. Three thousand four hundred and ninety.
12. Seven thousand and twenty-five.
13. Nine thousand and seven.
14. Forty thousand five hundred and sixty-three.
15. Seventy thousand and twenty.
16. Eighty-four thousand and two.
17. Ninety thousand and nine.
18. Sixty thousand six hundred.
19. One hundred and sixty-four thousand three hundred and ninety-four.
20. Two hundred and seven thousand four hundred and one.
21. Five hundred thousand and thirty.
22. Six hundred and forty thousand and forty.
23. Eight hundred and two thousand two hundred.
24. Nine hundred thousand nine hundred.
25. Seven hundred thousand and seven.
26. One million four hundred and twenty-one thousand six hundred and eighty-five.
27. Six millions sixty thousand and sixty.
28. Nine millions nine thousand and nine.
29. Seven millions and seventy.
30. Twenty-three millions six hundred and forty-five thousand nine hundred and seventy-one.
31. Ten millions one hundred thousand and ten.
32. Ninety millions ninety thousand and ninety.
33. Eighty-eight millions seventy thousand and five.
34. Sixty millions seven hundred and five thousand.
35. One hundred millions.
36. Two hundred and fifty millions three hundred and three thousand and twenty-six.
37. Eight hundred and seven millions forty thousand and thirty-one.
38. Seven hundred millions seventy thousand and seven.
39. Two billions and twenty millions.
40. Thirty trillions thirty millions thirty thousand and thirty.
41. Nineteen quadrillions twenty trillions and five hundred billions.
42. Ten quadrillions four hundred and three trillions ninety billions and six hundred millions.
43. Eighty octillions sixty sextillions three hundred and twenty-five quintillions and thirty-three billions.
44. Nine hundred decillions seventy nonillions six octillions forty septillions fifty quadrillions two hundred and four trillions ten millions forty thousand and sixty.

REVIEW.—36. What is the rule for notation? What is the proof?

ENGLISH METHOD OF NUMERATION.

ART. **37**. Although now but little used, the learner should understand this system of Numeration. The first six orders have the same names as in the French method, and these constitute the period of Units. The next or Millions period, consists of *six* orders. Each succeeding period consists of six orders, and their names are Billions, Trillions, Quadrillions, Quintillions, and so on. The following table illustrates this method:

of Trillions.						of Billions.						of Millions.						of Units.					
Hundreds of Th.	Tens of Th.	Thousands	Hundreds	Tens	Units	Hundreds of Th.	Tens of Th.	Thousands	Hundreds	Tens	Units	Hundreds of Th.	Tens of Th.	Thousands	Hundreds	Tens	Units	Hundreds of Th.	Tens of Th.	Thousands	Hundreds	Tens	Units
4	3	2	1	0	9	8	7	6	5	4	3	2	1	0	9	8	7	6	5	4	3	2	1

By this system the first twelve figures would be read, two hundred and ten thousand nine hundred and eighty-seven millions, six hundred and fifty-four thousand three hundred and twenty-one. By the French method they would be read, two hundred and ten billions, nine hundred and eighty-seven millions, six hundred and fifty-four thousand, three hundred and twenty-one.

ROMAN NOTATION.

ART. **38**. In the Roman Notation, numbers are represented by *letters*. The letter I represents *one;* V, *five;* X, *ten;* L, *fifty*; C, *one hundred;* D, *five hundred;* and M, *one thousand.* The other numbers are represented according to the following principles:

1st. Every time a letter is repeated, its value is repeated. Thus, II denotes two; XX denotes twenty.

REVIEW.—37. Explain the English method of Numeration. Give an example of its application. 38. How are numbers represented in the Roman Notation? What numbers are represented by I, V, X, C, D, M?

2d. Where a letter of *less* value is placed *before* one of *greater* value, the less is taken from the greater. If placed *after* it, the less is added to the greater. Thus, IV denotes four, while VI denotes six; IX denotes nine, while XI denotes eleven.

3d. A bar, —, placed over a letter increases its value a *thousand times*. Thus, $\overline{\text{V}}$ denotes *five thousand;* $\overline{\text{M}}$ denotes *one million.*

ROMAN TABLE.

I	One.	XXX	Thirty.
II	Two.	XL	Forty.
III	Three.	L	Fifty.
IV	Four.	LX	Sixty.
V	Five.	XC	Ninety.
VI	Six.	C	One hundred.
IX	Nine.	CCCC	Four hundred.
X	Ten.	D	Five hundred.
XI	Eleven.	DC	Six hundred.
XIV	Fourteen.	DCC	Seven hundred.
XV	Fifteen.	DCCC	Eight hundred.
XVI	Sixteen.	DCCCC	Nine hundred.
XIX	Nineteen.	M	One thousand.
XX	Twenty.	MM	Two thousand.
XXI	Twenty-one.	MDCCCLVI	1856.

III. ADDITION.

ART. **39.** ADDITION is the process of collecting two or more numbers into one sum.

Sum or *Amount* is the result obtained by Addition.

ART. **40.** Since a number is a collection of units of the *same* kind, *two or more numbers can be united into one sum, only when their units are of the same kind. Two apples* and 3 *apples* are 5 *apples;* but 2 *apples* and 3 *peaches* can not be united into *one* number, either of *apples* or of *peaches.*

REVIEW.—38. What is the effect of repeating a letter? Of placing a letter of less value before another of greater value? Of placing one of less value after one of greater value? Of placing a bar over a letter? 39. What is addition? Sum or amount?

Nevertheless, numbers of different names may be added together, if they can be brought under a *common denomination.* Thus, 2 *men* and 3 *women* are 5 *persons.* Two *horses*, 3 *sheep*, and 4 *cows*, are 9 *animals.*

RULE FOR ADDING SIMPLE NUMBERS.

1. *Write the numbers to be added, so that figures of the same order may stand in column, units under units, tens under tens, hundreds under hundreds, &c.*

2. *Begin at the right, and add each column separately, placing the units of each sum under the column added, and carrying the tens to the next column. At the last column, set down the whole amount.*

What is the sum of 639, 82, and 543?

```
 639
  82
 543
----
1264
```

SOLUTION.—Writing the numbers as in the margin, say, 3 and 2 are 5, and 9 are 14 *units*, which are 1 ten and 4 units. Write the 4 units beneath, and carry the 1 ten to the next column. Then 1 and 4 are 5, and 8 are 13, and 3 are 16 *tens*, which are 6 tens to be written beneath, and 1 hundred to be carried to the next column. Lastly, 1 and 5 are 6, and 6 are 12 hundreds, which is set beneath, there being no other columns to carry to or add.

DEMONSTRATION.—1. Figures of the same order are written under each other for *convenience*, since none but units of the same name can be added. (Art. 40.)

2. Commence at the right to add, so that when the sum of any column is greater than nine, the *tens* may be carried to the next column, and, thereby, units of the *same* name added together.

3. Carry *one* for every ten, since ten units of each order make *one* unit of the order next higher. (Art. 30.)

METHODS OF PROOF.

ART. **41.** 1. Add the figures *downward* instead of *upward;* or

REVIEW.—40. What is necessary in order that numbers may be added into one sum? Why? What is the rule for adding simple numbers? Explain the example, and give the reasons for the rule. 41. What is the first method of proof?

2. Separate the numbers into two or more divisions; find the sum of the numbers in each division, and then add the several sums together; or

3. Commence at the left; add each column separately; place each sum under that previously obtained, but extending one figure further to the right, and then add them together.

In all these methods the result should be the same as when the numbers are added upward.

NOTE.—For proof by casting out the 9's, see page 68.

EXAMPLES FOR PRACTICE.

Find the sum,

1. Of 76767; 7654; 50121; 775. *Ans.* 135317.
2. Of 97674; 686; 7676; 9017. *Ans.* 115053.
3. Of 971; 7430; 97476; 76734. *Ans.* 182611.
4. Of four hundred and three; 5025; sixty thousand and seven; eighty-seven thousand; two thousand and ninety, and 100. *Ans.* 154625.
5. Of 999; 3400; 73; 47; 452; 11000; 193; 97; 9903; 42, and 5100. *Ans.* 31306.
6. Of 20050; three hundred and seventy thousand two hundred; four millions and five; two millions ninety thousand seven hundred and eighty; one hundred thousand and seventy; 98002; seven millions five thousand and one; and 70070. *Ans.* 13754178.
7. Of 609505; 90070; 90300420; 9890655; 789; 37599; 19962401; 5278; 2109350; 41236; 722; 8764; 29753, and 370247. *Ans.* 123456789.
8. Of two hundred thousand two hundred; three hundred millions six thousand and thirty; seventy millions seventy thousand and seventy; nine hundred and four millions nine thousand and forty; eighty thousand; ninety millions nine thousand; six hundred thousand and sixty; five thousand seven hundred; four millions twenty thousand and twenty. *Ans.* 1369000120.

REVIEW.—What is the second method of proof? The third?

In each of the 7 following examples, find the sum of the numbers from A to B, including these numbers.

	A.		B.		
9.	119	. . .	131	. . . *Ans.*	1625.
10.	987	. . .	1001	. . . *Ans.*	14910.
11.	3267	. . .	3281	. . . *Ans.*	49110.
12.	4197	. . .	4211	. . . *Ans.*	63060.
13.	5397	. . .	5416	. . . *Ans.*	108130.
14.	7815	. . .	7831	. . . *Ans.*	132991.
15.	31989	. . .	32028	. . . *Ans.*	1280340.

16. Paid for coffee, $375; for tea, $280; for sugar, $564; for molasses, $119; and for spices, $75: what did the whole amount to? *Ans.* $1413.

17. Bought three pieces of cloth: the first cost $87; the second, $25 more than the first; and the third, $47 more than the second: what did all cost? *Ans.* $358.

18. Bought three bales of cotton. The first cost $325; the second, $16 more than the first; and the third, as much as both the others: what did the three bales cost? *Ans.* $1332.

19. A has $75; B has $19 more than A; C has as much as A and B, and $23 more; and D has as much as A, B, and C together: what sum do they all possess? *Ans.* $722.

IV. SUBTRACTION.

ART. **42**. SUBTRACTION is the process of finding the difference between two numbers.

The larger number is the *Minuend;* the less, the *Subtrahend;* the number left, the *Difference* or *Remainder.*

Minuend means to be *diminished;* subtrahend, *to be subtracted.*

ART. **43**. Subtraction is the reverse of Addition, and since none but numbers of the same kind can be added

REVIEW.—42. What is Subtraction? Minuend? Subtrahend? Remainder? What does minuend mean? What does subtrahend mean?

together, (Art. 40), it follows, therefore, that a number can be subtracted only from another of the same kind: 2 *cents* can not be taken from 5 *apples*, nor 3 *cows* from 8 *horses*.

ART. **44**. RULE FOR SUBTRACTING SIMPLE NUMBERS.

1. *Write the less number under the greater, placing units under units, tens under tens, &c.*
2. *Begin at the right, subtract each figure from the one above it, placing the remainder beneath.*
3. *If any figure exceeds the one above it, add ten to the upper, subtract the lower from the sum, and carry one to the next lower figure.*

From 827 dollars take 534 dollars.

Dollars.	
827	
534	
293	*Rem.*

SOLUTION.—After writing figures of the same order under each other, say, 4 units from 7 units leaves 3 units. Then, as 3 tens can not be taken from 2 tens, add 10 tens to the 2 tens, which makes 12 tens, and 3 tens from 12 tens leaves 9 tens. To compensate for the 10 tens added to the 2 tens, add one hundred (10 tens) to the 5 hundreds, and say, 6 hundreds from 8 hundreds leaves 2 hundreds; and the whole remainder is 2 hundreds 9 tens and 3 units, or 293.

DEMONSTRATION.—1. Since a number can be subtracted only from another of the same kind, (Art. 43), figures of the same name are placed in column, to be convenient to each other, the less number being placed below as a matter of custom.

2. Commence at the right to subtract, so that if any figure is greater than the one above it, the upper may be increased by 10, and the next lower figure by 1; both numbers being equally increased, since 10 in any place is equal to 1 in the next higher place.

The difference between two numbers is the same as the difference between those numbers increased equally. Thus, the difference between 2 and 5, is the same as the difference between 2+10 and 5+10, or 12 and 15.

There is another method of performing the operation

REVIEW.—43. What is subtraction the reverse of? Why? What is necessary in order that one number may be subtracted from another? Why? 44. What is the rule for subtracting simple numbers? Explain the example, and give the reasons for the rule.

when the lower figure is greater than the upper, called *borrowing*. To explain this, take 26 from 75.

SOLUTION.—Since 6 units can not be taken from 5 units, borrow 1 ten from the 7 tens, and add it to the 5, which will make 15 units. Then, 6 units from 15 units leaves 9 units; and 2 tens from 6 tens, (the number left in tens' place) leaves 4 tens: and the whole remainder is 4 tens and 9 units, or 49.

```
 75
 26
---
 49
```

PROOF.—Add the remainder to the less number. The sum will be equal to the greater.

NOTE.—For proof by casting out the 9's, see page 69.

EXAMPLES FOR PRACTICE.

1. From 30020037 take 50009.

When there are no figures in the lower number to correspond with those in the upper, consider the vacant places occupied by zeroes.

```
30020037
   50009
--------
29970028 Rem.
```

	FROM	TAKE	REM.
2.	79685	30253	49432.
3.	1145906	39876	1106030.
4.	7150490	92367	7058123.
5.	2900000	777888	2122112.
6.	71086540	64179730	6906810.
7.	10143072	9247568	895504.
8.	20286144	18495136	1791008.
9.	25047361	7140851	17906510.
10.	101067800	100259063	808737.

In each of the following examples to the 18th, subtract the first number from the second, then from the remainder, and so on, till a remainder is found less than the first number.

		LAST REM.
11.	7359 and 36800	5.
12.	7598 and 35768	5376.
13.	59649 and 248697	10101.
14.	730928 and 2500000	307216.
15.	3956847 and 21018802	1234567.
16.	9080907 and 60000000	5514558.
17.	1234567 and 10000000	123464.
18.	938979 and 9389790	0.

19. A merchant has $3050; what sum will he have, after paying $2364? *Ans.* $686.

20. How many years from the discovery of America in 1492, to its Independence in 1776? *Ans.* 284.

21. A has $5000, and owes $2705; what will be left after paying his debts? *Ans.* $2295.

22. A farm that cost $7253 was sold at a loss of $395; for how much was it sold? *Ans.* $6858.

23. What number must be added to 6428 that the sum shall be 8350? *Ans.* 1922.

24. The difference of two numbers is 19034, and the greater is 75421; what is the less? *Ans.* 56387.

25. How many times can 285 be subtracted from 1425? *Ans.* 5.

26. Which is the nearer number to 920736; 1816045 or 25427? *Ans.* Neither. Why?

27. Of the numbers 23452 and 60000, how much is one nearer to 41386 than the other? *Ans.* 680.

28. A merchant owing $5000, paid $2175, and $1850; how much was unpaid? *Ans.* $975.

29. From a tract of land containing 10000 acres, the owner sold to A 4750 acres; and to B 875 acres less than A: how many acres had he left. *Ans.* 1375.

30. A, B, C, D, are 4 places in order in a straight line. From A to D is 1463 miles; from A to C, 728 miles; and from B to D, 1317 miles. How far is it from A to B, from B to C, and from C to D?

Ans. A to B, 146 miles; B to C, 582; C to D, 735.

31. From Pittsburgh to New Orleans is 1999 miles; from Pittsburgh to Memphis is 1260 miles; and from Cincinnati to New Orleans is 1523 miles: how far is it from Cincinnati to Pittsburgh, from Cincinnati to Memphis, and from Memphis to New Orleans?

Ans. From C. to P. 476 miles; from C. to M. 784; from M. to N. O. 739 miles.

Art. 45. Examples in Addition and Subtraction.

1. Add 37452 to 19576, and from the result subtract the sum of 19914 and 19715. *Ans.* 17399.

Review.—44. What is the method of borrowing? What is the proof for subtraction?

2. From the sum of 28374 and 56132, subtract the difference of 100000 and 84506. *Ans.* 69012.

3. To the difference of 95603 and 44571, add the difference of 8632 and 4179. *Ans.* 55485.

4. From the difference of 7325 and 3149 subtract the difference of 8645 and 4702. *Ans.* 233.

5. From what number must 256 be subtracted to leave the remainder 144? *Ans.* 400.

6. To what number must 1327 be added, to make the sum 5000? *Ans.* 3673.

7. From what number must 27 be subtracted 3 times, to leave 38? *Ans.* 119.

8. To what number must 65 be added 4 times, to make the sum 297? *Ans.* 37.

9. The distance by railroad from Cincinnati to Cleveland is 275 miles. When one locomotive is 60 miles from Cincinnati, and another 118 miles from Cleveland, how far are they apart? How far apart, when one has traveled 174 miles, and the other 183 miles?

Ans. 97 miles and 82 miles.

10. A was 37 years old when B was born; how old was A, when B was 25 years of age? and how old was B, when A was 73 years of age? *Ans.* 62 and 36.

11. Which is the nearer number to 356908; 713866 or 7849, and how much? *Ans.* The last, by 7899.

V. MULTIPLICATION.

ART. **46.** MULTIPLICATION is taking one number as many times as there are units in another; or

Multiplication is a short method of adding numbers that are equal.

The number to be taken, is called the *Multiplicand;* the other number, the *Multiplier;* and the result obtained, the *Product.*

REVIEW.—46. What is Multiplication? What is the Multiplicand?

How many trees in 3 rows, each containing 42 trees.

SOLUTION.—Since 3 rows contain 3 *times* as many trees as *one* row, take 42 *three* times. This may be done by writing 42 three times, and then *adding*. This gives 126 for the whole number of trees.

First row, 42 trees.
Second row, 42 trees.
Third row, 42 trees.
126 trees.

Instead, however, of writing 42 *three* times, write it *once;* then placing under it the figure 3, the *number of times* it is to be taken, say, 3 times 2 are 6, and 3 times 4 are 12. This process is *Multiplication.*

$$\begin{array}{r} 42 \text{ trees.} \\ 3 \\ \hline 126 \text{ trees.} \end{array}$$

The Multiplicand and Multiplier are together called *Factors* (makers), because they *make* the product.

MULTIPLICATION TABLE.

1	2	3	4	5	6	7	8	9	10	11	12	13	14	15	16	17	18	19	20
2	4	6	8	10	12	14	16	18	20	22	24	26	28	30	32	34	36	38	40
3	6	9	12	15	18	21	24	27	30	33	36	39	42	45	48	51	54	57	60
4	8	12	16	20	24	28	32	36	40	44	48	52	56	60	64	68	72	76	80
5	10	15	20	25	30	35	40	45	50	55	60	65	70	75	80	85	90	95	100
6	12	18	24	30	36	42	48	54	60	66	72	78	84	90	96	102	108	114	120
7	14	21	28	35	42	49	56	63	70	77	84	91	98	105	112	119	126	133	140
8	16	24	32	40	48	56	64	72	80	88	96	104	112	120	128	136	144	152	160
9	18	27	36	45	54	63	72	81	90	99	108	117	126	135	144	153	162	171	180
10	20	30	40	50	60	70	80	90	100	110	120	130	140	150	160	170	180	190	200
11	22	33	44	55	66	77	88	99	110	121	132	143	154	165	176	187	198	209	220
12	24	36	48	60	72	84	96	108	120	132	144	156	168	180	192	204	216	228	240
13	26	39	52	65	78	91	104	117	130	143	156	169	182	195	208	221	234	247	260
14	28	42	56	70	84	98	112	126	140	154	168	182	196	210	224	238	252	266	280
15	30	45	60	75	90	105	120	135	150	165	180	195	210	225	240	255	270	285	300
16	32	48	64	80	96	112	128	144	160	176	192	208	224	240	256	272	288	304	320
17	34	51	68	85	102	119	136	153	170	187	204	221	238	255	272	289	306	323	340
18	36	54	72	90	108	126	144	162	180	198	216	234	252	270	288	306	324	342	360
19	38	57	76	95	114	133	152	171	190	209	228	247	266	285	304	323	342	361	380
20	40	60	80	100	120	140	160	180	200	220	240	260	280	300	320	340	360	380	400

ART. **47**. THEOREM.—*The product of two numbers is the same, whichever factor is the multiplier.*

DEMONSTRATION.—To find the number of stars in this diagram, multiply the *number of stars in a row*, by the *number of rows*. Counting *up and down*, there are 4 in a row, and 3 rows; hence, there must be 3 times 4, = 12 stars. But *across*, there are 3 in a row, and 4 rows; hence, there must be 4 times 3, = 12 stars. The two products, 3 times 4 and 4 times 3, are equal, since each gives the whole number of stars in the diagram. The same may be shown of any two numbers.

```
* * *
* * *
* * *
* * *
```

ART. **48**. THEOREM.—*The product is always of the same name as the multiplicand.*

DEMONSTRATION.—The *nature* of a number is not changed by repeating it. Thus, 5 dollars multiplied by 2, that is, taken twice, must be 10 *dollars*.

ART. **49**. THEOREM.—*The multiplier must always be an abstract number.*

DEMONSTRATION.—The multiplier shows the *number of times* the multiplicand is to be taken; hence, it can not be yards, bushels, or any concrete number. We can attach no idea to taking any thing so many *dollars times*, or so many *inches times*. It may then be asked, how explain the solution of such questions as "What will 3 yards of cloth cost at 5 dollars a yard?" *Ans.* By the following or a similar *Analysis:* Three yards will cost *three times* as much as *one* yard; therefore, if *one* yard cost 5 dollars, 3 yards will cost 3 *times* 5 dollars, = $15.

REMARK.—Hence, it is absurd to talk of multiplying dollars by dollars. It would be as rational to propose to find the product of 3 apples by 2 turnips.

ART. **50**. Multiplication is divided into two cases:

1. When the multiplier does not exceed 12.
2. When the multiplier exceeds 12.

RULE WHEN THE MULTIPLIER DOES NOT EXCEED 12.

1. *Write the multiplicand; place the multiplier under it, and draw a line beneath.*
2. *Begin with units; multiply each figure of the multiplicand by the multiplier, setting down and carrying as in Addition.*

At the rate of 53 miles an hour, how far will a railroad car run in four hours?

SOLUTION.—Here say, 4 times 3 (units) are 12 (units); write the 2 in units' place, and carry the 1 (ten); then, 4 times 5 are 20, and 1 carried makes 21 (tens), and the work is complete.

```
 53 miles.
  4
---
212 miles.
```

DEMONSTRATION.—The multiplier being written under the multiplicand for convenience, begin with units, so that if the product should contain tens, they may be carried to the tens; and so on for each successive order.

Since every figure of the multiplicand is multiplied, therefore, the *whole* multiplicand is multiplied.

PROOF.—Separate the multiplier into any two parts; multiply by these separately. The sum of the products must be equal to the first product.

EXAMPLES FOR PRACTICE.

	MULTIPLICAND.	MULTIPLIER.	PRODUCT.
1.	195	3	585.
2.	3823	4	15292.
3.	8765	5	43825.
4.	98374	6	590244.
5.	64382	7	450674.
6.	58765	8	470120.
7.	837941	9	7541469.
8.	645703	10	6457030.
9.	407649	11	4484139.

10. If 4 men can perform a certain piece of work in 15 days, how long will it require 1 man?

SOLUTION.—One man will be four times as long as four men. $4 \times 15 = 60$ days.

11. How many pages in a half-dozen books, each containing 336 pages? *Ans.* 2016.

12. How far can an ocean steamer travel in a week, at the rate of 245 miles a day? *Ans.* 1715 miles.

REVIEW.—46. Repeat the Multiplication Table. 47. Prove the theorem. 48. How is the denomination of the product known? Why? 49. What kind of a number is the multiplier? Why? 50. How is multiplication divided? What is the rule when the multiplier does not exceed 12? Give the reason. What is the proof?

13. What is the yearly expense of a cotton-mill, if $32053 are paid out every month? *Ans.* $384636.

Art. 51. Rule when the Multiplier exceeds 12.

1. *Write the multiplier under the multiplicand, placing figures of the same order under each other.*

2. *Multiply by each figure of the multiplier successively; first by the units' figure, then by the tens' figure, &c.; placing the right hand figure of each product under that figure of the multiplier which produces it.*

3. *Add the several partial products together; their sum will be the required product.*

Multiply 246 by 235.

```
  246
  235
 1230  product by   5
 738   product by  30
492    product by 200
57810  product by 235
```

Solution.—First multiply by 5 (units), and place the first figure of the product, 1230, under the 5 (units). Then multiply by 3 (tens), and place the first figure of the product, 738, under the 3 (tens). Lastly, multiply by 2 (hundreds), and place the first figure of the product, 492, under the 2 (hundreds). Then add these several products for the entire product.

Analysis.—The 0 of the first product, 1230, is *units*, Art. 50. The 8 of the 2d product, 738, is *tens*, because 3 (tens) times 6 = 6 times 3 (tens) = 18 (tens); giving 8 (tens) to be written in the tens' column. The 2 of the 3d product, 492, is hundreds, because 2 (hundreds) times 6 = 6 times 2 (hundreds) = 12 (hundreds), giving 2 (hundreds) to be written in the hundreds' column. The right hand figure of each product being in its proper column, the other figures will fall in their proper columns; and each line being the product of the multiplicand by a *part* of the multiplier, their sum will be the product by *all the parts* or the *whole* of the multiplier.

METHODS OF PROOF.

1. Multiply the multiplier by the multiplicand; this product must be the same as the first product.

2. The same as when the multiplier does not exceed 12.

Remark.—For proof by casting out the 9's, see Art. 100.

Review.—What is the rule when the multiplier exceeds 12?

ART. 52. Although it is customary to use the figures of the multiplier in regular order beginning with units, it will give the same product to use them in any order, observing that *the right-hand figure of each partial product must be placed under the figure of the multiplier which produces it.*

246		
235		
738	product by	30
492	product by	200
1230	product by	5
57810	product by	235

EXAMPLES FOR PRACTICE.

1. 7198 × 216 . . . . = 1554768.
2. 8862 × 189 . . . . = 1674918.
3. 7575 × 7575 . . . . . = 57380625.
4. 15607 × 3094 . . . . . = 48288058.
5. 93186 × 4455 . . . . . = 415143630.
6. 135790 × 24680 . . . . = 3351297200.
7. 3523725 × 2583 . . . = 9101781675.
8. 4687319 × 1987 . . . = 9313702853.
9. 9264397 × 9584 . . . . = 88789980848.
10. 9507340 × 7071 . . . . = 67226401140.
11. 1644405 × 7749 . . . . = 12742494345.
12. 1389294 × 8900 . . . . = 12364716600.
13. 2778588 × 9867 . . . . = 27416327796.
14. 204265 × 562402 . . . = 114879044530.
15. 39948123 × 6007 . . . = 239968374861.
16. 57902468 × 5008 . . . = 289975559744.
17. 57902468 × 5080 . . . = 294144537440.
18. 3081025 × 2008036 . = 6186809116900.
19. 860389657 × 96795 . . = 83281416849315.

CONTRACTIONS IN MULTIPLICATION.

CASE I.

WHEN THE MULTIPLIER IS A COMPOSITE NUMBER.

ART. 53. A composite number is the product of two or more whole numbers, each greater than 1, called its *factors.* Thus, 10 is a composite number, whose factors are 2 and 5: and 30 is one whose factors are 2, 3 and 5.

RULE.—*Separate the multiplier into two or more factors. Multiply first by one of the factors, then this product by another factor, and so on till each factor has been used as a multiplier. The last product will be the entire product required.*

At 7 cents a piece, what will 6 melons cost?

ANALYSIS.—Three times 2 times are 6 times. Hence, it is the same to take 2 times 7, and then take this product 3 times, as to take 6 times 7. And the same may be shown of any other composite number.

OPERATION.

```
 7 cents, cost of 1 melon.
 2
——
14 cents, cost of 2 melons.
 3
——
42 cents, cost of 6 melons.
```

EXAMPLES FOR PRACTICE.

1. At the rate of 37 miles a day, how far will a man walk in 28 days? *Ans.* 1036 miles.

2. Sound moves about 1130 feet per second: how far will it move in 54 seconds? *Ans.* 61020 feet.

3. Multiply 9765 by 35. *Ans.* 341775.

CASE II.

WHEN THE MULTIPLIER IS ONE WITH CIPHERS ANNEXED, AS 10, 100, 1000, &c.

ART. **54.** RULE.—*Annex to the multiplicand as many ciphers as there are in the multiplier; the result will be the required product.*

DEMONSTRATION.—By the principles of Notation, (Art. 26), placing *one* cipher on the right of a number, changes the units into tens, the tens into hundreds, and so on, and therefore, *multiplies the number by* 10.

Annexing *two* ciphers to a number changes the units into hundreds, the tens into thousands, and so on, and, therefore, *multiplies the number by* 100. Annexing *three* ciphers multiplies the number by 1000, &c.

EXAMPLES FOR PRACTICE.

1. Multiply 375 by 100 . . . . . . *Ans.* 37500.
2. Multiply 207 by 1000 . . . . . *Ans.* 207000.

REVIEW.—52. Explain the Example. 53. What is a composite number? What are its factors? What is the rule for multiplying by a composite number?

CASE III.

WHEN CIPHERS ARE AT THE RIGHT OF ONE OR BOTH FACTORS.

ART. 55. RULE.—*Multiply as if there were no ciphers at the right of the numbers; then annex to the product as many ciphers as there are at the right of both the factors.*

Find the product of 5400 by 130.

```
  5400
  130
 ─────
 162
 54
 ──────
 702000
```

SOLUTION.—Find the product of 54 by 13, and then annex *three* ciphers, the number at the right of both factors.

ANALYSIS.—Since 13 times 54 = 702, it follows that 13 times 54 hundreds (5400) = 702 hundreds (70200); and 130 times 5400 = 10 times 13 times 5400 = 10 times 70200 = 702000.

OR THUS: The operation is the same as that in the Rule for Multiplication, (Art. 51), except that the ciphers on the right are omitted until the sum of the partial products is found. This is evident from the operation in the margin.

```
   5400
    130
 ──────
 162000
 5400
 ──────
 702000
```

EXAMPLES FOR PRACTICE.

1. 15460 × 3200 . . . . . = 49472000.
2. 30700 × 5904000 . . = 181252800000.
3. 67051800 × 37800 . . = 2534558040000.
4. 9730000 × 645100 . = 6276823000000.
5. 46218000 × 757000 . = 34987026000000.
6. 9800100 × 6514000 . . = 63837851400000.

CASE IV.

WHEN THE MULTIPLIER IS A LITTLE LESS THAN 10, 100, 1000, &c.

ART. 56. RULE.—*Annex to the multiplicand as many ciphers as there are figures in the multiplier, and from the result subtract the product of the multiplicand by the number the multiplier wants of being* 10, 100, 1000, &c.

REVIEW.—54. What is the rule for multiplying by 1 with ciphers annexed? Give the reasons. 55. What is the rule for multiplying when ciphers are at the right of one or both factors? Explain the example.

Multiply 3046 by 997.

OPERATION.
3046
997
3046000
9138
3036862

ANALYSIS.—Since 997 is the same as 1000 diminished by 3, to multiply by it is the same as to multiply by 1000, (that is, to annex 3 ciphers), and by 3, and take the difference of the products, which agrees with the rule; and the same can be shown in any similar case.

EXAMPLES FOR PRACTICE.

1. 7023 × 99 . . . . = 695277.
2. 16642 × 996 . . . . = 16575432.
3. 372051 × 9998 . . . . . . = 3719765898.

NOTE.—If there are ciphers at the right of one or both the factors, the rule may still be applied, (Art. 55.)

4. 642438 × 93 . . . . = 59746734.
5. 28714 × 995 . . . . . . = 28570430.
6. 99999 × 99991 . . . . . =9999000009.
7. 525734 × 9994 . . . . = 5254185596.
8. 4127093 × 9989 . . . . . = 41225531977.
9. 2634527 × 9980 . . . . . = 26292579460.
10. 372918 × 9999600 . =3729030832800.
11. 203800 × 9997000 . = 2037388600000.
12. 811876 × 99988 . . . =81177857488.
13. 3606253 × 9999990 . =36062493937470.
14. 2055416 × 992 . . . . =2038972672.

CASE V.

ART. **57**. Another mode of contraction, of frequent use, is comprised in the following

RULE.—*Derive the partial products, when possible, from each other, commencing with that figure of the multiplier which is most convenient for the purpose; making use of two or more figures of the multiplier at once, when it can be done, and setting the right-hand figure of each partial product under the right-hand figure of the multiplier in use at the time.*

REVIEW.—56. What is the rule for multiplying by a number that wants but little of 100, 1000, &c.? Explain the example. If there are ciphers at the right of one or both factors, what may be done? 57. What other method is there of contracting multiplication?

Multiply 387295 by 216324.

OPERATION.
387295
216324
1161885
9295080
83655720
83781203580

SOLUTION.—Commence with the 3 of the multiplier, and obtain the first partial product, 1161885; then multiply this product by 8, which gives the product of the multiplicand by 24 at once, (since 8 times 3 times any number make 24 times it.) Set the right-hand figure under the right-hand figure 4 of the multiplier in use. Multiply the second partial product by 9, which gives the product of the multiplicand by 216, (since 9 times 24 times a number make 216 times that number.) Set the right-hand figure of this partial product under the 6 of the multiplicand; and, finally, add to obtain the total product.

EXAMPLES FOR PRACTICE.

1. 38057 × 48618 . . . . . . =1850255226.
2. 267388 × 14982 . . . . . . =4006007016.
3. 481063 × 63721 . . . . =30653815423.
4. 66917 × 849612 . . . =56853486204.
5. 102735 × 273162 . . . =28063298070.
6. 536712 × 729981 . . . =391789562472.
7. 750764 × 315135 . . . =236592013140.
8. 930079 × 251255 . . . =233686999145.

VI. DIVISION.

ART. 58. DIVISION is the process of finding how many times one number is contained in another.

Also, Division is the process of finding one of the factors of a given product, when the other is known.

The number contained in the other is the *Divisor*, the other number is the *Dividend*, and the result obtained is the *Quotient*.

The *Remainder* is the number which is sometimes left after dividing.

Thus, in the question, How often is 2 contained in 7? the divisor is 2, the dividend 7, the quotient 3, and the remainder 1.

NOTE.—Dividend signifies *to be divided.* Quotient is derived from the Latin word *quoties*, which signifies *how often.*

ART. 59. The divisor and quotient in Division, correspond to the factors in Multiplication, and the dividend corresponds to the product. Thus:

	FACTORS.	PRODUCT.	
MULTIPLICATION. .	3×5	$= 15.$	
	DIVIDEND.	DIVISOR.	QUOTIENT.
DIVISION. . . .	15 divided	by 3	= 5
	or, 15 divided	by 5	= 3.

There are three methods of expressing division; thus,

$12 \div 3, \qquad \frac{12}{3} \qquad \text{or} \qquad 3)12.$

Each indicates that 12 is to be divided by 3.

ART. 60 THEOREM.—*Division is a short method of making several subtractions of the same number.*

DEMONSTRATION.—To prove this, find how many times 2 cents is contained in 7 cents.

Two cents from 7 cents leaves 5 cents; 2 cents from 5 cents leaves 3 cents; 2 cents from 3 cents leaves 1 cent.

Here, 2 cents is taken 3 times from (*out of*) 7 cents, and 1 cent remains; hence, 2 cents is *contained in* 7 cents 3 times, with a remainder of 1 cent.

7 cents.
2 cents contained once.
5 cents.
2 cents contained 2 times.
3 cents.
2 cents contained 3 times.
1 cent left.

COROLLARIES.

COR. 1. *The Dividend and Divisor must always be of the same denomination.*

COR. 2. *The quotient is always an abstract number; merely showing how many times the divisor is contained in the dividend.*

COR. 3. *The remainder is always of the same name as the dividend.*

Multiplication is a short method of making several *additions* of the same number; Division is a short method of making several *subtractions* of the same number: hence, Division is the reverse of Multiplication.

REVIEW.—57. Explain the example. 58. What is Division? What is the Divisor? Dividend? Quotient? Remainder? What does dividend mean? What does quotient mean? 59. If the dividend is the product, what are the factors? If the divisor and quotient are factors, what is the product? What are the signs of Division?

The next step is to learn the Division Table, which gives the quotient in all cases where the divisor and quotient are both 12 or less, and there is no remainder. It is made from the Multiplication Table. Thus, 4 in 12 is contained 3 times, because 3 times 4 are 12.

ART. **61.** When the divisor does not exceed 12, the operation is called *Short Division.*

RULE FOR SHORT DIVISION.

1. *Write the divisor on the left of the dividend with a curved line between them. Begin at the left, divide successively each figure of the dividend by the divisor, and set the quotient beneath the figure divided.*
2. *Whenever a remainder occurs, prefix it to the figure in the next lower order, and divide as before.*
3. *If the figure in any order does not contain the divisor, place a cipher beneath it, prefix it to the figure in the next lower order, and divide as before.*
4. *If there is a remainder after dividing the last figure, place the divisor under it and annex it to the quotient.*

How often is 2 cents contained in 652 cents?

$$\begin{array}{r} 2\overline{)652} \\ \hline 326 \end{array}$$

SOLUTION.—Two in 6 (hundreds) is contained 3 (hundreds) times; 2 in 5 (tens) is contained 2 (tens) times, with a remainder of 1 (ten); lastly, 1 (ten) prefixed to 2 makes 12, and 2 in 12 (units) is contained 6 times, making the entire quotient 326.

DEMONSTRATION.—Commence at the left to divide, so that if there is a remainder it may be carried to the next lower order.

By the operation of the rule, the dividend is separated into parts corresponding to the different orders. Having found the number of times the divisor is contained in each of these parts, the *sum* of these must give the number of times the divisor is contained in the *whole* dividend. Analyze the preceding dividend thus:

$$652 = 600 + 40 + 12$$

	2 in	600	is contained	300	times.
	2 in	40	is contained	20	..
	2 in	12	is contained	6	..
Hence,	2 in	652	is contained	326	times.

REVIEW.—60. Division is a short method of performing what operation? Explain the example. What is division the reverse of? why?

METHODS OF PROOF.

1. Multiply the quotient by the divisor, and add in the remainder, if any; the result should equal the dividend; or,

2. Subtract the remainder, if any, from the dividend; the result divided by the quotient, should give the first divisor.

Art. **62.** It has been shown, (Art. 60), that the divisor and dividend must be of the same denomination, and that the quotient is always an abstract number. It is now to be shown how, according to these principles, the solution of questions in Division can be explained.

All questions in Division belong to two classes:

I. *To divide a number into parts each containing a certain number of units, and to find the number of parts.*

II. *To divide a number into a certain number of equal parts, and to find the number of units in each.*

EXAMPLES OF THE FIRST CLASS.

1. At $3 each, how many hats can I buy for $15?

Solution.—Since 1 hat costs $3, I can buy a hat as often as $3 is contained in $15; $3 in $15 is contained 5 times; therefore, I can buy 5 hats.

2. At $5 a yard, how much cloth can I buy for $35?

3. In a school of 60 pupils, how many classes of 10 pupils each?

4. I can walk 6 miles in 1 hour: in how many hours can I walk 54 miles?

EXAMPLES OF THE SECOND CLASS.

5. A man has 20 apples to divide equally among 4 boys: how many must he give to each?

Solution.—To give each boy 1 apple will require 4 apples; hence, each boy will receive one apple as often as 4 apples are contained in 20 apples; 4 apples are contained in 20 apples 5 times; therefore, each boy will receive 5 apples.

Review.—60. Of what denomination must the divisor be? the remainder? What kind of number is the quotient? why? 61. What is Short Division? Give the rule. Explain the example. What is the proof? 2d method? How is the remainder written? 62. What two classes of questions are performed by division?

REMARK.—The solution of this question evidently requires that the 20 apples should be divided into 4 equal parts, and since the answer is obtained by dividing the 20 apples by 4, it follows that to divide any number into 4 equal parts, that is, *to get one fourth* ($\frac{1}{4}$) *of a number, divide it by* 4.

In like manner, to divide any number into 3 equal parts, that is, *to get one third of a number, divide it by* 3; to divide any number into 2 equal parts, that is, *to get one half of a number, divide it by* 2, &c.

6. If 7 yards of cloth cost $35, what is 1 yard worth?

7. A school of 60 pupils is to be divided into 6 classes: how many will there be in each class?

8. I walked 54 miles in 9 hours: how many miles per hour did I go?

EXAMPLES FOR PRACTICE.

9. Divide 75191 by 3 . . . . *Ans.* $25063\frac{2}{3}$
10. Divide 171654 by 4 . . . . *Ans.* $42913\frac{2}{4}$
11. Divide 512653 by 5 . . . . *Ans.* $102530\frac{3}{5}$
12. Divide 534959 by 7 . . . . *Ans.* $76422\frac{5}{7}$
13. Divide 986028 by 8 . . . . *Ans.* $123253\frac{4}{8}$
14. Divide 986974 by 11 . . . . *Ans.* $89724\frac{10}{11}$

In the following examples, find the continued quotient of the given number, divided by 2, 4, 6, 8, 10, and 12; that is, divide the given number by 2, then the resulting quotient by 4, and so on.

EXAMPLES.	ANS.	EXAMPLES.	ANS.
15. 138240	3.	19. 783360	17.
16. 230400	5.	20. 1336320	29.
17. 322560	7.	21. 1658880	36.
18. 506880	11.	22. 2488320	54.

23. At $6 a head, how many sheep can be bought for $222? *Ans.* 37.

24. At $5 a barrel, how many barrels of flour can be bought for $895? *Ans.* 179.

25. If 8 pints make a gallon, how many gallons are there in 2520 pints? *Ans.* 315.

REVIEW.—62. Give an example of the first class. Explain it. Give an example of the second class. Explain it. How is the half, third, &c., of any number obtained?

26. At the rate of 9 gallons an hour, how long will it take to fill a cistern of 2115 gallons? *Ans.* 235 hours.

27. What number multiplied by 11 gives 638?
Ans. 58.

28. If three acres of land cost $741, what is the cost of 1 acre? *Ans.* $247.

29. If $1715 be divided into 7 equal parts, how many dollars will there be in each part? *Ans.* $245.

30. How many full weeks in 13 days? How many in 670 days? *Ans.* 1 in 13, and 95 in 670 days.

31. How many times can 9 be subtracted from 155?
Ans. 17 and 2 left.

ART. 63. *Short Division* is performed mentally, and only the result is written; but when all the work is written, it is called *Long Division.*

Short Division is generally used when the divisor does not exceed 12; Long Division, when the divisor exceeds 12. To show the difference between them, find how often 5 is contained in 820.

SHORT DIVISION.

```
5)820
------
  164 Quotient.
```

Both operations are performed on the same principle. In the first, the subtraction is performed mentally; in the second, it is written.

LONG DIVISION.

```
5)820(164 Quotient.
  5
  --
  32 tens.
  30
  --
   20 units.
   20
   --
```

ART. 64. RULE FOR LONG DIVISION.

1. *Draw curved lines on the right and left of the dividend, placing the divisor on the left.*

2. *Find how often the divisor is contained in the left-hand figure, or figures, of the dividend, and write the number in the quotient at the right of the dividend.*

3. *Multiply the divisor by this quotient figure, and write the product under that part of the dividend from which it was obtained.*

REVIEW.—63. Explain the difference between Short Division and Long Division. When is the latter used?

4. *Subtract this product from the figures above it; to the remainder bring down the next figure of the dividend, and divide as before, until all the figures of the dividend are brought down.*

5. *If at any time after a figure is brought down, the number thus formed is too small to contain the divisor, a cipher must be placed in the quotient, and another figure brought down, after which divide as before.*

PROOF.—The same as in Short Division.

Divide $4225 equally among 13 men.

```
     thous. hund. tens. units.   hund. tens. units.
13)  4      2     2     5      ( 3     2     5
     3      9  hun.
     ----
        3   2
        2   6  tens.
        ------
            6   5
            6   5  units.
            -----
```

SOLUTION.—As 13 is not contained in 4 (thousands), therefore, the quotient has no thousands. Next, take 42 (hundreds) as a partial dividend; 13 is contained in it 3 (hundreds) times; after multiplying and subtracting, there are 3 hundreds left. Then bring down 2 tens, and 32 tens is the next partial dividend. In this 13 is contained 2 (tens) times, with a remainder 6 tens. Lastly, bringing down the 5 units, 13 is contained in 65 (units) exactly 5 (units) times. The entire quotient is 3 hundreds 2 tens and 5 units, or 325.

DEMONSTRATION.—This operation involves no principle not explained in Short Division, (Art. 61). This may be further shown by decomposing the dividend into parts, each exactly divisible by 13, as follows:

```
DIVISOR.     DIVIDEND.            QUOTIENT.
  13) 3900 + 260 + 65 (300 + 20 + 5
      3900
      ----
           + 260
           + 260
           -----
                 + 65
                 + 65
                 ----
```

NOTES.—1. The product must never be greater than the partial dividend from which it is to be subtracted; if so, the quotient figure is too large, and must be diminished.

2. The remainder after each subtraction must be *less* than the divisor; if not, the last quotient figure is too *small*, and must be increased.

3. The order of each quotient figure is the same as the lowest order in the partial dividend from which it was derived.

ART. **65**. In the *Italian* method of Long Division, the multiplication and subtraction are both performed in the mind at the same time. It is illustrated thus:

Divide 20877514 by 3256.

```
OPERATION.                 Quot.
3256)20877514(6412
       13415
        3911
         6554
           42 Rem.
```

SOLUTION.—The first figure of the quotient being 6, multiply the divisor by it, and subtract the product from the dividend, saying 6 times 6 are 36, and as 36 can not be taken from 7, add 3 tens or 30 to the figure of the dividend, making 37, and then 36 from 37 leaves 1; to compensate for the 3 tens added to the upper figure, carry 3 to the next product when it is formed, saying 6 times 5 are 30 and 3 are 33, and 33 from 37 leaves 4, there being 3 to carry: then, 6 times 2 are 12, and 3 are 15, and 15 from 18 leaves 3, there being 1 to carry; 6 times 3 are 18, and 1 are 19, and 19 from 20 leaves 1, with 2 to carry, and 2 from 2 leaves 0. So proceed until the figures of the dividend are all used.

EXAMPLES FOR PRACTICE.

1. Divide 139180 by 453.

```
453)139180(307              PROOF.
    1359                     307 Quotient.
    ----                     453 Divisor.
     3280                   ----
     3171                     921
     ----                   1535
      109                  1228
                           ------
                           139071
                              109 Rem.
                           ------
                           139180
```

2. 1004835 ÷ 33 . . . . . $= 30449\frac{18}{33}$
3. 5484888 ÷ 67 . . . . . . $= 81864.$
4. 4326422 ÷ 961 . . . . . $= 4502.$
5. 1457924651 ÷ 1204. . . $= 1210900\frac{1051}{1204}$
6. 65358547823 ÷ 2789. . $= 23434402\frac{645}{2789}$
7. 33333333333 ÷ 5299. . . $= 6290495\frac{328}{5299}$
8. 245379633477 ÷ 1263. . $= 194283161\frac{1134}{1263}$
9. 555555555555 ÷ 123456. $= 4500028\frac{98787}{123456}$
10. 555555555555 ÷ 654321. $= 849056\frac{384579}{654321}$

In the following, multiply A by itself, also B by itself: divide the difference of the products by the sum of A and B.

	A.	B.		
11.	918	1902	. . . . *Ans.*	984.
12.	8473	9437	. . . . *Ans.*	964.
13.	2856	3765	. . . . *Ans.*	909.
14.	33698	42856	. . . . *Ans.*	9158.
15.	47932	152604	. . . . . . *Ans.*	104672.

In the following, multiply A by itself, also B by itself: divide the difference of the products by the difference of A and B.

	A.	B.		
16.	4986	5369	. . . . *Ans.*	10355.
17.	3973	4308	. . . . *Ans.*	8281.
18.	23798	59635	. . . . *Ans.*	83433.
19.	47329	65931	. . . . *Ans.*	113260.

20. If 25 acres produce 1825 bushels of wheat, how much is that per acre? *Ans.* 73 bushels.

21. How many times 1024 in 1048576? *Ans.* 1024.

22. How many sacks, each containing 55 pounds, can be filled by 2035 pounds of flour? *Ans.* 37.

23. How many pages in a book of 7359 lines, each page containing 37 lines? *Ans.* $198\frac{33}{37}$

24. In what time will a vat of 10878 gallons be filled, at the rate of 37 gallons an hour? *Ans.* 294 hours.

25. In what time will a vat of 3354 gallons be emptied, at the rate of 43 gallons an hour? *Ans.* 78 hours.

26. The product of two numbers is 212492745; one is 1035; what is the other? *Ans.* 205307.

27. What number multiplied by 109, and 98 added to the product, will give 106700? *Ans.* 978.

CONTRACTIONS IN DIVISION.

CASE I.—WHEN THE DIVISOR IS A COMPOSITE NUMBER.

ART. **66**. RULE.—1. *Divide the dividend by one of the factors of the divisor, and this quotient by another factor, and so on, until all the factors have been used. The last quotient will be the one required.*

REVIEW.—64. Give the rule; the proof. Explain the example. When is the quotient figure too large? When too small? How is the order of each quotient figure known?

2. *To find the true remainder: Multiply each remainder by all the preceding divisors, except that which produced it, and to the sum of the products, add the remainder from the first division*

Divide 217 by 15.

$$\begin{array}{l} 3)\underline{217} \\ 5)\underline{72} \text{ and 1 rem.} \\ 14 \text{ and 2 rem.} \end{array}$$

SOLUTION.—Since $15=3\times 5$, divide by 3 and then by 5. The true remainder is $2\times 3+1=7$. Hence, the quotient is 14, remainder 7.

DEMONSTRATION.—Dividing by 3 and then by 5, is the same as dividing by 15; for, in the former case, the quotient must be multiplied by 5 and then by 3, and, in the latter case, by 15, to produce the given dividend; and, since multiplying by 5 and then by 3, is the same as multiplying by 15, (Art. 53), the quotients on which these operations are performed must be alike, or they could not both produce the same dividend.

To prove the rule for finding the remainder; dividing 217 by 3, the quotient is 72 *threes*, and 1 *unit* remainder. Dividing by 5, the quotient is 14 (fifteens), and a remainder of 2 *threes*. Hence, the whole remainder is 2 *threes* + 1, or 7.

EXAMPLES FOR PRACTICE.

1. $1036\div 28$. $=37$.
2. $3640\div 35$. $=104$.
3. $2332\div 54$. $=43\frac{10}{54}$
4. $3800\div 56$. $=67\frac{48}{56}$
5. $34855\div 168$ . . $=207\frac{79}{168}$
6. $620314\div 231$ . . $=2685\frac{79}{231}$

CASE II.—WHEN THE DIVISOR IS ONE WITH CIPHERS ANNEXED, AS 10, 100, 1000, &c.

ART. **67**. RULE.—*Cut off as many figures from the right of the dividend as there are ciphers in the divisor; the figures cut off will be the remainder, the other figures will be the quotient.*

Divide 23543 by 100.

$$\begin{array}{l} 1|00)\underline{235|43} \\ 235 \text{ Quotient, 43 Remainder.} \end{array}$$

ANALYSIS.—To divide 23543 by 100, is to find how many *hundreds* it contains. This is done by mere inspection, the part on the right of the line, 43 (units), being the remainder, since it is less than 100: while that on the left, 235, is the quotient, being the *number of hundreds* in the given number.

REVIEW.—65. What is the Italian method of long division? Explain the example. 66. What is the rule for dividing by a composite number?

1. $4567 \div 100$ . . . . $= 45\frac{67}{100}$
2. $8325 \div 1000$ . . . . $= 8\frac{325}{1000}$
3. $95043 \div 100$ . . . . $= 950\frac{43}{100}$
4. $730015 \div 10000$ . . . . $= 73\frac{15}{10000}$

CASE III.—WHEN CIPHERS ARE ON THE RIGHT OF THE DIVISOR.

ART. 68. RULE.—1. *Cut off the ciphers at the right of the divisor, and as many figures from the right of the dividend.*

2. *Divide the remaining part of the dividend by the remaining part of the divisor.*

3. *Annex the figures cut off to the remainder, and it will give the true remainder.*

Divide 3846 by 400.

4|00)38|46

9 Quotient, $200 + 46 = 246$ Rem.

DEMONSTRATION.—To divide by 400 is the same as to divide by 100 and then by 4, (Art. 66). Dividing by 100 gives 38, and 46 remainder, (Art. 67); then dividing by 4 gives 9, and 2 remainder· the true remainder is $2 \times 100 + 46 = 246$, (Art. 66.)

EXAMPLES FOR PRACTICE.

1. $34500 \div 130$ . . . . $= 265\frac{50}{130}$
2. $82500 \div 1100$ . . . . $= 75.$
3. $60000 \div 1700$ . . . . $= 35\frac{500}{1700}$
4. $1896000 \div 24000$ . . . . $= 79.$

CASE IV.—WHEN THE DIVISOR WANTS BUT LITTLE OF BEING 100, 1000, 10000, &c.

ART. 69. RULE.—*Cut off from the right of the dividend, by a vertical line, as many figures as the divisor contains; multiply the part on the left of the line by what the divisor wants of being* 100, 1000, &c., *and set the product under the dividend, commencing at units' place. Multiply the part of this product on the left of the*

REVIEW.—66. Explain the rule. How is the true remainder found? 67. What is the rule for dividing by 1 with ciphers annexed? Explain the example. 68. What is the rule for dividing by a number that has ciphers on its right? Explain the example.

line by the same multiplier, and set down as before. Continue so until no figure falls to the left of the line; then add the several results, and for every 1 *carried across the line, add to the sum already obtained on the right of the line, the number used as a multiplier. This will be the true remainder, and the part on the left, the quotient.*

NOTE.—If the remainder is larger than the divisor, carry one to the quotient, and the excess will be the true remainder.

Divide 3289662 by 95.

OPERATION.

```
95)32896|62
    1644|80
      82|20
       4|10
        |20
   -----|--
   34627|92
       1| 5
   -----|--
   34628|97
   Quot. 95
         --
   Rem.   2
```

ANALYSIS.—The reason of the rule can be seen in the explanation of this example. The divisor is assumed to be 100, and the quotient, on that supposition, is 32896, with a remainder 62. If this quotient were multiplied by the *true* divisor 95, which is 5 less than the one assumed, the product, after the remainder 62 was added in, would be less than the dividend by 5 times the quotient 32896. Therefore, 32896 and 62 are the true quotient and remainder for *only a part* of the dividend; the surplus remaining to be divided is 5 times 32896 or 164480. This surplus is divided like the first dividend, the quotient and remainder going to increase those already obtained. There is a surplus also in this operation, which is treated like the first, and so on, until the quotient is nothing, when the successive divisions must cease.

The sum of the remainders is 192; from which, after setting down 92, we carry 1 to the first column of the quotients. This 1 is 100, and as 1 must be carried to the quotient for every 95 in the remainder, 5 out of this 100 should continue in the remainder, and be added to the 92 already set down, making 97; but 97, being large enough to contain the divisor 95, must furnish another unit to the quotient, making it 34628, and the excess 2 is the true remainder.

EXAMPLES FOR PRACTICE.

1. $75076822 \div 99$ . . . $= 758351\frac{73}{99}$
2. $24206778 \div 989$ . . . $= 24476\frac{14}{989}$
3. $680571293 \div 9996$ . . . $= 68084\frac{3629}{9996}$
4. $8662309749 \div 9994$ . . . $= 866751\frac{255}{9994}$
5. $52351346504 \div 99930$ . . $= 523880\frac{18104}{99930}$
6. $42104678535 \div 99800$ . . $= 421890\frac{56535}{99800}$

GENERAL PRINCIPLES OF MULTIPLICATION AND DIVISION.

ART. 70. THEOREM I.—*Multiplying either factor of a product multiplies the product by the same number.*

DEMONSTRATION.—9 multiplied by 5 gives a product 45, and 9 multiplied by 10 (2 times 5) gives a product 90, which is 2 times 45, and so in all cases; for, when a multiplier is made 2, 3, 4, &c. times as large as at first, the multiplicand must be taken 2, 3, 4, &c. times as often as before, and, therefore the product will be 2, 3, 4, &c. times as large as the first product. And since this is true of the multiplier, it is also true of the multiplicand, because it can be used *as* the multiplier. (Art. 47.)

ART. 71. THEOREM II.—*Dividing either factor of a product divides the product by the same number.*

DEMONSTRATION.—9 multiplied by 10 gives a product 90, and 9 multiplied by 5 ($\frac{1}{2}$ of 10) gives a product 45, which is $\frac{1}{2}$ of 90, and so in all cases; for when a multiplier is made $\frac{1}{2}$, $\frac{1}{3}$, $\frac{1}{4}$, &c., as large as at first, the multiplicand must be taken $\frac{1}{2}$, $\frac{1}{3}$, $\frac{1}{4}$, &c., as often as before, and, therefore, the product will be $\frac{1}{2}$, $\frac{1}{3}$, $\frac{1}{4}$, &c., as large as the first product. And since this is true of the multiplier, it is also true of the multiplicand, because it can be used *as* the multiplier. (Art. 47.)

ART. 72. THEOREM III.—*Multiplying one factor of a product, and dividing the other factor by the same number, does not alter the product.*

DEMONSTRATION.—Multiplying either factor by 2, 3, 4, &c., multiplies the product by 2, 3, 4, &c., while dividing the other factor by 2, 3, 4, &c., divides the product by 2, 3, 4, &c. When these operations are performed together, the product is both multiplied and divided by the same number, and must, therefore, remain unchanged.

ART. 73. THEOREM IV.—*Multiplying the dividend, or dividing the divisor, by any number, multiplies the quotient by that number.*

DEMONSTRATION.—If any divisor is contained in a given dividend a certain number of times, it must be contained in twice that dividend, twice as often; in 3 times that dividend, 3 times as often;

REVIEW.—69. What is the rule for dividing by a number that wants but little of being 100, 1000, &c.? If the remainder is larger than the divisor, what must be done? Explain example.

and so on. Thus, 3 is contained in 12 *four* times, and in twice 12 it is contained twice 4, = 8 times. Again, in any given dividend, *half* the divisor will be contained *twice* as often as the whole divisor; *one third* of the divisor, 3 times as often, and so on; thus, 4 is contained in 12 *three* times, and one half of 4, which is 2, is contained in 12, twice 3, = 6 times.

ART. **74.** THEOREM V.—*Dividing the dividend, or multiplying the divisor by any number, divides the quotient by that number.*

DEMONSTRATION.—If any divisor is contained in a given dividend a certain number of times, it must be contained in half that dividend, half as many times; in one third of that dividend, one third as many times, and so on. Thus, 3 is contained in 12 *four* times, and in one half of 12, which is six, it is contained one half of 4 = 2 times.

Again, in any given dividend, *twice* the divisor will be contained *half* as often as the divisor itself; *three times* the divisor, *one third* as often, and so on; thus, 3 is contained in 24 *eight* times, and 2 times 3, which is 6, is contained in 24 one half of 8 = 4 times.

ART. **75.** THEOREM VI.—*Multiplying or dividing both dividend and divisor by the same number, does not change the quotient.*

DEMONSTRATION.—This follows from the two preceding theorems; for the effects produced by multiplying or dividing both dividend and divisor exactly balance each other; thus, 6 is contained in 24 4 times; and twice 6 is contained in twice 24, just 4 times; also, one half of 6 is contained in one half of 24, just 4 times; and so of any other dividend and divisor.

CONTRACTIONS IN MULTIPLICATION AND DIVISION.

ART. **76.** To multiply by any simple part of 100, 1000, &c.,

RULE.—*Multiply by* 100, 1000, &c., *and take such a part of the result as the multiplier is of* 100, 1000, &c.

NOTE.—To get *two* thirds ($\frac{2}{3}$) of a number, divide it by 3, which gives one third ($\frac{1}{3}$) of it, (Art. 62), and multiply the quotient by 2; or, if it is more convenient, multiply the number by 2 first, and divide the product by 3. In like manner, to get three fourths ($\frac{3}{4}$) of a number, multiply it by 3, and divide the product by 4, and so on.

PARTS OF 100.	PARTS OF 1000.
$12\frac{1}{2} = \frac{1}{8}$ of 100.	$125 = \frac{1}{8}$ of 1000.
$16\frac{2}{3} = \frac{1}{6}$ of 100.	$166\frac{2}{3} = \frac{1}{6}$ of 1000.
$25 = \frac{1}{4}$ of 100.	$250 = \frac{1}{4}$ of 1000.
$33\frac{1}{3} = \frac{1}{3}$ of 100.	$333\frac{1}{3} = \frac{1}{3}$ of 1000.
$37\frac{1}{2} = \frac{3}{8}$ of 100.	$375 = \frac{3}{8}$ of 1000.
$62\frac{1}{2} = \frac{5}{8}$ of 100.	$625 = \frac{5}{8}$ of 1000.
$66\frac{2}{3} = \frac{2}{3}$ of 100.	$666\frac{2}{3} = \frac{2}{3}$ of 1000.
$75 = \frac{3}{4}$ of 100.	$750 = \frac{3}{4}$ of 1000.
$83\frac{1}{3} = \frac{5}{6}$ of 100.	$833\frac{1}{3} = \frac{5}{6}$ of 1000.
$87\frac{1}{2} = \frac{7}{8}$ of 100.	$875 = \frac{7}{8}$ of 1000.

Multiply 246 by $87\frac{1}{2}$.

SOLUTION.—Since $87\frac{1}{2}$ is $\frac{7}{8}$ of 100, annex two ciphers to the multiplicand, which multiplies it by 100, (Art. 54), and then take $\frac{7}{8}$ of the result. (See Note).

$$\begin{array}{r} 24600 \\ 7 \\ \hline 8)172200 \\ \hline \textit{Ans. } 21525 \end{array}$$

EXAMPLES FOR PRACTICE.

1. $422 \times 33\frac{1}{3}$ . . . $= 14066\frac{2}{3}$
2. 3708×25 . . . $= 92700.$
3. $6564 \times 62\frac{1}{2}$ . . . $= 410250.$
4. $10724 \times 16\frac{2}{3}$ . . . $= 178733\frac{2}{6}$
5. 8740×75 . . . $= 655500.$
6. 53840×125 . . . $= 6730000.$
7. $4456 \times 666\frac{2}{3}$ . . . $= 2970666\frac{2}{3}$
8. $7293 \times 833\frac{1}{3}$ . . . $= 6077500.$
9. 852×875 . . . $= 745500.$
10. 64082×375 . . . $= 24030750.$

ART. **77**. To multiply by any number whose digits are all alike,

RULE.—*Multiply as if the digits were* **9**'s, (Art. 56), *and take such a part of the product as the digit is of* **9**.

Multiply 592643 by 66666.

SOLUTION.—Multiply 592643 by 99999, (Art. 56), the product is 59263707357; take $\frac{2}{3}$ of this product, since 6 is $\frac{2}{3}$ of 9; the result is 39509138238.

REVIEW.—70–75. State and prove the theorems.

5

1. 451402 × 3333 . . . . . = 1504522866.
2. 281257 × 555555 . . . = 156253732635.
3. 630224 × 4444000 . . . = 2800715456000.

ART. 78. To divide by a number ending in any simple part of 100, 1000, &c.,

RULE.—*Multiply both dividend and divisor by such a number* (3, 4, 6, *or* 8,) *as will convert the final figures of the divisor into ciphers, and then divide the former product by the latter.*

NOTE.—1. If there be a remainder, it should be divided by the multiplier, to get the true remainder.

2. The multiplier is 3, 4, 6, or 8, according as the final portion of the divisor is *thirds*, *fourths*, *sixths*, or *eighths* of 100, 1000.

3. Several successive multiplications may sometimes be made before dividing.

Divide 6903141128 by 21875. *Ans.* $315572\frac{3628}{21875}$

SOLUTION.—Multiply both by 8 and 4 successively. The divisor becomes 700000, and the dividend 220900516096, while the quotient remains the same, (Art. 75). Performing the division as in Art. 68, the quotient is 315572, and remainder 116096. The remainder being a part of the dividend, has been made too large by the multiplication by 8 and 4, and is, therefore, reduced to its true dimensions by dividing by 8 and 4. This gives 3628 for the true remainder.

EXAMPLES FOR PRACTICE.

1. 300521761 ÷ 225 . . . = $1335652\frac{61}{225}$
2. 1510337264 ÷ 43750 . . = $34521\frac{43514}{43750}$
3. 22500712361 ÷ 1406250 . = $16000\frac{712361}{1406250}$
4. 620712480 ÷ $20833\frac{1}{3}$ = 29794, Rem. $4146\frac{2}{3}$
5. 742851692 ÷ $2916\frac{2}{3}$ = 254692 .. $25\frac{1}{3}$

ART. 79. To divide by any number whose digits are all alike,

RULE.—*Treat the dividend and divisor alike by multiplication, and division, if necessary, until the digits of the divisor are 9's; then divide by Rule in* Art. 69.

REVIEW.—76. What is the rule for multiplying by any simple part of 100, 1000, &c.? 77. For multiplying by any number whose digits are all alike? 78. For dividing by any number ending in any simple part of 100, 1000, &c?

NOTE.—1. If the divisor ends in ciphers, divide first as in Art. 68.
2. If a remainder occurs, reverse upon it the operations performed on the dividend and divisor, to get the *true* remainder.

Divide 420565342 by 666.

SOLUTION.—Divide both numbers by 2, and multiply the results by 3; this makes the divisor 999, and does not alter the quotient, (Art. 75), which is found, (by Art. 69), to be 631479, with a remainder 492; the latter being a part of the dividend, differs from the true remainder, by reason of the division by 2, and the multiplication by 3, and is brought to its true dimensions by *multiplying* by 2 and *dividing* by 3.

EXAMPLES FOR PRACTICE.

1. $13641096 \div 2222 \ldots = 6139\frac{238}{2222}$
2. $376802902 \div 7770 \ldots = 48494\frac{4522}{7770}$
3. $8811857528 \div 3333 \ldots = 2643821\frac{2135}{3333}$
4. $5606258492 \div 88888 \ldots = 63071\frac{3444}{88888}$

SUMMARY OF GENERAL PRINCIPLES.

ART. **80.** Notation and Numeration, Addition, Subtraction, Multiplication, and Division, are called the fundamental rules of Arithmetic, because all the various operations of Arithmetic are performed by them.

ART. **81.** NOTATION is the method of expressing numbers in *figures*. NUMERATION is the method of expressing numbers in *words*.

ART. **82.** ADDITION is the process of finding the *aggregate* or *sum* of two or more numbers, (Art. 40).

ART. **83.** SUBTRACTION is the process of finding the *difference* between two numbers, (Art. 43).

ART. **84.** MULTIPLICATION is the process of finding what is *produced* by taking one number as *many times* as there are units in another, (Art. 46).

REVIEW.—79. What is the rule for dividing by a number whose digits are all alike? How is the true remainder obtained? 80. Which are the fundamental rules of Arithmetic? Why are they so called? 81. What is Notation? Numeration? 82. Addition? 83. Subtraction?

ART. **85**. DIVISION is the process of finding how *many times* one number is contained in another, (Art. 58).

GENERAL PROBLEMS.

1. When the separate cost of several things is given, how is the entire cost found?

2. When the sum of two numbers and one of them are given, how is the other found?

3. When the less of two numbers and the difference between them are given, how is the greater found?

4. When the greater of two numbers and the difference between them are given, how is the less found?

5. When the cost of one article is given, how do you find the cost of any number at the same price?

6. When the divisor and quotient are given, how do you find the dividend?

7. How do you divide a number into parts each containing a certain number of units?

8. How do you divide a number into a given number of equal parts?

9. If the product of two numbers and one of them is given, how do you find the other?

10. If the dividend and quotient are given, how do you find the divisor?

11. If you have the product of three numbers, and two of them are given, how do find the third?

12. If the divisor, quotient, and remainder are given, how do you find the dividend?

13. If the dividend, quotient, and remainder are given, how do you find the divisor?

ART. **86**. PARTICULAR EXAMPLES.

1. I bought 3 horses for $165; two cows for $37; and 7 sheep for $29: what did all cost? *Ans.* $231.

2. The sum of two numbers is 664, and one of them is 369: what is the other? *Ans.* 295.

3. The difference between two numbers is 168, and the less number is 289: what is the greater? *Ans.* 457.

REVIEW.—84. What is Multiplication? 85. Division?

4. The greater of two numbers is 753, and their difference 457: what is the less? *Ans.* 296.

5. At $7 a yard, what will 5 yards of cloth cost?

SOLUTION.—5 yards are 5 *times* 1 yard. If 1 yard cost $7, 5 yards will cost 5 *times* $7, which are $35, *Ans.*

6. The divisor is 753, and the quotient 245: what is the dividend? *Ans.* 184485.

7. How many classes of 9 pupils can be formed in a school containing 54 pupils?

SOLUTION.—Since it takes 9 pupils to form *one* class, there will be as many classes as 9 pupils are contained in 54 pupils; 9 in 54, 6 times. *Ans.* 6.

8. If you divide 28 apples equally among 4 boys, what will be the share of each?

SOLUTION.—To give each boy 1 apple, will require 4 apples; hence, each boy will receive 1 apple as often as 4 apples are contained in 28 apples; 4 in 28, 7 times. *Ans.* 7 apples.

9. The product of two numbers is 612451, and one of them is 203: what is the other? *Ans.* 3017.

10. The dividend is 395631145 and the quotient 4007: what is the divisor? *Ans.* 98735.

11. The product of three numbers is 195318005; two are 307 and 703: what is the third? *Ans.* 905.

12. What number, divided by 473, will give the quotient 8061 and remainder 365? *Ans.* 3813218.

13. The dividend is 7781174, the quotient 8216, the remainder 622: what is the divisor? *Ans.* 947.

ART. **87.** MISCELLANEOUS EXERCISES.

1. A grocer gave 153 barrels of flour, worth $6 a barrel, for 54 barrels of sugar: what did the sugar cost per barrel? *Ans.* $17.

2. When the divisor is 35, quotient 217, and remainder 25, what is the dividend? *Ans.* 7620.

3. What number besides 41 will divide 4879 without a remainder? *Ans.* 119.

4. Of what number is 103 both divisor and quotient? *Ans.* 10609.

5. What is the nearest number to 53815, that can be divided by 375 without a remainder? *Ans.* 54000.

6. A farmer bought 25 acres of land for $2675: what did 19 acres of it cost? *Ans.* $2033.

7. Bought 15 horses at $75 a head: at how much per head must I sell them to gain $210? *Ans.* $89.

8. A locomotive has 391 miles to run in 11 hours: after running 139 miles in 4 hours, at what rate per hour must the remaining distance be run? *Ans.* 36 miles.

9. A merchant bought 235 yards of cloth at $5 per yard: after reserving 12 yards, what will he gain by selling the remainder at $7 per yard? *Ans.* $386.

10. A grocer bought 135 barrels of pork for $2295; he sold 83 barrels at the same rate at which he purchased, and the remainder at an advance of $2 per barrel: how much did he gain? *Ans.* $104.

11. A drover bought 5 horses at $75 each, and 12 at $68 each; he sold them all at $73 each: what did he gain? *Ans.* $50.

At what price per head must he have sold them to have gained $118? *Ans.* $77.

12. A merchant bought 3 pieces of cloth of equal lengths at $4 a yard; he gained $24 on the whole, by selling 2 pieces for $240: how many yards were there in each piece? *Ans.* 18.

13. If 18 men can do a piece of work in 15 days, in how many days will one man do it?

Solution.—It will require 1 man 18 times as long as 18 men. Eighteen times 15 days are 270 days. *Ans.*

14. If 13 men can build a wall in 15 days, in how many days can it be done if 8 men leave? *Ans.* 39.

15. If 14 men can perform a job of work in 24 days, in how many days can they perform it with the assistance of 7 more men? *Ans.* 16 days.

16. A company of 45 men have provisions for 30 days: how many men must depart, that the provisions may last the remainder 50 days? *Ans.* 18 men.

17. A horse worth $85, and 3 cows at $18 each, were exchanged for 14 sheep and $41 in money: at how much each were the sheep valued? *Ans.* $7.

18. A drover bought an equal number of sheep and hogs for $1482; he gave $7 for a sheep, and $6 for a hog: what number of each did he buy? *Ans.* 114.

Suggestion.—1 sheep and 1 hog cost $7 + $6 = $13.

19. A trader bought a lot of horses and oxen for $1260; the horses cost $50, and the oxen $17 a head; there were twice as many oxen as horses: how many were there of each? *Ans.* 15 horses and 30 oxen.

20. In a lot of silver change worth 1050 cents, one-seventh of the value is in 25 cent pieces; the rest is made up of 10 cent, 5 cent, and 3 cent pieces, of each an equal number: how many of each coin are there?
Ans. of 25 cent pieces, 6; of the others, 50 each.

21. A speculator had 140 acres of land, which he might have sold at $210 an acre, and gained $6300, but after holding, he sold at a loss of $5600: how much an acre did it cost him, and how much an acre did he sell it for?
Ans. $165, cost; and $125, sold for.

VII. PROPERTIES OF NUMBERS.

ART. 88. DEFINITIONS.

1. An integer is a whole number; as, 1, 2, 3, &c.

2. Whole numbers are divided into two classes—*prime* numbers and *composite* numbers.

3. A *prime* number is one that can be exactly divided by no other whole number but itself and unity, (1); as, 1, 2, 3, 5, 7, 11, &c.

4. A *composite* number is one that can be exactly divided by some other whole number besides itself and unity; as, 4, 6, 8, 9, 10, &c.

REMARK.—Every composite number is the product of two or more other numbers.

5. Two numbers are *prime to each other*, when unity is the only number that will exactly divide both; as, 4 and 5.

REMARK.—Two prime numbers are always prime to each other: sometimes, also, two composite numbers: as, 4 and 9.

6. An *even* number is one which can be divided by 2 without a remainder; as, 2, 4, 6, 8, &c.

REVIEW.—88. What is an integer? Into what classes are integers divided? What is a prime number? a composite number? When are numbers prime to each other? What kind of numbers must be prime to each other? What kind may be? What is an even number?

7. An *odd* number is one which can not be divided by 2 without a remainder; as, 1, 3, 5, 7, &c.

REMARK.—All even numbers except 2 are composite: the odd numbers are partly prime and partly composite.

8. A *divisor* or *measure* of a number, is a number that will divide it without a remainder: 2 is a divisor of 4; 5 of 10, &c.

9. One number is *divisible* by another when it contains that other without a remainder; 8 is divisible by 2.

10. A *multiple* of a number is the product obtained by taking it a certain number of times; 15 is a multiple of 5, being equal to 5 taken 3 times; hence,

1st. *A multiple of a number can always be divided by it without a remainder.*

2d. *Every multiple is a composite number.*

11. Since every composite number is the product of factors, (Def. 4), each factor must divide it exactly; hence, every *factor* of a number is a divisor of it.

12. A *prime factor* of a number is a prime number that will exactly divide it: 5 is a *prime* factor of 20; while 4 is a *factor* of 20, not a prime factor; hence,

1st. The prime factors of a number are all the prime numbers that will exactly divide it; 1, 2, 3, and 5, are the prime factors of 30.

2d. Every composite number is equal to the product of all its prime factors. All the prime factors of 15 are 1, 3, and 5; and $1 \times 3 \times 5 = 15$.

13. Any factor of a number is called an *aliquot* part of it; 1, 2, 3, 4, and 6, are aliquot parts of 12.

FACTORING.

ART. **89.** Factoring depends on the following PRINCIPLES and PROPOSITIONS.

PRINCIPLE 1. *A factor of a number is a factor of any multiple of that number.*

DEMONSTRATION.—Since $6 = 2 \times 3$, therefore, any multiple of $6 = 2 \times 3 \times$ some number; hence, every factor of 6 is also a factor of the multiple.

PRINCIPLE 2. *A factor of any two numbers is also a factor of their sum.*

DEMONSTRATION.—Since each of the numbers contains the factor a certain number of times, their sum must contain it as often as both the numbers; 2, which is a factor of 6 and 10, must be a factor of their sum, for 6 is 3 *twos*, and 10 is 5 *twos*, and thei sum is 3 *twos* + 5 *twos* = 8 *twos*.

ART. **90**. From these principles are derived six

PROPOSITIONS.

PROP. I.—*Every number ending with* 0, 2, 4, 6, *or* 8, *is divisible by* 2.

DEMONSTRATION.—Every number ending with a 0, is either 10 or some number of *tens;* and since 10 is divisible by 2, therefore, by Principle 1st, (Art. 89), any number of tens is divisible by 2.

Again, any number ending with 2, 4, 6, or 8, may be considered as a certain number of tens plus the figure in the units' place; and since each of the two parts of the number is divisible by 2, therefore, by Principle 2d, (Art. 89), the number itself is divisible by 2; thus, 36 = 30 + 6 = 3 tens + 6; each part is divisible by 2, hence, 36 is divisible by 2.

Conversely, *no number is divisible by* 2, *unless it ends with* 0, 2, 4, 6, *or* 8.

PROP. II.—*A number is divisible by* 4, *when the number denoted by its two right hand digits, is divisible by* 4.

DEMONSTRATION.—Since 100 is divisible by 4, any number of hundreds will be divisible by 4, (Art. 89, Principle 1st); and any number consisting of more than two places may be regarded as a certain number of hundreds plus the number expressed by the digits in tens' and units' places, (thus, 384 is equal to 3 hundreds + 84); then, if the latter part (84) is divisible by 4, both parts, or the number itself, will be divisible by 4, (Art. 89, Prin. 2d).

Conversely, *no number is divisible by* 4, *unless the number denoted by its two right-hand digits is divisible by* 4.

REVIEW.—88. What is an odd number? Are the even numbers prime or composite? Are odd numbers prime or composite? What is a divisor of a number? When is one number divisible by another? What is a multiple of a number? What two axioms concerning multiples? What is a factor of a number? A prime factor? What two axioms concerning prime factors? What is an aliquot part of a number? Give examples illustrating the definitions.

PROP. III.—*A number ending in* 0 *or* 5 *is divisible by* 5.

DEMONSTRATION.—Ten is divisible by 5, and every number of two or more figures, is a certain number of tens, plus the right hand digit; if this is 5, both parts of the number are divisible by 5, and, hence, the number itself is divisible by 5, (Art. 89, Prin. 2d).

Conversely, *no number is divisible by* 5, *unless it ends in* 0 *or* 5.

PROP. IV.—*Every number ending in* 0, 00, &c., *is divisible by* 10, 100, &c.

DEMONSTRATION.—If the number ends in 0, it is either 10 or a multiple of 10; if it ends in 00, it is either 100, or a multiple of 100, and so on; hence, by Prin. 1st, Art. 89, the proposition is true.

PROP. V.—*A composite number is divisible by the product of any two or more of its prime factors.*

DEMONSTRATION.—Since $2 \times 3 \times 5 = 30$, it follows that 2×3 taken 5 times, makes 30; hence, 30 contains 2×3 (6) exactly 5 times. In like manner, 30 contains 3×5 (15) exactly 2 times, and 2×5 (10), exactly 3 times.

Hence, *if any even number is divisible by* 3, *it is also divisible by* 6.

DEMONSTRATION.—An even number is divisible by 2; and if also by 3, it must be divisible by their product 2×3, or 6.

PROP. VI.—*Every prime number, except* 2 *and* 5, *ends with* 1, 3, 7, *or* 9.

DEMONSTRATION.—This is in consequence of Prop. I and III.

ART. **91.** To find the prime factors of a composite number,

RULE.—*Divide the given number by any prime number that will exactly divide it; divide the quotient in like manner, and so continue until the quotient is a prime number; the last quotient and the several divisors are the prime factors.*

REVIEW.—89. What is the first principle used in factoring? Prove it. What is the second principle used in factoring? Prove it. 90. When is a number divisible by 2, and when not? Why? When by 4, and when not? Why? When by 5, and when not? Why? When by 10, 100, &c., and when not? Why? What is every composite number divisible by? Why? If an even number is divisible by 3, what else must divide it? Why? How do the prime numbers end? Why? 91. What is the rule for finding the prime factors of a number?

REMARKS.—1. Divide first by the smallest prime factor.

2. The *least* divisor of any number is a *prime* number; for, if it were a composite number, its factors, which are less than itself, would also be divisors, (Art. 89), and then it would not be the *least* divisor. Therefore, the prime factors of any number may be found by dividing it first by the *least* number that will exactly divide it, then dividing this quotient in like manner, and so on.

3. Since 1 is a factor of every number either prime or composite, it is not usually specified as a factor.

Find the prime factors of 42.

DEMONSTRATION.—By trial, 2 is found to be a factor of 42. Also, 3 is found by trial to be a factor of 21, and consequently a factor of 42, which is a multiple of 21, (Art. 89). In like manner, 7 being a factor of 21, must be a factor of 42; and since $2 \times 3 \times 7 = 42$, there can be no other factors of 42 besides 2, 3, and 7.

```
2)42
3)21
   7
```

SEPARATE INTO PRIME FACTORS,

1.	45	. . *Ans.* 3, 3, 5.	8.	72	*Ans.* 2, 2, 2, 3, 3.
2.	48	*Ans.* 2, 2, 2, 2, 3.	9.	75	. . *Ans.* 3, 5, 5.
3.	50	. . *Ans.* 2, 5, 5.	10.	80	*Ans.* 2, 2, 2, 2, 5.
4.	54	. *Ans.* 2, 3, 3, 3.	11.	84	. *Ans.* 2, 2, 3, 7.
5.	56	. *Ans.* 2, 2, 2, 7.	12.	96	*Ans.* 2, 2, 2, 2, 2, 3.
6.	60	. *Ans.* 2, 2, 3, 5.	13.	98	. . *Ans.* 2, 7, 7.
7.	63	. . *Ans.* 3, 3, 7.	14.	99	. *Ans.* 3, 3, 11.

THE PUPIL who desires to be an expert Arithmetician, should be able to give the prime factors of all numbers under 100 by mere inspection, the operation being performed mentally.

15. Factor 210. . . . . *Ans.* 2, 3, 5, 7.
16. Factor 1155. . . . . . . *Ans.* 3, 5, 7, 11.
17. Factor 10010. . . . . *Ans.* 2, 5, 7, 11, 13.
18. Factor 36414. . . . *Ans.* 2, 3, 3, 7, 17, 17.
19. Factor 58425. . . . . *Ans.* 3, 5, 5, 19, 41.

The prime factors common to several numbers may be found by resolving each into its prime factors, then taking the prime factors alike in all.

REVIEW.—91. Which number is it most convenient to divide by first? What is said of the least divisor of a number? Of 1, as a factor? Explain the example, and show the reason of the rule.

Find the prime factors common to

20.	42 and 98	. . . .	*Ans.* 2, 7.
21.	45 and 105	. . . .	*Ans.* 3, 5.
22.	90 and 210	. . . .	*Ans.* 2, 3, 5.
23.	210 and 315	. . . .	*Ans.* 3, 5, 7.

ART. **92.** Since any composite number is divisible, not only by each of its prime factors, but also by the product of any two or more of them, (Art. 90, Prop. V.); hence,

To find all the divisors of any composite number,

RULE.—*Resolve the number into its prime factors; and then form from these factors all the different products of which they will admit; the prime factors and their products will be all the divisors of the given number.*

$42 = 2 \times 3 \times 7$; and all its divisors are 2, 3, 7, and 2×3, 2×7, and 3×7; or, 2, 3, 7, 6, 14, 21.

Find all the divisors

1.	Of 70	. .	*Ans.* 2, 5, 7 and 10, 14 and 35.
2.	Of 30	. .	*Ans.* 2, 3, 5 and 6, 10 and 15.
3.	Of 196	. .	*Ans.* 2, 7, 4, 14, 28, 49 and 98.
4.	Of 231	. .	*Ans.* 3, 7, 11 and 21, 33 and 77.

GREATEST COMMON DIVISOR.

ART. **93.** A *common* divisor of two or more numbers, is a number that will divide each of them without a remainder; 3 is a common divisor of 12 and 18.

The *greatest* common divisor of two or more numbers, is the *greatest* number that will divide each without a remainder; 6 is the greatest common divisor of 12 and 18.

REMARK.—1. Two numbers may have several common divisors, ut they can have only *one greatest* common divisor.

2. The greatest common divisor is often termed the *greatest common measure*, or *greatest common factor.*

REVIEW.—92. What is the rule for finding all the divisors of a number? Why can there be no other divisor than these? 93. What is a common divisor of several numbers? the greatest common divisor? Give examples.

ART. 94. To find the greatest common divisor of two numbers,

RULE I.—*Resolve the given numbers into their prime factors; the product of the factors common to both numbers will be the greatest common divisor sought.*

Find the greatest common divisor of 30 and 105.

$$\left.\begin{array}{r} 30 = 2 \times 3 \times 5. \\ 105 = 3 \times 5 \times 7. \end{array}\right\} \quad 3 \times 5 = 15, \text{ the greatest common divisor.}$$

DEMONSTRATION.—The product 3×5 is a divisor of both the numbers, since each contains it; and it is their greatest common divisor, since it contains all the factors common to both.

FIND THE GREATEST COMMON DIVISOR

1. Of 30 and 42 . . . . . *Ans.* $2 \times 3 = 6$.
2. Of 42 and 70 . . . . . *Ans.* $2 \times 7 = 14$.
3. Of 63 and 105 . . . . *Ans.* $3 \times 7 = 21$.
4. Of 66 and 165 . . . . *Ans.* $3 \times 11 = 33$.
5. Of 90 and 150 . . . *Ans.* $2 \times 3 \times 5 = 30$.

The greatest common divisor contains, as factors, all the other common divisors; thus, 30, which is $2 \times 3 \times 5$, contains 2, 3, 5, $2 \times 3 = 6$, $2 \times 5 = 10$, and $3 \times 5 = 15$, the only remaining common divisors of 90 and 150.

6. Of 60 and 84 . . . *Ans.* $2 \times 2 \times 3 = 12$.
7. Of 90 and 225 . . . *Ans.* $3 \times 3 \times 5 = 45$.
8. Of 112 and 140 . . . *Ans.* $2 \times 2 \times 7 = 28$.

The greatest common divisor of more than two numbers may be found, by resolving each into its prime factors, and taking the product of the factors common to all.

9. Of 30, 45, and 75 . . *Ans.* $3 \times 5 = 15$.
10. Of 84, 126, and 210 . *Ans.* $2 \times 3 \times 7 = 42$.

ART. 95. Rule 1 is generally used when the numbers are small; but when they are large, apply

RULE II.—*Divide the greater number by the less, and the divisor by the remainder, and so on; always dividing the last divisor by the last remainder, till nothing remains; the last divisor will be the greatest common divisor sought.*

Find the greatest common divisor of 24 and 66

SOLUTION.—Divide 66 by 24; the quotient is 2, and the remainder 18. Next, divide 24 by 18; the quotient is 1, and the remainder 6. Lastly, divide 18 by 6; there is no remainder; hence, 6 is the greatest common divisor of 24 and 66.

```
24)66(2
   48
   18)24(1
      18
       6)18(3
         18
```

Rule 2 depends on the following

PRINCIPLES.

1. *A divisor of a number is a divisor of any multiple of that number.* As shown in Art. 89, Principle 1st.

2. *A common divisor of two numbers is a divisor of their* SUM. As shown, Art. 89, Principle 2d.

3. *A common divisor of two numbers is a divisor of their* DIFFERENCE.

DEMONSTRATION.—Since each of the numbers contains the common divisor a certain number of times, their difference must contain it as many times as the larger contains it more times than the smaller; 2 being a divisor of 16 and 10, must be a divisor of their difference; for, 16 is 8 *twos*, and 10 is 5 *twos*, and their difference is 8 *twos* minus 5 *twos* = 3 *twos*.

4. *The greatest common divisor of two numbers is also the greatest common divisor of the smaller, and their remainder after division.*

OBSERVE, that in the following demonstration, G. C. D. signifies greatest common divisor.

DEMONSTRATION.—The G. C. D. of 24 and 66 divides 24, and must, therefore, divide 48, which is 2 times 24, (Principle 1st); as it divides both 66 and 48, it must divide their difference 18, (Principle 3d); therefore, the G. C. D. of 24 and 66, is *a common divisor* of 24 and 18.

```
24)66(2
   48
   18)24(1
      18
       6)18(3
         18
```

It now remains to show that it is their *greatest* common divisor.

If there could be any greater common divisor of 24 and 18, as it would divide 24, it would also divide 48 (Principle 1st); and as it

REVIEW.—94. What is Rule 1 for finding the greatest common divisor of two numbers? Prove it. What are contained in the greatest common divisor? How is the rule applied to more than two numbers?

would divide both 48 and 18, it would divide their sum 66, (Principle 2d), and would, consequently, be *a common divisor* of 24 and 66. We should then have a common divisor of 24 and 66 greater than their greatest common divisor, which is absurd. Hence, the G. C. D. of 24 and 66 is not only a com. divisor of 24 and 18, but is their G. C. D.; and the proposition is proved.

DEMONSTRATION OF RULE II.

The reason of the rule follows immediately from Principle 4th; for, as the G. C. D. of the two given numbers is the same as the G. C. D. of the smaller and their remainder after division, and as this is the same as the G. C. D. of the smaller of those two and their remainder after division, and so on; it follows, that when we get the G. C. D. of any remainder and the previous divisor (which is always the smaller of the two numbers used in the division), this will also be the G. C. D. of the original two numbers; but, whenever a remainder is exactly contained in the previous divisor, it will necessarily be the G. C. D. of those two, since no number can have a greater divisor than itself; and, therefore, whenever this exact division takes place, the divisor will be the G. C. D. not only of the two numbers last used, but also of the two numbers first given.

NOTES.—1. To find the G. C. D. of more than two numbers, first find the G. C. D. of any two; then of that G. C. D. and one of the remaining numbers, and so on for all the numbers; the last C. D. will be the G. C. D. of all the numbers.

2. If two given numbers be divided by their G. C. D. the quotients will be prime to each other.

FIND THE GREATEST COMMON DIVISOR OF

		ANS.			ANS.
1.	85 and 120	5.	7.	597 and 897	3.
2.	91 and 133	7.	8.	825 and 1287	33.
3.	357 and 525	21.	9.	423 and 2313	9.
4.	425 and 493	17.	10.	18607 and 24587	23.
5.	324 and 1161	27.	11.	105, 231 and 1001	7.
6.	589 and 899	31.	12.	165, 231 and 385	11.

13. 816, 1360, 2040 and 4080 . . . . . *Ans.* 136.

14. 1274, 2002, 2366, 7007 and 13013 . *Ans.* 91.

REVIEW.—95. When is Rule 1 generally used? What is Rule 2? Illustrate. What are the 1st and 2d principles on which the demonstration of the rule depends? The 3d? Prove it. The 4th? Prove it. Demonstrate the rule.

LEAST COMMON MULTIPLE.

ART. 96. A *common* multiple of two or more numbers, is a number that can be divided by *each* of them without a remainder; 24 is a common multiple of 3 and 4.

The *least* common multiple of two or more numbers, is the least number that is divisible by each of them. 12 is the least common multiple of 3 and 4.

REMARKS.—1. Since the common multiple of two or more numbers contains each of them as a factor, *every common multiple is a composite number.*

2. Since the continued product of two or more numbers is divisible by each of them, a common multiple of two or more numbers may always be found by taking their continued product; and since any multiple of this product will be divisible by each of the numbers, (Art. 89, Principle 1st), an unlimited number of common multiples may be found for the same numbers.

TO FIND THE LEAST COMMON MULTIPLE.

ART. 97. RULE I.—*Separate the given numbers into their prime factors; then multiply together such, and only such, of those factors, as are necessary to form a product that will contain all the prime factors of each number.*

Find the least common multiple of 10, 12, 15.

$$10 = \not{2} \times \not{5}$$
$$12 = 2 \times 2 \times \not{3}$$
$$15 = 3 \times 5$$
$$2 \times 2 \times 3 \times 5 = 60. \textit{ Ans.}$$

SOLUTION.—Factor the numbers; the prime factor 2 occurs *once* in 10 and *twice* in 12; strike out the first 2 and reserve 2×2 as factors of the least common multiple. The factor 5 occurs once in 10 and once in 15; strike out the first 5, and reserve the second as a factor of the least common multiple. The factor 3 occurs once in 12 and once in 15; strike out the first 3, and reserve the second 3 as a factor of the least common multiple. Lastly, take the product of the reserved prime factors $2 \times 2 \times 3 \times 5 = 60$, for the least common multiple.

DEMONSTRATION.—The product $2 \times 2 \times 3 \times 5 = 60$, is *a* common multiple, because it contains all the factors of each of the

REVIEW.—96. What is a common multiple of several numbers? What is the least common multiple? Give examples. How many common multiples may be found for several numbers? Why? 97. Give Rule 1. Prove it. How often must a factor be taken in the least common multiple? Why?

numbers; it is the *least* common multiple, because it does not contain any prime factor not found in some one of the numbers.

REMARKS.—Each factor must be taken in the least common multiple the greatest number of times it occurs in either of the numbers. In the preceding solution, 2 must be taken *twice*, because it occurs twice in 12, the number containing it most.

2. To avoid mistakes, after resolving the numbers into their prime factors, strike out the needless factors.

FIND THE LEAST COMMON MULTIPLE OF

		ANS.			ANS.
2.	8, 10, 15	120.	5.	8, 14, 21, 28	168.
3.	6, 9, 12	36.	6.	10, 15, 20, 30	60.
4.	12, 18, 24	72.	7.	15, 30, 70, 105	210.

ART. 98. RULE II.—1. *Place the numbers in a line; strike out any that will exactly divide any of the others; divide by any prime number that will divide two or more of them without a remainder; set the quotients and undivided numbers in a line beneath.*

2. *Proceed with this line as before, and continue the operation till no number greater than* 1 *will exactly divide two or more of the numbers.*

3. *Multiply together the divisors and the numbers in the last line; their product will be the least common multiple required.*

Find the least common multiple of, 10, 20, 25 and 30.

2)	~~10~~	20,	25,	30
5)		10,	25,	15
		2	5	3

$2 \times 5 \times 2 \times 5 \times 3 = 300$ *Ans.*

SOLUTION.—Strike out the 10, because it is exactly contained in 20; then divide by 2, because it is a factor of two of the numbers, 20 and 30. Next divide by 5, because it is a factor of more than one of the remaining numbers. Lastly, multiply together the divisors 2 and 5, and the remaining numbers, 2, 5 and 3; their product, 300, is the least com. multiple.

DEMONSTRATION.—The 10 is marked out to shorten the operation. The least common multiple of 20, 25 and 30, contains 20, and therefore contains 10, which is a factor of 20, and will be the least

REVIEW.—98. Give Rule 2. Prove it. What kind of numbers must be used as divisors? Why? When the numbers are prime to each other, how is the least common multiple found?

common multiple of *all* the numbers. The rest of the demonstration is similar to the previous one; for the division by the prime numbers serves to strike out the needless factors, and those divisors with the factors remaining in the last line, are evidently the reserved factors necessary to constitute the least common multiple. For a full analysis, see Ray's Arithmetic, Third Book, Art. 120.

NOTE.—In dividing to cancel the needless factors, divide by a *prime* number. Dividing by a *composite* number would not, in all cases, cancel all the needless factors; in the preceding example, if we divide first by 10, the 5 in the number 25 would not be canceled.

ART. 99. RULE III.—1. *Set the numbers in a line, striking out such as are contained in any of the others, and separate any convenient one, generally the largest, from the others, by a curved line.*

2. *Divide each of the remaining numbers by the greatest divisor common to it and the number cut off, and set the quotients in a line beneath.*

3. *Proceed with the second line exactly as with the first, and continue so until all the quotients are observed to be prime to each other; the continued product of these quotients and the numbers cut off, will be the least common multiple required.*

NOTES.—1. If the number cut off is found to be prime to all the rest in the same line, cut off another, and proceed with it, reserving the first as a factor of the least common multiple.

2. When the given numbers contain no common factor, it is evident their product will be the least common multiple required; the least common multiple of

$$4,\ 9 \text{ and } 25, \text{ is } 4 \times 9 \times 25 = 900.$$

3. The least common multiple of several numbers is equal to the product of their greatest common divisor, by those factors of each number not found in the others.

4. If the least common multiple of several numbers be divided by either of them, the quotient will be the product of all the factors of the others not found in the divisor.

Resume the example under Rule 2.

OPERATION.

$$\begin{array}{lll} \not{1}\not{0}, & 20, & 25(30 \\ \hline & 2 & 5 \end{array}$$

$$2 \times 5 \times 30 = 300 \ \textit{Ans.}$$

SOLUTION.—After striking out the 10, cut off 30 from the rest by a curve, and then divide 20 by 10, which is the greatest divisor common to it and 30; also divide 25 by 5, the greatest divisor common to it and 30. The quotients are set beneath, and being prime to each other, further cutting off and dividing are unnecessary, and the product of 2, 5 and 30, gives 300, the least common multiple required.

DEMONSTRATION.—The least common multiple must contain the number cut off (30), and such factors of the rest as are not found in the 30; on this account, divide each of them by the greatest factor common to it and the number cut off, thus getting rid of the needless factors; in like manner, the least common multiple must contain the number cut off in the second line, and such factors of the rest as are not found in the one cut off; therefore, we repeat the process for each line, until all the numbers in the same line are found to be prime to each other; when this is the case, there will be no more needless factors, and the least common multiple will be the continued product of the numbers cut off and those in the last line.

FIND THE LEAST COMMON MULTIPLE OF

		ANS.			ANS.
1.	4,6,15 . . .	60.	5.	35,45,63,70	. 630.
2.	6,9,20 . . .	180.	6.	10,14,20,35	. 140.
3.	15,20,30 . .	60.	7.	8,15,20,25,30	. 600.
4.	7,11,13,5 .	5005.	8.	15,24,40,140	. 840.

9. 30,45,48,80,120,135 . . . . *Ans.* 2160.
10. 77,91,105,143,165,195 . . . *Ans.* 15015.
11. 174,485,4611,14065,15423 . *Ans.* 4472670.
12. 498,85988,235803,490546. *Ans.* 244291908.

PROOF OF THE ELEMENTARY RULES

BY CASTING OUT THE NINES.

ART. **100.** To cast the 9's out of any number, is to divide the sum of its digits by 9, and find the excess. To do this, begin at the left, add the digits together, and when the sum is nine or more, drop the 9, and carry the excess to the next digit, and so on.

For example, to cast the 9's out of 768945, say 7 and 6 are 13, which is 4 above 9; drop the 9, and carry the 4;

REVIEW.—99. Give Rule 3. If a number cut off is prime to the rest in the same line, what must be done? Prove the rule. 100. What is meant by casting the 9's out of any number?

4 and 8 are 12, which is 3 above 9; then, 3 (the digit 9 being passed over) and 4 are 7 and 5 are 12, which is 3 above 9; consequently, the excess in this instance, is 3

All the methods of proof are founded on this

PRINCIPLE.—*Any number divided by 9 will leave the same remainder as the sum of its digits divided by 9.*

For example, take 3456.

$$3456=\begin{cases}3000=3(1000)=3\times(999+1)=3\times999+3\\ \ \ 400=\ 4(100)=\ 4\times(99+1)=\ 4\times99+4\\ \ \ \ \ 50=\ \ 5(10)=\ \ 5\times(9+1)=\ \ 5\times9+5\\ \ \ \ \ \ \ 6=\ .\ .\ .\ \ .\ .\ .\ .\ .\ .\ .\ .\ 6\end{cases}$$

DEMONSTRATION.—Observe that 3000 is a certain number of 9's, with a remainder 3; 400, a certain number of 9's, with a remainder 4; 50, a certain number of 9's, with a remainder 5; and 6, being less than 9, may be termed a remainder. The remainders, 3, 4, 5, 6, are the digits of the number 3456; hence, the excess of 9's in the number 3456 is the same as that in its digits, 3, 4, 5 and 6.

PROOF OF ADDITION.

Cast the 9's (or 11's) out of each of the numbers added, also out of their sum; the last excess must equal the sum of the others after dropping all 9's (or 11's).*

* To cast the 11's out of any number, see page 70.

The excesses in the numbers are 8, 2, 4 and 3, and the excess in the sum of these excesses is 8. The excess in the sum of the numbers is 8, the two excesses being the same, as they ought to be when the work is correct.

	EXCESS.
7352	8
5834	2
6241	4
7302	3
26729	8

NOTE.—In proving Addition by this method, it is not necessary to stop at each number and write the excess, but regard all the numbers as forming one horizontal line.

DEMONSTRATION.—Divide each number by 9, and add up the remainders, omitting the 9's, if any; the result must be the same as the remainder obtained by dividing the sum of the numbers by 9; and as these remainders are most easily obtained by casting the 9's out of each number, (see Principle), the reason of the proof is apparent.

REVIEW.—100. On what principle do the proofs by casting out the 9's depend? Prove it. What is the proof of addition?

PROOF OF SUBTRACTION.

Cast the 9's (or 11's) out of the subtrahend, the remainder, and the minuend; the last excess will be equal to the sum of the other two, after dropping all 9's (or 11's).

EXAMPLE.—Proceeding as directed in the note to last proof, we find the excess in the subtrahend and remainder together to be 8; the excess in the minuend is also 8, as it should be when the work is right.

Minuend,	7640
Subtrahend,	1234
Remainder,	6406

DEMONSTRATION.—As the minuend is the sum of the subtrahend and remainder, the reason of this proof is seen from that of Addition.

PROOF OF MULTIPLICATION.

Cast the 9's (or 11's) out of the multiplicand and multiplier; multiply the two excesses together; cast the 9's (or 11's) out of the result, and also out of the product; when the work is correct, the last two excesses will agree.

Multiply 835 by 76; the product is 63460. The excess in the multiplicand is 7, in the multiplier 4, and in the product 1; the two former multiplied give 28, and the excess in 28 is also 1, as it should be.

$$\begin{array}{r} 835 = 92\times 9+7 \\ 76 = 8\times 9+4 \\ \hline 368\times 9+28 \\ 736\times 81 + 56\times 9 \\ \hline 736\times 81+424\times 9+28 \end{array}$$

DEMONSTRATION.—Since each of the numbers to be multiplied contains a certain number of 9's and an excess, their product, as seen from the work, will consist of *three* parts, two of which are divisible by 9; therefore, the excess of 9's in the product must be the same as the excess of 9's in the 3d part, (28), which is the product of the excesses in the multiplicand and multiplier; and since these excesses are most easily obtained by casting the 9's out of the digits of each number, (see Principle), the reason of the proof is apparent.

PROOF OF DIVISION.

Cast the 9's (or 11's) out of the divisor, dividend, quotient, and remainder. Multiply together the excesses in the divisor and quotient, and cast the 9's (or 11's) out of the result; then, to this excess, add the excess in the remainder, and cast the 9's (or 11's) out of the result; when the work is correct, this excess will be the same as the excess in the dividend.

REVIEW.—What is the proof of subtraction? Of multiplication?

Divide 8915 by 25; the quotient is 356 and the remainder 15.

Excess of 9's in the divisor	7.
Excess of 9's in the quotient	5.
$7 \times 5 = 35$. Excess of 9's in 35	8.
Excess of 9's in the remainder	6.
$6 + 8 = 14$. Excess of 9's in 14	5.
Excess of 9's in the dividend is also	5.

DEMONSTRATION.—Since the dividend is the product of the divisor and quotient, with the remainder added, the reason of this proof is seen from those of Multiplication and Addition.

PROOF BY CASTING OUT THE ELEVENS.

To cast the 11's out of any number, add its alternate figures, commencing at the right, and dropping 11 when the sum exceeds that number; then do the same with the remaining figures, and subtract the last result from the former, increased by 11 if necessary.

For example, to cast the 11's out of 30752486, say 6 and 4 are 10 and 5 are 15; drop 11 and 4 remains, then 4 and 0 are 4; also, 8 and 2 are 10 and 7 are 17, drop 11 and 6 remains, and 6 and 3 are 9; 9 from 4 can not be taken, but 4 increased by 11 is 15, and 9 from 15 leaves 6, which is the excess required.

The Proofs by casting out the 11's depend on this

PRINCIPLE.—*Any number divided by 11 leaves the same remainder as the excess obtained by casting out its 11's.*

This principle, and the proofs to which it leads, can be demonstrated in a manner similar to those regarding the 9's, and, by putting 11 for 9 in the proofs by 9's, we have the proofs by 11's. (See Proofs.)

REMARK.—These methods of proof are liable to this objection; two figures may be substituted for each other, or the correct figures may be replaced by wrong ones having the same sum, and the work appear to prove when it is wrong. These, however, occur so rarely as not to detract much from the merits of the methods.

REVIEW.—What is the proof of division? Illustrate and give the reason for these proofs. 100. Explain the proofs by casting out the 11's. When will these proofs fail to detect an error?

CANCELLATION.

ART. **101.** CANCELLATION is a short method of obtaining results by omitting equal factors from a dividend and divisor. Its use is seen in the following examples.

Exchanged 24 hats at $5 each, for coats at $24 each: how many coats did I get?

SOLUTION.—To solve this question, multiply 24 by 5, and divide the product by 24. Instead of performing this work, indicate it thus, $\frac{\not{2}\not{4}\times 5}{\not{2}\not{4}}$; then, since dividing either factor of a product divides the product, (Art. 71), the result is $1\times 5=5$; the same as would be got by canceling the 24 from both dividend and divisor.

Multiply 105 by 18, and divide the product by 30.

$$\frac{{}^{21}\not{1}\not{0}\not{5}\times \not{1}\not{8}^{3}}{\not{3}\not{0}_{\not{5}}} \qquad 21\times 3=63.$$

SOLUTION.—Indicate the operations as in the margin. Divide both dividend and divisor by 6; this gives 105×3 (Art. 71,) above, and 5 below, and does not alter the quotient, (Art. 75). The quotient is found, as in last example, to be 21×3 (Art. 71) or 63. As each factor is used in canceling, it is crossed to indicate that there is no further use of it; and each quotient is placed beside the number from which it is obtained. The 21 and 3 being the factors left of the dividend after cancellation, are multiplied together; their product is 63, and as no factor of the divisor is left, the 63 is not to be divided, and is, therefore, the quotient required.

Multiply 75, 153 and 28 together, and divide by the product of 63 and 36.

$$\frac{\overset{25}{\not{7}\not{5}}\times\overset{17}{\not{1}\not{5}\not{3}}\times\not{2}\not{8}^{\not{7}}}{\underset{\not{9}\,3}{\not{6}\not{3}}\times\not{3}\not{6}_{\not{9}}}$$

$$\frac{25\times 17}{3}=\frac{425}{3}=141\tfrac{2}{3}$$

SOLUTION.—Indicate the operations as in the margin. Cancel 4 out of 28 and 36, leaving 7 above, and 9 below. Cancel this 7 out of the dividend and out of the 63 in the divisor, leaving 9 below. Cancel a 9 out of the divisor, and out of 153 in the dividend, leaving 17 above. Cancel 3 out of 9 and 75, leaving 25 above and 3 below. No further canceling is possible; the factors remaining in the dividend are 25 and 17, whose product, 425, divided by the 3 in the divisor, gives $141\frac{2}{3}$.

ART. **102**. If the dividend or divisor, or both, is the product of several factors, the quotient can often be easily obtained by this

RULE FOR CANCELLATION.

1. *Indicate the multiplications which produce the dividend, and those, if any, which produce the divisor.*

2. *Cancel equal factors from dividend and divisor; multiply together the factors remaining in the dividend, and divide the product by the product of the factors left in the divisor.*

NOTES.—1. If no factor remains in the divisor, the product of the factors remaining in the dividend will be the quotient; if only one factor is left in the dividend, it will be the answer.

2. Cancellation can only take place between a factor of the dividend and a factor of the divisor; not between two factors of the dividend, or two factors of the divisor.

3. Canceling equal factors is dividing both dividend and divisor by the same number, which does not affect the value of the quotient, (Art. 75.)

EXAMPLES FOR PRACTICE.

1. How many cows, worth $24 each, can I get for 9 horses worth $80 each? *Ans.* 30.

2. I exchanged 8 barrels of molasses, each containing 33 gallons, at 40 cents a gallon, for 10 chests of tea, each containing 24 pounds: how much a pound did the tea cost me? *Ans.* 44 cents.

3. How many bales of cotton, of 400 pounds each, at 12 cents a pound, are equivalent to 6 hogsheads of sugar, 900 pounds each, at 8 cents a pound? *Ans.* 9.

4. Divide $15 \times 24 \times 112 \times 40 \times 10$ by $25 \times 36 \times 56 \times 90$. *Ans.* $3\frac{5}{9}$

VIII. COMMON FRACTIONS.

ART. **103**. A fraction is a part of a unit or whole thing, when it is divided into *equal* parts. If an apple is divided into 3 equal parts, each part is a fraction of the apple.

NOTE.—*Fraction* is derived from the Latin *fractus, broken.*

REVIEW.—101. What is Cancellation? Illustrate it. 102. In what case may cancellation be employed? Give the rule.

ART. **104**. *The name of a fraction, and the size of its parts, depend on the number of parts into which the unit is divided;* thus, when the unit is divided into 2, 3, or 4 equal parts, the fractions are named *halves*, *thirds* and *fourths* (Art. 62): moreover, a *half* is evidently larger than a *third*, a *third* than a *fourth*, and so on.

ART. **105**. Fractions are divided into two classes, *Common* and *Decimal*.

A common fraction is expressed by two numbers, one above the other, with a horizontal line between them; one-half is expressed by $\frac{1}{2}$, two-thirds by $\frac{2}{3}$.

The number below the line is called the *denominator*, because it *denominates* or gives *name* to the fraction: it shows into how many parts the *unit* is divided.

The number above the line is called the *numerator*, because it *numbers* the parts: it shows how many parts the *fraction* contains; in the fraction $\frac{4}{7}$, the denominator 7, shows that the *unit* contains *seven equal* parts; the numerator 4, shows that the *fraction* contains 4 of those parts.

The *terms* of a fraction are the numerator and denominator; the terms of $\frac{5}{8}$ are 5 and 8.

ART. **106**. There are two methods of considering a fraction whose numerator is greater than 1; thus, to divide 2 apples of the same size, equally among 3 boys, divide each apple into *three equal* parts, making 6 parts in all, and then give to each boy two of the parts, expressed by $\frac{2}{3}$. The 2 parts may be taken from *one* apple, or 1 part from each of the *two* apples; hence, $\frac{2}{3}$ expresses either 2 thirds of *one* thing, or 1 third of *two* things. Also, $\frac{4}{5}$ expresses either 4 fifths of one thing or 1 fifth of four things; therefore,

The numerator of a fraction may be regarded as showing the number of units to be divided; the denominator, the number of parts into which the numerator is to be divided: the fraction itself being the value of one of those parts.

REVIEW.—102. If no factor remains in the divisor, what is the quotient? Between what factors only can cancellation take place? What does canceling equal factors amount to? 103. What is a Fraction? Why so called? 104. How are fractions named? What does the value of a fractional part depend on? 105. How are fractions divided? How are common fractions expressed? What is the denominator? What is the numerator? Why are they called so? What are both together called?

Hence, a fraction may be considered as an expression of *unexecuted* division, in which

The DIVIDEND is called the NUMERATOR;
The DIVISOR is called the DENOMINATOR;
The QUOTIENT is called the FRACTION itself.

$\frac{1}{4}$ is the quotient of 1 (num'tor) divided by 4 (denom.)

$\frac{3}{5}$ is the quotient of 3 (num'tor) divided by 5 (denom.)

$\frac{7}{6}$ is the quotient of 7 (num'tor) divided by 6 (denom.)

NUMERATION AND NOTATION.

ART. **107**. Since Fractions arise from Division, one of the *signs* of Division, (Art. 59), is used in expressing them; that is, the numerator is written above, and the denominator below a horizontal line; hence,

TO READ COMMON FRACTIONS,

RULE.—*Read the number of parts taken as expressed by the numerator, and then the size of the parts as expressed by the denominator.*

$\frac{7}{9}$ is read *seven ninths.*

REMARK.—Seven ninths, ($\frac{7}{9}$), signifies either 7 ninths of one, or $\frac{1}{9}$ of 7, or 7 divided by 9, (Art. 106).

Fractions to be read; that is, expressed in words:

$\frac{5}{7}$, $\frac{3}{8}$, $\frac{7}{10}$, $\frac{9}{11}$, $\frac{13}{17}$, $\frac{7}{27}$, $\frac{14}{41}$, $\frac{9}{65}$, $\frac{13}{90}$, $\frac{7}{100}$, $\frac{15}{230}$, $\frac{8}{1345}$.

Of these fractions, which expresses parts of the largest size? Which, the smallest size? Which, the least number of parts? Which, the greatest number of parts? Which, the same number of parts? Which, parts of the same size?

TO WRITE COMMON FRACTIONS,

ART. **108**. RULE.—*Write the number of parts; place a horizontal line below it, under which write the number which indicates the size of the parts.*

REVIEW.—106. How many methods of considering a fraction? Illustrate them. What operation is expressed by a fraction? What is the dividend? the divisor? the quotient?

Fractions to be expressed in figures:

Seven *eighths.* Four *elevenths.* Five *thirteenths.* One *seventeenth.* Three *twenty-ninths.* Eight *twenty-oneths.* Nine *forty-twoths.* Nineteen *ninety-thirds.* Thirteen *one hundredths.* Twenty-four *one hundred and fifteenths.*

To express a whole number in the form of a fraction, write 1 *below* it for a denominator; thus, $2 = \frac{2}{1}$, $7 = \frac{7}{1}$, &c.

PROPOSITIONS.

ART. 109. PROP. 1.— *When the numerator of a fraction is less than the denominator, the fraction is less than* 1.

PROP. 2.— *When the numerator of a fraction is equal to the denominator, the fraction is equal to* 1.

PROP. 3.— *When the numerator of a fraction is greater than the denominator, the fraction is greater than* 1.

DEMONSTRATION.—Prop. 1 is true, because, in that case, the number of parts in the fraction is *less* than the number of parts in a unit.

Prop. 2 is true, because, in that case, the number of parts in the fraction is the *same* as the number of parts in a unit.

Prop. 3 is true, because, in that case, the number of parts in the fraction is *greater* than the number of parts in a unit.

ART. 110. DEFINITIONS.—1. A *proper fraction* is one whose numerator is *less* than the denominator; as, $\frac{1}{2}$; $\frac{6}{7}$.

2. An *improper fraction* is one whose numerator is *equal to*, or *greater* than, the denominator; as, $\frac{3}{3}$ and $\frac{6}{5}$.

REMARK.—The word fraction primarily signifies a part. A *proper* fraction is *properly* a fraction expressing a value less than the whole. An *improper* fraction is *not properly* a fraction, the value expressed being *equal to*, or *greater* than, the whole.

3. A *simple fraction* is a single fraction, proper or improper; as, $\frac{1}{2}$, $\frac{2}{3}$, or $\frac{5}{4}$.

REVIEW.—107. What is the rule for reading fractions? for writing them? 108. What two other methods of reading a fraction are there? How may a whole number be expressed as a fraction? 109. When is a fraction less than 1? equal to 1? greater than 1? Give the reasons. 110. What is a proper fraction? an improper fraction? Why so called? What is a simple fraction?

4. A *compound fraction* is a fraction of a fraction, or several fractions connected by *of*; as, $\frac{2}{3}$ of $\frac{3}{4}$ of $\frac{4}{5}$.

5. A *mixed number* is a fraction joined with a whole number; as, $1\frac{1}{2}$ and $2\frac{3}{4}$.

6. A *complex fraction* is one having a fraction either in one or both of its terms; as, $\frac{2\frac{1}{3}}{4}$, $\frac{1}{3\frac{1}{2}}$, $\frac{\frac{2}{3}}{\frac{4}{5}}$, and $\frac{1\frac{1}{2}}{3\frac{1}{4}}$.

ART. **111.** To show that $\frac{1}{2}$ of $\frac{1}{2} = \frac{1}{4}$; that $\frac{1}{3}$ of $\frac{1}{2} = \frac{1}{6}$; that $\frac{1}{3}$ of $\frac{1}{4} = \frac{1}{12}$, and so on.

DEMONSTRATION.—Divide a line as A B into two equal parts; one of the parts, as A C, is termed one half ($\frac{1}{2}$): divide a half into 3 equal parts, as in the figure; one of the parts is called one third of one half, which is expressed by figures thus, $\frac{1}{3}$ of $\frac{1}{2}$. This is evidently one sixth of the whole line, that is, $\frac{1}{3}$ of $\frac{1}{2} = \frac{1}{6}$. In like manner, $\frac{1}{2}$ of $\frac{1}{2} = \frac{1}{4}$; $\frac{1}{2}$ of $\frac{1}{3} = \frac{1}{6}$; $\frac{1}{3}$ of $\frac{1}{4} = \frac{1}{12}$, and so on.

A——.——.——.——————B
C

What is $\frac{1}{2}$ of $\frac{1}{3}$? Why? What is $\frac{1}{2}$ of $\frac{1}{4}$? $\frac{1}{3}$ of $\frac{1}{5}$? What is $\frac{1}{4}$ of $\frac{1}{5}$?

PROPOSITIONS.

ART. **112.** PROP. I.—*Multiplying the numerator of a fraction likewise multiplies the fraction.* Multiply the numerator of the fraction $\frac{2}{5}$ by 3; the result, $\frac{6}{5}$, is *three times* as great as $\frac{2}{5}$.

ART. **113.** PROP. II.—*Dividing the numerator of a fraction likewise divides the fraction.* Divide the numerator of the fraction $\frac{8}{5}$ by 2; the result, $\frac{4}{5}$, is only *one half* as great as $\frac{8}{5}$.

ART. **114.** PROP. III.—*Multiplying the denominator of a fraction, on the contrary, divides the fraction.* Multiply the denominator of the fraction $\frac{12}{2}$ by 3; the result, $\frac{12}{6}$, is *one third* as great as $\frac{12}{2}$.

ART. **115.** PROP. IV.—*Dividing the denominator of a fraction, on the contrary, multiplies the fraction.* Divide the denominator of the fraction, $\frac{12}{6}$, by 2; the result, $\frac{12}{3}$, is *twice* as great as $\frac{12}{6}$.

REVIEW.—What is a compound fraction? a mixed number? a complex fraction? 112. What is the effect of multiplying the numerator of a fraction? 113. Of dividing the numerator? 114. Of multiplying the denominator? 115. Of dividing the denominator?

ART. 116. PROP. V.—*Multiplying both terms of a fraction by the same number, changes its form but does not alter its value.* Multiply both terms of the fraction $\frac{2}{3}$ by 3; the result, $\frac{6}{9}$, has the *same* value as $\frac{2}{3}$.

ART. 117. PROP. VI.—*Dividing both terms of a fraction by the same number, changes its form, but does not alter its value.* Divide both terms of the fraction $\frac{4}{8}$ by 4; the result, $\frac{1}{2}$, has the *same* value as $\frac{4}{8}$.

DEMONSTRATION.—Observe that the numerator of a fraction is a dividend of which the denominator is the divisor, and the value of the fraction the quotient. (Art. 106).

Prop. 1 is true, because, (Art. 73), multiplying the dividend by any number, multiplies the quotient by the same number.

Prop. 2 is true, because, (Art. 74), dividing the dividend by any number, divides the quotient by the same.

Prop. 3 is true, because, (Art. 74), multiplying the divisor by any number, divides the quotient by the same.

Prop. 4 is true, because, (Art. 73), dividing the divisor by any number, multiplies the quotient by the same.

Prop. 5 and 6 are true, because, (Art. 75), multiplying or dividing both dividend and divisor by the same number, the quotient remains the same.

REMARK.—Hence, there are *two* ways of multiplying a fraction, *two* ways of dividing a fraction, and *two* ways of changing its form without altering its value; that is, *multiplying or dividing the numerator does the same to the fraction; but multiplying or dividing the denominator does the opposite to the fraction; while multiplying or dividing both terms alike makes no change in its value.*

REDUCTION OF FRACTIONS.

ART. 118. Reduction of Fractions consists in changing their forms without altering their values.

CASE I.

TO REDUCE A FRACTION TO ITS LOWEST TERMS.

ART. 119. A fraction is in its lowest terms when the numerator and denominator are prime to each other, (Art. 88, Def. 5); as, $\frac{2}{3}$, but not $\frac{4}{6}$.

REVIEW.—116. What is the effect of multiplying both terms alike? 117. Of dividing both terms alike? Illustrate and prove these propositions.

RULE.—*Divide both terms by any common factor, do the same to the resulting fraction, and so on, till both terms are prime to each other.*

Or, *divide both terms of the fraction by their greatest common divisor.*

Reduce $\frac{20}{30}$ to its lowest terms.

SOLUTION.—Dividing both terms by the common factor 2, the result is $\frac{10}{15}$; dividing this by 5, the result is $\frac{2}{3}$, which can not be reduced lower.

$$2)\frac{20}{30}=5)\frac{10}{15}=\frac{2}{3}\ \textit{Ans.}$$

Or, dividing at once by 10, the greatest common divisor of both terms, the result is $\frac{2}{3}$, as before.

$$10)\frac{20}{30}=\frac{2}{3}\ \textit{Ans.}$$

DEMONSTRATION.—The value of the fraction is not changed, because both terms are divided by the same number (Art. 117).

REDUCE TO THEIR LOWEST TERMS,

		ANS.			ANS.			ANS.
1.	$\frac{30}{45}$	$\frac{2}{3}$	4.	$\frac{105}{195}$	$\frac{7}{13}$	7.	$\frac{253}{414}$	$\frac{11}{18}$
2.	$\frac{32}{56}$	$\frac{4}{7}$	5.	$\frac{154}{210}$	$\frac{11}{15}$	8.	$\frac{667}{783}$	$\frac{23}{27}$
3.	$\frac{42}{189}$	$\frac{2}{9}$	6.	$\frac{156}{221}$	$\frac{12}{17}$	9.	$\frac{1767}{4557}$	$\frac{19}{49}$

EXPRESS IN THEIR SIMPLEST FORMS,

10.	$923\div1491$	$=\frac{13}{21}$	12.	$2261\div4123$	$=\frac{17}{31}$
11.	$890\div1691$	$=\frac{10}{19}$	13.	$6160\div40480$	$=\frac{7}{46}$

CASE II.

TO REDUCE AN IMPROPER FRACTION TO A WHOLE OR MIXED NUMBER.

ART. **120.** RULE.—*Divide the numerator by the denominator; the quotient will be the whole or mixed number.*

NOTE.—If there be a fraction in the answer, reduce it to its lowest terms.

To reduce $\frac{13}{5}$ of a dollar to dollars, divide 13 by 5, making $2\frac{3}{5}$ dollars.

DEMONSTRATION.—Since 5 *fifths* make 1 dollar, there will be as many dollars in 13 fifths as 5 fifths are contained times in 13 fifths that is, $2\frac{3}{5}$ dollars; and so in all such cases.

REVIEW.—118. What is Reduction of Fractions? 119. When is a fraction in its lowest terms?

1. In $\frac{37}{8}$ of a dollar, how many dollars? *Ans.* $4\frac{5}{8}$
2. In $\frac{137}{4}$ of a bushel, how many bushels? *Ans.* $34\frac{1}{4}$
3. In $\frac{785}{60}$ of an hour, how many hours? *Ans.* $13\frac{1}{12}$

REDUCE TO WHOLE OR MIXED NUMBERS,

4. $\frac{23}{23}$ . . *Ans.* 1.		7. $\frac{1162}{11}$ . . *Ans.* $105\frac{7}{11}$
5. $\frac{1295}{37}$ . . *Ans.* 35.		8. $\frac{4260}{13}$ . . *Ans.* $327\frac{9}{13}$
6. $\frac{800}{9}$ . . *Ans.* $88\frac{8}{9}$		9. $\frac{15780}{31}$ . . *Ans.* $509\frac{1}{31}$

CASE III.

TO REDUCE A WHOLE OR MIXED NUMBER TO AN IMPROPER FRACTION.

ART. 121. RULE.—*Multiply together the whole number and the denominator of the fraction; to the product add the numerator, and write the sum over the denominator.*

Reduce $3\frac{3}{4}$ to an improper fraction; to *fourths*.

DEMONSTRATION.—In 1 (unit), there are 4 fourths; in 3 (units), there are 3 times 4 fourths, = 12 fourths: and 12 fourths + 3 fourths = 15 fourths.

$3\frac{3}{4}$
4
12 = fourths in 3
3 = fourths in fraction.
15 = fourths in $3\frac{3}{4}$
Ans. $\frac{15}{4}$

REMARK.—1. This demonstration shows that the whole number is really the multiplier, and the denominator the multiplicand; but, multiplying by the denominator is more convenient, and gives the same result, (Art. 47).

2. This, and the preceding case, are the reverse of, and mutually prove each other.

1. In $\$7\frac{3}{8}$, how many 8ths of a dollar? . *Ans.* $\frac{59}{8}$
2. In $19\frac{3}{4}$ gallons, how many fourths? . . *Ans.* $\frac{79}{4}$
3. In $13\frac{37}{60}$ hours, how many sixtieths? . *Ans.* $\frac{817}{60}$

REDUCE TO IMPROPER FRACTIONS,

4. $11\frac{2}{3}$ . . . *Ans.* $\frac{35}{3}$	7. $109\frac{9}{19}$ . . *Ans.* $\frac{2080}{19}$
5. $15\frac{8}{11}$ . . . *Ans.* $\frac{173}{11}$	8. $5\frac{207}{211}$ . . *Ans.* $\frac{1262}{211}$
6. $127\frac{11}{17}$. *Ans.* $\frac{2170}{17}$	9. $13\frac{51}{73}$ . . *Ans.* $\frac{1000}{73}$

REVIEW.—119. What is the rule for reducing a fraction to its lowest terms? Prove the rule. 120. Give the rule for reducing an improper fraction to a whole or mixed number. Prove it.

ART. **122**. To reduce a whole number to a fraction having a given denominator, is merely an example of the preceding case, the numerator of the fractional part being zero, (0). It is done by multiplying together the whole number and the denominator, and writing the product over the denominator.

To reduce 4 to a fraction whose denominator is 5, is the same as to reduce $4\frac{0}{5}$ to an improper fraction.

1. Reduce 7 to 4ths. . . . . *Ans.* $\frac{28}{4}$
2. Reduce 9 to sevenths. . . . . . . *Ans.* $\frac{63}{7}$
3. Reduce 23 to twenty-thirds. . . . . *Ans.* $\frac{529}{23}$
4. Reduce 19 to a fraction whose denominator is 29. *Ans.* $\frac{551}{29}$

CASE IV.

TO REDUCE COMPOUND TO SIMPLE FRACTIONS.

ART. **123**. RULE.—*Multiply all the numerators together for a new numerator, and all the denominators together for a new denominator.*

Reduce $\frac{2}{3}$ of $\frac{4}{5}$ to a simple fraction.

$$\frac{2}{3} \text{ of } \frac{4}{5} = \frac{2 \times 4}{3 \times 5} = \frac{8}{15} \ \textit{Ans.}$$

DEMONSTRATION.—$\frac{2}{3}$ of $\frac{4}{5}$ = 2 times $\frac{1}{3}$ of $\frac{4}{5}$ = 2 times $\frac{1}{3}$ of $\frac{1}{5}$ of 4, (since $\frac{1}{5}$ of 4 is the same as $\frac{4}{5}$, by Art. 106,) = 2 times $\frac{1}{15}$ of 4, (since $\frac{1}{3}$ of $\frac{1}{5}$ is the same as $\frac{1}{15}$, by Art. 111,) = 2 times $\frac{4}{15}$, (since $\frac{1}{15}$ of 4 is the same as $\frac{4}{15}$, by Art. 106,) = $\frac{8}{15}$ (since multiplying the numerator multiplies the fraction, by Art. 112.)

NOTE.—Before applying the rule, reduce mixed or whole numbers to a fractional form.

1. Reduce $\frac{1}{3}$ of $\frac{2}{5}$ of $3\frac{1}{7}$ to a simple fraction.

   SOLUTION.—$3\frac{1}{7} = \frac{22}{7}$, and $\frac{1}{3}$ of $\frac{2}{5}$ of $\frac{22}{7} = \frac{44}{105}$ *Ans.*

2. $\frac{3}{7}$ of $\frac{5}{11}$ to a simple fraction. . . . . *Ans.* $\frac{15}{77}$
3. $\frac{2}{3}$ of $\frac{5}{9}$ of $2\frac{3}{7}$ to a simple fraction. . . *Ans.* $\frac{170}{189}$
4. $\frac{3}{4}$ of $\frac{1}{2}$ of $3\frac{4}{5}$ to a simple fraction. . . *Ans.* $1\frac{17}{40}$

NOTE.—Equal factors may be canceled out of the numerator and denominator, as out of any other dividend and divisor, (Art. 101), before the multiplications are performed.

REVIEW.—121. Give the rule for reducing a whole or mixed number to an improper fraction. Prove it. Which number is really the multiplier?

5. Reduce $\frac{2}{3}$ of $\frac{9}{10}$ of $\frac{7}{12}$ to a simple fraction.

SOLUTION.—The factors 2, 3, and 3 are common to both terms. Canceling them, and multiplying together the remaining factors, the result is 7 twentieths.

$$\frac{\not{2} \times \overset{3}{\not{9}} \times 7}{\not{3} \times \underset{5}{\not{10}} \times \underset{4}{\not{12}}} = \frac{7}{20}\ \textit{Ans.}$$

REDUCE TO SIMPLE FRACTIONS,

6. $\frac{1}{3}$ of $\frac{3}{4}$ of $\frac{4}{7}$. . *Ans.* $\frac{1}{7}$.
7. $\frac{2}{5}$ of $\frac{4}{7}$ of $2\frac{5}{8}$. *Ans.* $\frac{3}{5}$.
8. $\frac{4}{5}$ of $\frac{15}{16}$ of $2\frac{2}{3}$. *Ans.* 2.
9. $\frac{1}{2}$ of $\frac{4}{5}$ of $3\frac{3}{4}$. *Ans.* $1\frac{1}{2}$
10. $\frac{3}{4}$ of $\frac{8}{9}$ of $\frac{4}{7}$ of $8\frac{3}{4}$. . . . . *Ans.* $3\frac{1}{3}$
11. $\frac{1}{3}$ of $\frac{3}{5}$ of $\frac{6}{7}$ of $\frac{3}{4}$ of $4\frac{2}{3}$. . . . . . . *Ans.* $\frac{3}{5}$
12. $\frac{8}{11}$ of $\frac{3}{7}$ of $\frac{4}{19}$ of $\frac{77}{24}$ of $7\frac{1}{8}$. . . . . *Ans.* $1\frac{1}{2}$
13. $\frac{12}{13}$ of $\frac{9}{16}$ of $\frac{7}{18}$ of $\frac{10}{21}$ of $1\frac{4}{35}$. . . . *Ans.* $\frac{3}{28}$

NOTE.—To reduce complex to simple fractions, see Art. 132.

CASE V.

ART. **124.** To reduce fractions of different denominators to equivalent fractions of a common denominator,

RULE.—*Multiply both terms of each fraction by the product of all the denominators except its own.*

DEMONSTRATION.—Multiplying both terms of each fraction by the same number, does not alter its value; the new denominator of each fraction will be the *same*, since it will be the product of the same numbers; viz., of all the denominators.

NOTE.–Reduce compound to simple fractions, and whole or mixed numbers to improper fractions, before applying the rule.

Reduce $\frac{1}{2}$, $\frac{2}{3}$ and $\frac{3}{4}$ to a common denominator.

Both terms of the first fraction are multiplied by $3 \times 4 = 12$; of the second, by $2 \times 4 = 8$; and of the third, by $2 \times 3 = 6$.

$$\frac{1 \times 3 \times 4}{2 \times 3 \times 4} = \frac{12}{24} \begin{array}{l}\text{new num.}\\ \text{new denom.}\end{array}$$

$$\frac{2 \times 2 \times 4}{3 \times 2 \times 4} = \frac{16}{24} \begin{array}{l}\text{new num.}\\ \text{new denom.}\end{array}$$

$$\frac{3 \times 2 \times 3}{4 \times 2 \times 3} = \frac{18}{24} \begin{array}{l}\text{new num.}\\ \text{new denom.}\end{array}$$

Since the denominator of each new fraction is the product of the same numbers; viz., all the denominators of the given fractions, it

REVIEW.—122. How is a whole number reduced to a fraction having a given denominator? 123. How are compound fractions reduced to simple ones? Prove the rule.

is unnecessary to find this product more than once. The operation is generally performed as in the following example.

Reduce $\frac{1}{2}$, $\frac{3}{5}$ and $\frac{6}{7}$ to a common denominator.

$1 \times 5 \times 7 = 35$ 1st num.
$3 \times 2 \times 7 = 42$ 2d num.
$6 \times 2 \times 5 = 60$ 3d num.
$2 \times 5 \times 7 = 70$ com. denom.

Ans. $\frac{35}{70}, \frac{42}{70}, \frac{60}{70}$.

REDUCE TO A COMMON DENOMINATOR,

1. $\frac{1}{2}, \frac{2}{3}, \frac{3}{5}$ . . . . *Ans.* $\frac{15}{30}, \frac{20}{30}, \frac{18}{30}$
2. $\frac{1}{4}, \frac{1}{5}, \frac{1}{6}$ . . . . *Ans.* $\frac{30}{120}, \frac{24}{120}, \frac{20}{120}$
3. $\frac{2}{3}, \frac{3}{7}, \frac{5}{8}$ . . . . *Ans.* $\frac{112}{168}, \frac{72}{168}, \frac{105}{168}$
4. $\frac{1}{2}, \frac{3}{5}, \frac{5}{6}, \frac{7}{8}$ . . . . . . *Ans.* $\frac{240}{480}, \frac{288}{480}, \frac{400}{480}, \frac{420}{480}$
5. $\frac{2}{3}$, $\frac{1}{2}$ of $3\frac{1}{2}$, $\frac{2}{3}$ of $\frac{3}{5}$ . . . . *Ans.* $\frac{40}{60}, \frac{105}{60}, \frac{24}{60}$
6. $\frac{2}{3}$ of $\frac{6}{7}$, $\frac{3}{4}$ of $\frac{8}{9}$, $\frac{1}{2}$ of $\frac{4}{5}$ of $\frac{3}{7}$ of $2\frac{5}{8}$. *Ans.* $\frac{240}{420}, \frac{280}{420}, \frac{189}{420}$

ART. **125**. When the terms of the fractions are small, and one denominator is a multiple of the others, reduce the fractions to a common denominator, by multiplying both terms of each by such a number as will render its denominator the same as the largest denominator. *This number will be found by dividing the largest denominator by the denominator of the fraction to be reduced.*

Reduce $\frac{1}{3}$ and $\frac{5}{6}$ to a common denominator.

$$\frac{1 \times 2}{3 \times 2} = \frac{2}{6}$$
$$\frac{5}{6} = \frac{5}{6}$$

SOLUTION—The largest denominator, 6, is a multiple of 3; therefore, if we multiply both terms of $\frac{1}{3}$ by 6 divided by 3, which is 2, it is reduced to $\frac{2}{6}$.

REDUCE TO A COMMON DENOMINATOR,

1. $\frac{1}{2}$, $\frac{3}{4}$ and $\frac{5}{8}$. . . . . *Ans.* $\frac{4}{8}, \frac{6}{8}, \frac{5}{8}$
2. $\frac{2}{3}$, $\frac{5}{6}$ and $\frac{7}{12}$. . . . . *Ans.* $\frac{8}{12}, \frac{10}{12}, \frac{7}{12}$
3. $\frac{3}{4}$, $\frac{4}{5}$, $\frac{9}{10}$ and $\frac{11}{20}$. . . . . *Ans.* $\frac{15}{20}, \frac{16}{20}, \frac{18}{20}, \frac{11}{20}$

REVIEW.—123. If there are any mixed numbers, what must be done? What may be done before multiplying? 124. How are fractions of different denominators brought to a common denominator? Prove the rule. What must be done with compound fractions? With whole or mixed numbers?

CASE VI.

ART. **126.** To reduce fractions of different denominators, to equivalent fractions of the least common denominator.

RULE.—*Find the least common multiple of the given denominators; multiply both terms of each fraction by the quotient obtained by dividing this least common multiple by the denominator of the fraction.*

Reduce $\frac{1}{2}$, $\frac{3}{4}$ and $\frac{5}{6}$ to the least common denominator.

$$\begin{array}{r} 2)\not{2}\;\;4\;\;6 \\ \hline 2\;\;3 \end{array} \quad \begin{array}{r} 2)12 \\ \hline 6 \end{array} \quad \begin{array}{r} 4)12 \\ \hline 3 \end{array} \quad \begin{array}{r} 6)12 \\ \hline 2 \end{array}$$

$2 \times 2 \times 3 = 12$ $\frac{1 \times 6}{2 \times 6} = \frac{6}{12}$, $\frac{3 \times 3}{4 \times 3} = \frac{9}{12}$, $\frac{5 \times 2}{6 \times 2} = \frac{10}{12}$

least com. mul. *Ans.* $\frac{6}{12}$, $\frac{9}{12}$ and $\frac{10}{12}$

DEMONSTRATION.—Since multiplying both terms of a fraction by the same number does not alter its value, (Art. 116), each of the given fractions may be reduced to an equivalent fraction, whose denominator is *any multiple* of its own; and they may all be reduced to equivalent fractions of *the least common* denominator, by taking for that denominator the *least common multiple* of the given denominators.

After getting the least common denominator, both terms of each fraction are multiplied by the quotient of the least common denominator divided by its own denominator, as in Art. 125.

NOTES.—1. Before commencing, each fraction must be in its lowest terms.

2. Reduce compound to simple fractions, and whole or mixed numbers to improper fractions.

3. When the pupil is acquainted with the principles of the operation, the multiplication of the denominators may be omitted, as the new denominator of each fraction will be equal to the least common multiple.

4. The object of reducing fractions to a common denominator, is to prepare them for Addition or Subtraction.

REVIEW.—125. When one denominator is a multiple of the others, how can the fractions be reduced to a common denominator? 126. How are fractions of different denominators reduced to a least common denominator? What must be done if a fraction is not in its lowest terms? If there are compound fractions? If there are whole or mixed numbers? What will the least common denominator be?

REDUCE TO LEAST COMMON DENOMINATOR,

1. $\frac{1}{3}, \frac{3}{4}, \frac{5}{6}$ . . . . *Ans.* $\frac{4}{12}, \frac{9}{12}, \frac{10}{12}$
2. $\frac{1}{2}, \frac{3}{5}, \frac{9}{10}, \frac{3}{4}$ . . . . *Ans.* $\frac{10}{20}, \frac{12}{20}, \frac{18}{20}, \frac{15}{20}$
3. $\frac{3}{7}, \frac{5}{8}, \frac{11}{14}$ . . . . *Ans.* $\frac{24}{56}, \frac{35}{56}, \frac{44}{56}$
4. $\frac{3}{4}, \frac{6}{8}, \frac{9}{12}, \frac{15}{20}$ * . . . . *Ans.* $\frac{3}{4}, \frac{3}{4}, \frac{3}{4}, \frac{3}{4}$
5. $\frac{6}{9}, \frac{9}{12}, \frac{12}{20}, \frac{7}{10}$ . . . . . . *Ans.* $\frac{40}{60}, \frac{45}{60}, \frac{36}{60}, \frac{42}{60}$
6. $\frac{2}{3}, \frac{6}{10}, \frac{3}{4}$ of $\frac{14}{15}, \frac{34}{60}$ . . . . *Ans.* $\frac{20}{30}, \frac{18}{30}, \frac{21}{30}, \frac{17}{30}$
7. $1\frac{3}{4}, 3\frac{2}{3}$ and $\frac{3}{10}$ of $3\frac{4}{7}$ . . . . *Ans.* $\frac{147}{84}, \frac{308}{84}, \frac{90}{84}$

* See Note 1, preceding page.

ADDITION OF FRACTIONS.

ART. **127.** Addition of Fractions is the process of adding together two or more fractional numbers.

RULE.—*Reduce the fractions to a common denominator; add their numerators together, and place the sum over the common denominator.*

Add $\frac{3}{4}, \frac{5}{8}, \frac{7}{12}$

$$\begin{array}{l} \not{4} \quad 8 \quad (12 \\ \hline \quad 2 \\ 2 \times 12 = 24 \\ \text{least com. mul.} \end{array} \qquad \begin{array}{r|l} & 24 \\ \hline 4 & 6 \times 3 = 18 \\ 8 & 3 \times 5 = 15 \\ 12 & 2 \times 7 = 14 \end{array} \Big\} \text{new numerators.}$$

$$\tfrac{47}{24} = 1\tfrac{23}{24} \textit{ Ans.}$$

DEMONSTRATION.—When the denominators of two or more fractions are the same, they express parts of the *same* size; we can, therefore, add

1 *fourth*		1 *cent.*
2 *fourths*	as we would add	2 *cents.*
3 *fourths*		3 *cents.*

The sum being 6 fourths, ($\frac{6}{4}$), in one case, and 6 *cents* in the other.

That is, *to add fractions having a common denominator, find the sum of the numerators and write the result over the denominators.*

But, if the denominators are different, fractions do not express things of the *same unit value*, and can not, therefore, be added together (Art. 40); in the preceding example we can not add *fourths, eighths* and *twelfths*, but, by reducing them to *twenty-fourths*, they express things of the same denomination, and can then be added.

REVIEW.—127. Why are fractions reduced to a common denominator? What is Addition of Fractions? What is the rule? Prove it.

NOTES.—1. Before commencing the operation, each fraction should be in its lowest terms, and compound fractions must be reduced to simple ones.

2. Mixed numbers may be reduced to improper fractions, and then added; or, the fractions may be added, and then the whole numbers, and the results united.

3. After adding, reduce the result to its lowest terms.

1. Add $\frac{4}{37}$, $\frac{8}{37}$, $\frac{15}{37}$, $\frac{29}{37}$ and $\frac{18}{37}$ . . . *Ans.* 2.

2. $\frac{17}{96}$, $\frac{35}{96}$, $\frac{43}{96}$ and $\frac{25}{96}$ . . . . *Ans.* $1\frac{1}{4}$

3. $\frac{2}{5}$, $\frac{3}{8}$ and $\frac{7}{10}$ . . . . . . *Ans.* $1\frac{19}{40}$

4. $\frac{1}{6}$, $\frac{2}{9}$ and $\frac{5}{12}$ . . . . . . *Ans.* $\frac{29}{36}$

5. $\frac{2}{3}$, $\frac{3}{5}$, $\frac{7}{9}$ and $\frac{4}{15}$ . . . . . . *Ans.* $2\frac{14}{45}$

6. $1\frac{2}{3}$ and $2\frac{3}{5}$ . . . . *Ans.* $4\frac{4}{15}$

7. $2\frac{1}{4}$, $3\frac{2}{7}$ and $4\frac{5}{6}$ . . . . . . *Ans.* $10\frac{31}{84}$

8. $\frac{4}{6}$, $\frac{9}{15}$, $\frac{18}{20}$ and $1\frac{1}{4}$ . . . . . . *Ans.* $3\frac{5}{12}$

9. $\frac{6}{10}$, $\frac{4}{14}$, $\frac{8}{12}$ and $2\frac{1}{3}$ . . . . *Ans.* $3\frac{31}{35}$

10. $1\frac{1}{2}$, $2\frac{1}{3}$, $3\frac{1}{4}$ and $4\frac{1}{5}$ . . . . *Ans.* $11\frac{17}{60}$

11. $\frac{2}{3}$ of $\frac{4}{5}$, and $\frac{3}{7}$ of $\frac{5}{8}$ of $2\frac{1}{3}$ . . *Ans.* $1\frac{19}{120}$

12. What is $\frac{5}{8}+\frac{5}{14}+\frac{1}{6}$? . . . . . *Ans.* $1\frac{25}{168}$

13. $\frac{3}{8}+\frac{1}{22}+\frac{7}{24}+\frac{29}{88}$? . . . . *Ans.* $1\frac{1}{24}$

14. $\frac{3}{4}+1\frac{1}{3}+\frac{4}{7}+1\frac{3}{4}$? . . . . *Ans.* $4\frac{17}{42}$

15. $2\frac{3}{5}+4\frac{7}{8}+5\frac{3}{10}$? . . . . . *Ans.* $12\frac{31}{40}$

16. $\frac{1}{7}+\frac{2}{3}$ of $\frac{5}{9}$ of $6\frac{1}{2}$? . . . . *Ans.* $2\frac{104}{189}$

17. $\frac{7}{8}+\frac{11}{12}+\frac{17}{18}+\frac{23}{24}+\frac{20}{27}$? . . *Ans.* $4\frac{47}{108}$

18. $\frac{1}{9}$ of $6\frac{3}{4}+\frac{8}{15}$ of $\frac{6}{7}$ of $7\frac{1}{2}$? . . *Ans.* $4\frac{5}{28}$

19. $\frac{4}{7}$ of $96\frac{1}{4}+\frac{8}{9}$ of $\frac{11}{12}$ of $5\frac{1}{6}$? . *Ans.* $59\frac{17}{81}$

20. $\frac{2}{3}+\frac{8}{9}+\frac{26}{27}+\frac{80}{81}+\frac{242}{243}+\frac{728}{729}$? *Ans.* $5\frac{365}{729}$

SUBTRACTION OF FRACTIONS.

ART. **128.** Subtraction of Fractions is the process of finding the difference between two fractional numbers.

RULE.—*Reduce the fractions to a common denominator; find the difference of their numerators, and place it over the common denominator.*

REVIEW.—127. Before commencing, what should be done? What with mixed numbers? What should be done with the answer, when obtained?

Find the difference between $\frac{5}{6}$ and $\frac{8}{15}$.

SOLUTION.—The fractions, when reduced to the least common denominator, become $\frac{25}{30}$ and $\frac{16}{30}$; their difference is $\frac{9}{30} = \frac{3}{10}$.

DEMONSTRATION.—When the denominators of two fractions are the same, they express parts of the same size, and their difference can be found as in the case of whole numbers.

Thus	5 *sevenths*,	5 *cents*.
	3 *sevenths*,	3 *cents*.
Difference,	2 *sevenths* ($\frac{2}{7}$) in one case, and	2 *cents* in the other.

But, if the denominators are different, the fractions do not express things of the same kind; therefore, one can not be subtracted from the other, any more than 3 *cents* can be taken from 5 *apples*, (Art. 43); in the preceding example, *fifteenths* can not be taken from *sixths*; but by reducing them both to *thirtieths*, their difference can be found.

NOTES.—1. Before commencing, reduce compound to simple fractions, and see that each fraction is in its lowest terms.

2. After subtracting, reduce the result to its lowest terms.

	WHAT IS	ANS.		WHAT IS	ANS.
1.	$\frac{4}{5} - \frac{5}{12}$? . . .	$\frac{23}{60}$	5.	$\frac{11}{14} - \frac{4}{63}$? . . .	$\frac{13}{18}$
2.	$\frac{8}{11} - \frac{3}{17}$ of $\frac{1}{2}$? .	$\frac{239}{374}$	6.	$\frac{9}{55} - \frac{1}{15}$? . . .	$\frac{16}{165}$
3.	$\frac{11}{54} - \frac{3}{28}$ of $\frac{5}{6}$? .	$\frac{173}{1512}$	7.	$\frac{7}{45} - \frac{11}{75}$? . . .	$\frac{2}{225}$
4.	$\frac{5}{11} - \frac{1}{13}$ of 4? .	$\frac{21}{143}$	8.	$\frac{10}{39} - \frac{8}{65}$? . . .	$\frac{2}{15}$

ART. **129**. Mixed numbers may be brought to improper fractions, and then subtracted; or, the fractions may be subtracted, and then the whole numbers and the results united. In the latter method, if the lower fraction is the larger, increase the upper fraction by the number of parts in 1 unit, and carry 1 to the first figure of whole numbers in the lower line.

Thus, $6\frac{1}{8} - 4\frac{3}{4} = \frac{49}{8} - \frac{19}{4} = \frac{11}{8} = 1\frac{3}{8}$; or,

$$\begin{array}{r} 6\frac{1}{8} \\ 4\frac{3}{4} \\ \hline 1\frac{3}{8} \end{array} \qquad \begin{array}{r} \frac{1}{8} + \frac{8}{8} = \frac{9}{8} \\ \frac{3}{4} = \frac{6}{8} \end{array}$$

REVIEW.—128. What is Subtraction of Fractions? What is the rule? Prove it. What should be done before commencing? What should be done with the answer, when obtained? 129. How are mixed numbers subtracted?

WHAT IS	ANS.	WHAT IS	ANS.
9. $12\frac{3}{4}-10\frac{13}{16}$? .	$1\frac{15}{16}$	13. $15-\frac{3}{7}$? . . .	$14\frac{4}{7}$
10. $12\frac{23}{28}-9\frac{27}{35}$? .	$3\frac{1}{20}$	14. $18-5\frac{3}{8}$? . .	$12\frac{5}{8}$
11. $5\frac{23}{32}-2\frac{2}{7}$? . .	$3\frac{97}{224}$	15. $\frac{5}{3}$ of $2\frac{7}{9}-3\frac{17}{18}$?	$\frac{37}{54}$
12. $7\frac{5}{12}-3\frac{1}{2}$? . .	$3\frac{11}{12}$	16. $3\frac{1}{3}-\frac{4}{5}$ of $1\frac{7}{8}$?	$1\frac{5}{6}$

17. $\frac{16}{3}$ of $4\frac{1}{2}-\frac{13}{4}$ of $3\frac{1}{5}$? . . . . . . *Ans.* $13\frac{3}{5}$

18. $11\frac{2}{3}+8\frac{7}{9}-9\frac{19}{22}$? . . . . *Ans.* $10\frac{115}{198}$

19. A man owned $\frac{25}{38}$ of a ship, and sold $\frac{3}{5}$ of his share: how much had he left? *Ans.* $\frac{5}{19}$

20. After selling $\frac{4}{7}$ of $\frac{5}{8}+\frac{1}{5}$ of $\frac{3}{7}$ of a farm, what part of it remains? *Ans.* $\frac{39}{70}$

21. $3\frac{1}{4}+4\frac{2}{5}-5\frac{1}{2}+16\frac{5}{8}-7\frac{11}{24}+10-14\frac{5}{6}$, is equal to what? *Ans.* $6\frac{29}{60}$

22. $5\frac{1}{5}-2\frac{5}{6}+\frac{13}{2}-3\frac{3}{10}+3\frac{1}{12}+8\frac{1}{9}-16\frac{1}{4}=$ what? *Ans.* $\frac{23}{45}$

23. $1-\frac{3}{8}$ of $\frac{5}{6}-\frac{2}{3}$ of $\frac{3}{7}=$ how much? *Ans.* $\frac{45}{112}$

MULTIPLICATION OF FRACTIONS.

ART. **130.** Multiplication of Fractions is the process of multiplication, when one or both of the factors are fractional numbers. It embraces three operations:

1. To multiply a fraction by a whole number.
2. To multiply a whole number by a fraction.
3. To multiply one fraction by another.

Since any whole number may be expressed in the form of a fraction, (Art. 108), these 3 cases may all be performed by this

GENERAL RULE FOR MULTIPLYING FRACTIONS.

Multiply the numerators together for a new numerator, and the denominators together for a new denominator.

Multiply $\frac{4}{5}$ by $\frac{2}{3}$. . . . . $\frac{4}{5}\times\frac{2}{3}=\frac{8}{15}$ *Ans.*

DEMONSTRATION.—We can attach no other idea to the product of 4 fifths by 2 thirds, than that signified by 2 thirds of 4 fifths.

REVIEW.—130. What is Multiplication of Fractions? What does it embrace? What is the general rule? Prove it.

But, 2 thirds of 4 fifths is 8 fifteenths, (Art. 123); therefore, the product of 4 fifths by 2 thirds, is also 8 fifteenths. Hence, these

COROLLARIES.

I To Multiply a Fraction by a Whole Number.—*Multiply the numerator of the fraction by the whole number, and write the product over the denominator*, (Art. 112).

Or, *Divide the denominator of the fraction by the whole number, when it can be done without a remainder, and over the quotient write the numerator*, (Art. 115).

II. To Multiply a Whole Number by a Fraction.—*Multiply the whole number by the numerator of the fraction, and divide the product by the denominator.*

Remarks.—1. After indicating the operations, if the numerator and denominator contain common factors, cancel them before multiplying. The result will be in its lowest terms.

2. Multiplying one fraction by another is the same as reducing a compound to a simple fraction, (Art. 123).

EXAMPLES FOR PRACTICE.

1.	$\frac{10}{13} \times 12$ . .	$= 9\frac{3}{13}$	4.	$\frac{9}{16} \times 28$ . .	$= 15\frac{3}{4}$
2.	$\frac{11}{24} \times 18$ . .	$= 8\frac{1}{4}$	5.	$\frac{13}{15} \times 30$ . .	$= 26$
3.	$\frac{29}{48} \times 24$ . .	$= 14\frac{1}{2}$	6.	$3\frac{2}{3} \times 5$ . .	$= 18\frac{1}{3}$

Suggestion.—In multiplying a mixed number by a whole number, multiply the fraction and the whole number separately, and add the products; or, reduce the mixed number to an improper fraction, and multiply it; as,

$\frac{2}{3} \times 5 = 3\frac{1}{3}$, and $3 \times 5 = 15$; and $15 + 3\frac{1}{3} = 18\frac{1}{3}$.

Or, $3\frac{2}{3} = \frac{11}{3}$, and $\frac{11}{3} \times 5 = \frac{55}{3} = 18\frac{1}{3}$.

7.	$45 \times \frac{7}{9}$ . .	$= 35.$	11.	$28 \times 3\frac{2}{3}$.	$= 102\frac{2}{3}$
8.	$50 \times \frac{11}{14}$ . .	$= 39\frac{2}{7}$	12.	$\frac{12}{35} \times \frac{7}{16}$ . .	$= \frac{3}{20}$
9.	$25 \times \frac{3}{4}$ . .	$= 18\frac{3}{4}$	13.	$\frac{15}{16} \times \frac{24}{25}$ . .	$= \frac{9}{10}$
10.	$32 \times 2\frac{3}{8}$ . .	$= 76.$	14.	$\frac{42}{55} \times \frac{22}{35}$ . .	$= \frac{12}{25}$

15. What will $3\frac{1}{3}$ yards of cloth cost at $\$4\frac{1}{2}$ per yard?

Review.—130. How is a fraction multiplied by a whole number? How is a whole number multiplied by a fraction? How may the work be shortened? Multiplying fractions is equivalent to what case of reduction? How may a mixed number be multiplied by a whole number?

SUGGESTION.—In finding the product of two mixed numbers, it is generally best to reduce them to improper fractions; thus,

$4\frac{1}{2}=\frac{9}{2}$; $3\frac{1}{3}=\frac{10}{3}$; $\frac{9}{2}\times\frac{10}{3}=\frac{90}{6}=\$15$ *Ans.*

The operation may be performed without reducing to improper fractions; thus, 3 yards will cost $\$13\frac{1}{2}$, and $\frac{1}{3}$ of a yard will cost $\frac{1}{3}$ of $\$4\frac{1}{2}=\$1\frac{1}{2}$; hence, the whole will cost $15.

$$\begin{array}{r} 4\frac{1}{2} \\ \underline{3\frac{1}{3}} \\ 13\frac{1}{2} \\ \underline{1\frac{1}{2}} \\ 15 \end{array}$$

16. $6\frac{2}{3}\times4\frac{1}{2}$ . . $=30$.
17. $4\frac{4}{5}\times2\frac{2}{3}$ . . $=12\frac{4}{5}$
18. $12\frac{3}{8}\times3\frac{3}{11}$. $=40\frac{1}{2}$
19. $7\frac{11}{12}\times3\frac{7}{19}$. $=26\frac{2}{3}$
20. Multiply $\frac{1}{5}$ of 8 by $\frac{1}{4}$ of 10 . . . . *Ans.* 4.
21. Multiply $\frac{2}{3}$ of $5\frac{2}{5}$ by $\frac{3}{5}$ of $3\frac{1}{3}$ . . . *Ans.* $7\frac{1}{5}$
22. Multiply $\frac{3}{4}$ of $\frac{2}{3}$ of $5\frac{3}{5}$ by $\frac{3}{7}$ of $3\frac{3}{8}$ . . *Ans.* $4\frac{1}{20}$
23. Multiply 5, $4\frac{1}{4}$, $2\frac{1}{3}$ and $\frac{3}{7}$ of $4\frac{4}{9}$ . . *Ans.* $94\frac{4}{9}$
24. Multiply $\frac{4}{5}$, $\frac{3}{7}$, $\frac{5}{11}$, $\frac{1}{3}$ of $2\frac{1}{2}$, $\frac{4}{7}$ of $3\frac{1}{7}$. *Ans.* $\frac{80}{343}$
25. Multiply $\frac{3}{5}$, $\frac{7}{9}$, $\frac{9}{11}$, $3\frac{1}{3}$ and $3\frac{1}{7}$ . . . *Ans.* 4.
26. Multiply $3\frac{1}{2}$, $4\frac{2}{3}$, $5\frac{3}{5}$, $\frac{2}{9}$ of $\frac{5}{14}$, $6\frac{3}{4}$. *Ans.* 49.

27. At $\frac{7}{8}$ of a dollar per yard, what will 25 yards of cloth cost? *Ans.* $\$21\frac{7}{8}$

28. A quantity of provisions will last 25 men $12\frac{3}{4}$ days: how long will the same last one man? *Ans.* $318\frac{3}{4}$ days.

29. At $3\frac{1}{2}$ cents a yard, what will $2\frac{3}{4}$ yards of tape cost? *Ans.* $9\frac{5}{8}$ cts.

30. What must be paid for $\frac{3}{5}$ of $\frac{2}{3}$ of a lot of ground that cost $\$18\frac{3}{4}$? *Ans.* $\$7\frac{1}{2}$

31. K owns $\frac{5}{8}$ of a ship, and sells $\frac{3}{4}$ of his share to L: what part has he left? *Ans.* $\frac{5}{32}$

32. B bought $\frac{3}{5}$ of a farm of $219\frac{3}{8}$ acres, and sold $\frac{2}{9}$ of his part to C: what part of the whole, and how many acres, did he sell? *Ans.* $\frac{2}{15}$, and $29\frac{1}{4}$ acres.

DIVISION OF FRACTIONS.

ART. 131. Division of Fractions is the process of division, (Art. 58), when the dividend or divisor, or both, are fractional numbers. It embraces three operations:

1. To divide a fraction by a whole number.
2. To divide a whole number by a fraction.
3. To divide one fraction by another.

Since any whole number may be expressed in the form of a fraction, (Art. 108), these 3 cases may all be performed by this

GENERAL RULE FOR DIVIDING FRACTIONS.

Invert the divisor; then multiply the numerators together for a new numerator, and the denominators for a new denominator.

Divide $\frac{3}{4}$ by $\frac{2}{3}$.

$$\frac{2}{3} \text{ inverted} = \frac{3}{2}, \text{ and } \frac{3}{4} \times \frac{3}{2} = \frac{9}{8} = 1\frac{1}{8} \textit{ Ans.}$$

1st DEMONSTRATION.—Suppose the divisor were 2 instead of $\frac{2}{3}$; the quotient would be $\frac{3}{4} \times 2 = \frac{3}{8}$, (Art. 114). But, the real divisor, ($\frac{2}{3}$), is only *one third* as large as the supposed divisor (2); therefore, the real quotient must be *three times* as large as the supposed quotient, (Art. 73), and 3 times $\frac{3}{8} = \frac{9}{8}$, (Art. 112); which agrees with the rule.

2d DEM.—It has been shown, (Art. 60), that the divisor and dividend must be of the same denomination; hence, to find how often 2 *thirds* is contained in 3 *fourths*, reduce them to *twelfths*. But, $\frac{2}{3} = \frac{8}{12}$, and $\frac{3}{4} = \frac{9}{12}$, and 8 *twelfths* in 9 *twelfths* is the same as 8 in 9; that is, $\frac{9}{8} = 1\frac{1}{8}$ times.

Hence, the rule might have been expressed thus:

To divide one fraction by another, reduce them both to a common denominator, and divide the numerator of the dividend by the numerator of the divisor.

COROLLARIES.

I. TO DIVIDE A FRACTION BY A WHOLE NUMBER.—*Multiply the denominator of the fraction by the whole number, and over the product write the numerator.*

Or, *Divide the numerator of the fraction by the whole number, when it can be done without a remainder, and under the quotient write the denominator.*

II. TO DIVIDE A WHOLE NUMBER BY A FRACTION.—*Multiply the whole number by the denominator of the fraction, and divide the product by the numerator.*

REVIEW.—130. How may mixed numbers be multiplied together? 131. What is Division of Fractions? What does it embrace? What is the general rule? Illustrate it. What is the 1st demonstration? the 2d? How is a fraction divided by a whole number? How is a whole number divided by a fraction?

The rule may be simply demonstrated, as follows:

3d Dem.—Inverting the divisor, shows how often it is contained in a unit, and this multiplied by the dividend, shows how often *it* contains the divisor; in the example given, since $\frac{1}{3}$ is contained in a unit 3 times, $\frac{2}{3}$ is contained in a unit $\frac{3}{2}$ times, (Art. 74), and in $\frac{3}{4}$, it is contained $\frac{3}{2}\times\frac{3}{4}=\frac{9}{8}$ times, (Art. 130).

Notes.—1. Before commencing, reduce compound to simple fractions, and mixed numbers to improper fractions; also, express whole numbers in the form of fractions.

2. In all cases, reduce the result to its lowest terms.

EXAMPLES FOR PRACTICE.

1. $\frac{9}{16}\div 3$ . . . $=\frac{3}{16}$
2. $\frac{14}{23}\div 7$ . . . $=\frac{2}{23}$
3. $\frac{3}{5}\div 8$ . . . $=\frac{3}{40}$
4. $6\div\frac{2}{3}$ . . . $=9.$
5. $21\div\frac{9}{10}$ . . . $=23\frac{1}{3}$
6. $\frac{3}{4}\div\frac{1}{2}$ . . . $=1\frac{1}{2}$
7. $\frac{2}{3}\div\frac{1}{40}$ . . . $=26\frac{2}{3}$
8. $\frac{21}{25}\div\frac{14}{15}$ . . . $=\frac{9}{10}$
9. $\frac{12}{35}\div\frac{30}{77}$ . . . $=\frac{22}{25}$
10. $1\frac{3}{4}\div 5$ . . . $=\frac{7}{20}$
11. $8\frac{1}{6}\div\frac{2}{3}$ . . $=12\frac{1}{4}$
12. $19\frac{1}{2}\div 1\frac{7}{8}$. $=10\frac{2}{5}$
13. $73\frac{1}{2}\div 9\frac{4}{5}$. $=7\frac{1}{2}$
14. $54\frac{43}{48}\div 25\frac{5}{6}$. $=2\frac{1}{8}$
15. Divide $1\frac{1}{2}$ by $\frac{1}{2}$ of $\frac{3}{5}$ of $7\frac{1}{2}$ . . . . *Ans.* $\frac{2}{3}$
16. Divide $\frac{3}{10}$ of $\frac{3}{16}$ of $\frac{5}{12}$ by $\frac{7}{24}$ of $\frac{12}{35}$ . . *Ans.* $\frac{15}{64}$

Note.—After indicating the operations, if the numerator and denominator contain common factors, cancel them before multiplying. The result will be in its lowest terms.

17. Divide $\frac{1}{4}$ of $2\frac{1}{3}$ by $\frac{5}{8}$ of $1\frac{2}{5}$

Solution.—$2\frac{1}{3}=\frac{7}{3}$, and $1\frac{2}{5}=\frac{7}{5}$

$$\text{Then, } \frac{1}{\not 4}\times\frac{\not 7}{3}\times\frac{\overset{2}{\not 8}}{\not 5}\times\frac{\not 5}{\not 7}=\frac{2}{3}\ \textit{Ans.}$$

Suggestion.—In the division of fractions, or other operations by cancellation, it is often convenient to place the divisors on the left of a vertical line, and the multipliers on the right.

$$\begin{array}{c|l} \not 4 & 1 \\ 3 & \not 7 \ 2 \\ \not 5 & \not 8 \\ \not 7 & \not 5 \\ \hline 3 & 2 \end{array}$$

Review.—131. What is the 3d demonstration? Before commencing what must be done? How may the work be shortened?

18. Divide $\frac{3}{4}$ of $\frac{1}{2}$ by $\frac{1}{4}$ of $\frac{2}{3}$ . . . . *Ans.* $2\frac{1}{4}$
19. Divide $\frac{7}{9}$ of $3\frac{3}{5}$ by $\frac{13}{14}$ of 7 . . . . . . *Ans.* $\frac{28}{65}$
20. Divide $\frac{1}{3}$ of $\frac{4}{5} \times \frac{39}{320}$ by $\frac{2}{75}$ of $3\frac{1}{4}$ . . *Ans.* $\frac{3}{8}$
21. Divide $\frac{2}{7}$ of $5\frac{1}{2}$ by $\frac{2}{5}$ of $\frac{9}{14}$ of $3\frac{1}{3}$ . . . *Ans.* $1\frac{5}{6}$
22. Divide $\frac{1}{3}$ of $\frac{2}{7}$ of $\frac{4}{11}$ by $\frac{2}{5}$ of $\frac{1}{3}$ of $\frac{4}{7}$ . . *Ans.* $\frac{5}{11}$
23. Divide $1\frac{7}{8}$ times $4\frac{2}{3}$ by $1\frac{7}{11}$ times $3\frac{8}{9}$. *Ans.* $1\frac{3}{8}$
24. Divide $3\frac{1}{7}$ by $\frac{4}{9}$ of $8\frac{1}{4}$ times $\frac{5}{11}$ of $3\frac{3}{10}$. *Ans.* $\frac{4}{7}$
25. Divide $\frac{9}{11}$ of $\frac{2}{3}$ of $27\frac{1}{2}$ by $\frac{4}{9}$ of $\frac{3}{17}$ of $5\frac{1}{2}$ *Ans.* $34\frac{17}{22}$
26. What is $2\frac{5}{8} \times \frac{4}{5}$ of $19\frac{1}{3} \div 4\frac{5}{6} \times \frac{3}{10}$ of 8? *Ans.* $3\frac{1}{2}$

ART. **132.** To reduce complex to simple fractions.

RULE.—*Divide the numerator by the denominator as in Division of fractions,* (Art. 131).

Or, *Multiply both terms of the complex fractions by the least common multiple of the denominators of their fractional parts.*

Reduce $\dfrac{1\frac{5}{6}}{2\frac{3}{4}}$ to a simple fraction.

OPERATION. $1\frac{5}{6} = \frac{11}{6}$; $2\frac{3}{4} = \frac{11}{4}$; then, $\frac{11}{6} \div \frac{11}{4} = \frac{\cancel{11}}{\cancel{6}_3} \times \frac{\cancel{4}^2}{\cancel{11}} = \frac{2}{3}$ *Ans.*

The least common multiple of 6 and 4 is 12; hence,

2D OPERATION. $\dfrac{1\frac{5}{6} \times 12}{2\frac{3}{4} \times 12} = \dfrac{22}{33} = \dfrac{2}{3}$ *Ans.*

DEMONSTRATION.—Since every fraction indicates that the numerator is to be divided by the denominator, the 1st Rule needs no demonstration.

Since the value of a complex fraction is not changed when both terms are multiplied by the same number, (Art. 116), and since 12 is the most convenient multiplier that will cause the small fractions to disappear, the reason of the 2d Rule is evident.

REDUCE TO SIMPLE FRACTIONS,

1. $\dfrac{\frac{3}{8}}{\frac{11}{4}}$. *Ans.* $\frac{3}{22}$
2. $\dfrac{1\frac{3}{4}}{2\frac{4}{5}}$. *Ans.* $\frac{5}{8}$
3. $\dfrac{4\frac{2}{3}}{2\frac{11}{12}}$. *Ans.* $1\frac{3}{5}$
4. $\dfrac{2\frac{5}{14}}{2\frac{2}{21}}$. *Ans.* $1\frac{1}{8}$
5. $\dfrac{12\frac{3}{8}}{18}$ *Ans.* $\frac{11}{16}$
6. $\dfrac{62}{16\frac{10}{11}}$ *Ans.* $3\frac{2}{3}$

REVIEW.—132. What are the rules for reducing a complex to a simple fraction? Illustrate and prove them.

Complex fractions may be multiplied or divided, by reducing them to simple fractions. The operation may often be shortened by cancellation.

7.	$\frac{7}{10} \times \frac{4\frac{1}{4}}{8}$ . .	*Ans.* $\frac{119}{320}$	10.	$\frac{6\frac{1}{4}}{2\frac{2}{5}} \div \frac{7}{12\frac{1}{2}}$	*Ans.* $4\frac{437}{672}$
8.	$\frac{31}{25} \times \frac{8\frac{1}{2}}{10\frac{1}{5}}$.	*Ans.* $1\frac{1}{30}$	11.	$\frac{7\frac{3}{7}}{40\frac{5}{9}} \div \frac{17\frac{1}{3}}{73}$	*Ans.* $\frac{27}{35}$
9.	$\frac{5\frac{1}{3}}{9\frac{1}{5}} \times \frac{11\frac{1}{12}}{24\frac{1}{5}}$	*Ans.* $\frac{650}{2277}$	12.	$\frac{2\frac{4}{11}}{2\frac{3}{5}} \div \frac{2\frac{7}{11}}{8\frac{7}{10}}$	*Ans.* 3.

TO FIND THE GREATEST COM. DIVISOR OF FRACTIONS.

ART. **133.** RULE.—*Reduce the numbers to simple fractions, and to their lowest terms; find the greatest common divisor of the numerators, and divide it by the least common multiple of the denominators: the quotient will be the greatest common divisor of the fractions.*

NOTES.—1. If the numerators are prime to each other, and the denominators are also prime to each other, the greatest common divisor of the numbers will be, 1 divided by the product of the denominators.

2. The greatest common divisor of more than two fractions can be obtained by first finding the greatest common divisor of two of them, then of this and a third, and so on; the last divisor will be the greatest common divisor of all.

Find the greatest com. divisor of $26\frac{1}{4}$ and $1085\frac{5}{8}$

SOLUTION.—The numbers when reduced to the form of fractions, are $\frac{105}{4}$ and $\frac{8685}{8}$; the greatest common divisor of the numerators, 105 and 8685, is 15, (Art. 95), and the least common multiple of the denominators, 4 and 8, is 8, (Art. 99); hence, the greatest common divisor of the given numbers is $\frac{15}{8} = 1\frac{7}{8}$

DEMONSTRATION.—If the fractions, $\frac{105}{4}$ and $\frac{8685}{8}$, are divided by 15 the greatest common divisor of their numerators, the quotients are fractions, viz: $\frac{7}{4}$ and $\frac{579}{8}$. If the divisor 15 be divided by 8, the quotients must be multiplied by 8, (Art. 73), and become whole numbers, viz: 14 and 579. Now, 8 is the smallest possible number, which, used as a multiplier, will convert the quotients, $\frac{7}{4}$ and $\frac{579}{8}$,

REVIEW.—133. What is the rule for finding the greatest common divisor of fractions? Illustrate and prove it.

into whole numbers at the same time, since it is the least common multiple of the denominators 4 and 8; therefore, as $\frac{15}{8}$, when used as a divisor, gives the smallest possible whole numbers for quotients, it must be the greatest common divisor of the given numbers.

FIND THE GREATEST COMMON DIVISOR

1. Of $83\frac{1}{3}$, and $268\frac{3}{4}$ . . . . *Ans.* $2\frac{1}{12}$
2. Of $14\frac{7}{12}$ and $95\frac{3}{8}$ . . . . *Ans.* $\frac{7}{24}$
3. Of $59\frac{1}{9}$ and $735\frac{14}{15}$ . . . . *Ans.* $2\frac{43}{45}$
4. Of $23\frac{7}{16}$ and $213\frac{13}{24}$ . . . . *Ans.* $2\frac{29}{48}$
5. Of $418\frac{3}{5}$ and $1772\frac{1}{3}$. . . . . *Ans.* $\frac{13}{15}$
6. Of $237\frac{2}{3}$ and $1751\frac{1}{2}$. . . . . *Ans.* $5\frac{1}{6}$
7. Of $261\frac{13}{14}$ and $652\frac{11}{21}$ . . . . *Ans.* $4\frac{25}{42}$
8. Of $44\frac{4}{9}$, $546\frac{2}{3}$ and 3160 . . . . . *Ans.* $4\frac{4}{9}$
9. Of $137\frac{1}{2}$, $478\frac{1}{8}$ and $2093\frac{3}{4}$ . . . . . *Ans.* $3\frac{1}{8}$
10. Of $3977\frac{1}{6}$, $1022\frac{7}{10}$, $2954\frac{7}{15}$, $16801\frac{1}{2}$. *Ans.* $16\frac{7}{30}$

TO FIND THE LEAST COMMON MULTIPLE OF FRACTIONS.

ART. **134.** RULE.—*Reduce the numbers to simple fractions and to their lowest terms. Find the least common multiple of the numerators, and divide it by the greatest common divisor of the denominators; the quotient will be the least common multiple of the fractions.*

Find the least com. multiple of $3\frac{3}{4}$, $4\frac{1}{6}$, $1\frac{1}{8}$, $\frac{5}{12}$.

SOLUTION.—The numbers, when reduced to the form of simple fractions, become $\frac{15}{4}$, $\frac{25}{6}$, $\frac{9}{8}$ and $\frac{5}{12}$; the least common multiple of the numerators is 225; the greatest common divisor of their denominators is 2; the former divided by the latter gives $112\frac{1}{2}$, the least common multiple of the given numbers.

DEMONSTRATION.—If the given numbers were the whole numbers 15, 25, 9, 5, their least common multiple would be 225, which would contain them respectively 15, 9, 25 and 45 times; moreover, 225 contains $\frac{15}{4}$, $\frac{25}{6}$, $\frac{9}{8}$, $\frac{5}{12}$, respectively 4×15, 6×9, 8×25, 12×45 times, since dividing the divisor multiplies the quotient, (Art. 73); and $\frac{1}{2}$ of $225=112\frac{1}{2}$ will contain the same numbers half as many times respectively, viz: 2×15, 3×9, 4×25, 6×45 times; and since no other number will divide all these quotients exactly, $112\frac{1}{2}$ is the least number which will contain the fractions exactly, and is, therefore, their least common multiple.

REMARK.—The least common multiple of fractional numbers can also be found by Rule 3, Art. 99, taking care to obtain the greatest common divisors required, by the rule in the last article.

FIND THE LEAST COMMON MULTIPLE

1. Of $\frac{2}{3}$, $\frac{3}{4}$, $\frac{4}{5}$, $\frac{5}{6}$ and $\frac{6}{7}$ . . . . . . *Ans.* 60.
2. Of $4\frac{1}{2}$, $6\frac{3}{4}$, $5\frac{5}{6}$ and $10\frac{1}{2}$ . . . . . *Ans.* $472\frac{1}{2}$
3. Of $3\frac{1}{3}$, $4\frac{3}{8}$, $\frac{5}{12}$, $5\frac{5}{9}$ and $12\frac{1}{2}$ . . . *Ans.* 350.
4. Of $14\frac{2}{7}$, $9\frac{1}{11}$, $16\frac{2}{3}$ and 25 . . . *Ans.* 100.
5. Of $18\frac{3}{4}$, $66\frac{2}{3}$, $21\frac{3}{7}$ and 15 . . . *Ans.* 600.
6. Of $9\frac{3}{8}$, $22\frac{11}{12}$, $25\frac{5}{16}$ and $12\frac{3}{20}$ . . *Ans.* $16706\frac{1}{4}$
7. Of $8\frac{1}{6}$, $10\frac{8}{9}$, $5\frac{13}{15}$, $6\frac{5}{12}$ and $\frac{2}{3}$ . . *Ans.* $1437\frac{1}{3}$
8. Of $\frac{5}{12}$, $\frac{7}{18}$, $1\frac{19}{30}$, $3\frac{5}{24}$ and $3\frac{31}{48}$ . . *Ans.* $2245\frac{5}{6}$
9. Of $12\frac{1}{2}$, $13\frac{1}{3}$, $14\frac{1}{4}$, $15\frac{1}{5}$ and $17\frac{1}{7}$. . *Ans.* 11400.

ART. 135. PROMISCUOUS EXERCISES.

1. Add together $3\frac{1}{2}$, $4\frac{1}{3}$, $5\frac{1}{4}$, $\frac{3}{4}$ of $\frac{7}{8}$, and $\frac{1}{2}$ of $\frac{1}{3}$ of $\frac{5}{8}$. *Ans.* $13\frac{27}{32}$

2. The sum of $1\frac{1}{26}$ and $\dfrac{1}{1\frac{4}{9}}$ is equal to how many times their difference? *Ans.* 5 times.

3. What is $\left\{2\frac{3}{4}+\dfrac{5}{2} \text{ of } \dfrac{7}{3\frac{4}{5}}-\dfrac{1\frac{2}{3}}{2\frac{1}{2}}\right\}\div 1\frac{77}{228}$? *Ans.* 5.

4. Reduce $\dfrac{4\frac{4}{15} \text{ of } 2\frac{5}{8}}{5\frac{1}{5}-4\frac{1}{2}}$; and $\dfrac{5}{7}\times(100-\dfrac{200}{3}+\dfrac{7\frac{1}{3}}{2\frac{1}{4}})$ to their simplest forms. *Ans.* 16 and $26\frac{26}{189}$

5. What is $\frac{1}{4}$ of $5\frac{1}{4}-\frac{1}{3}$ of $3\frac{7}{8}$? *Ans.* $\frac{1}{48}$

6. What is $\frac{32}{51}\times\frac{85}{112}\times\frac{189}{207}\times\frac{23}{36}$ equal to? *Ans.* $\frac{5}{18}$

7. Also $\dfrac{1}{2}\times\dfrac{1-\frac{1}{2}}{2}\times\dfrac{2-\frac{1}{2}}{3}$? *Ans.* $\frac{1}{16}$

8. Also $\dfrac{1}{3}\times\dfrac{1-\frac{1}{3}}{2}\times\dfrac{2-\frac{1}{3}}{3}\times\dfrac{3-\frac{1}{3}}{5}\times\dfrac{4-\frac{1}{3}}{4}$? *Ans.* $\frac{22}{729}$

9. $\dfrac{(2+\frac{1}{5})\div(3+\frac{1}{7})}{(2-\frac{1}{3})\times(4-3\frac{3}{7})}=$ what? *Ans.* $\frac{147}{200}$

REVIEW.—134. What is the rule for finding the least common multiple of fractions? Illustrate and prove it.

10. $\dfrac{4\frac{1}{3} \times 4\frac{1}{3} \times 4\frac{1}{3} - 1}{4\frac{1}{3} \times 4\frac{1}{3} - 1} =$ what? *Ans.* $4\frac{25}{48}$

11. Add $\frac{2}{3}$ of $\frac{3}{4}$ of $\frac{7}{8}$, $\frac{1}{5} \times \frac{2}{3}$ of $1\frac{7}{8}$, and $\frac{1}{4}$ *Ans.* $\frac{15}{16}$

12. $\frac{3}{5}$ of $\frac{10}{9}$ of what number, diminished by $\dfrac{\frac{3}{10}}{2\frac{1}{6} + \frac{7}{30}}$, leaves $\frac{25}{64}$? *Ans.* $\frac{99}{128}$

13. Find the least common multiple of the numbers from 10 to 20 inclusive. *Ans.* 232792560.

14. K leaves L for N, (109 miles apart) at the same time that B leaves N for L. K travels $7\frac{1}{2}$ miles per hour, and B, $8\frac{1}{4}$ miles per hour: in how many hours will they meet, and how far will each have traveled?
Ans. $6\frac{58}{63}$ hours. K, $51\frac{19}{21}$ miles; B, $57\frac{2}{21}$ miles.

15. What number multiplied by $\frac{5}{9}$ of $\frac{3}{7}$ of $3\frac{11}{15}$ will produce $2\frac{1}{2}$? *Ans.* $2\frac{13}{16}$

16. What, divided by $1\frac{3}{5}$, gives $14\frac{3}{4}$? *Ans.* $23\frac{3}{5}$

17. What, added to $14\frac{5}{8}$, gives $29\frac{23}{35}$? *Ans.* $15\frac{9}{280}$

18. I spend $\frac{5}{9}$ of my income in board, $\frac{1}{6}$ in clothes, and save \$60 a year: what is my income? *Ans.* \$216.

19. Find the G. C. D. (greatest common divisor), of 96, 120, 160 and 200, and the least com. mul. of 13, 19, 57 and 65. *Ans.* 8 and 3705.

20. The least com. mul. of 10, 24, 35 and an unknown number prime to these three, is 9240. What is the unknown number? (See Rem. 1, Art. 97). *Ans.* 11.

21. What two numbers between 35 and 840, have the former for their G. C. D., and the latter for their least com. mul.? (See Note 3, Art. 99). *Ans.* 105 and 280.

22. Find a number between 697 and 731, which shall have with each, the same greatest common divisor that they have with each other. *Ans.* 714.

23. The G. C. D. of three numbers is 15, and their least com. mul. is 450. What are the numbers?
Ans. 30, 45 and 75.

24. $\frac{1}{2}$ is what part of $\frac{2}{3}$? *Ans.* $\frac{3}{4}$

25. Divide $\frac{7}{9}$ of $3\frac{3}{5}$ by $\frac{13}{14}$ of 7; and $\frac{9}{11}$ of $\frac{2}{3}$ of $27\frac{1}{2}$ by $\frac{4}{9}$ of $\frac{3}{17}$ of $5\frac{1}{2}$. *Ans.* $\frac{28}{65}$ and $34\frac{17}{22}$

26. Multiply $\frac{7}{11}$ of $2\frac{1}{2}$ by $\frac{3}{13}$ of $19\frac{1}{2}$; and divide $\frac{7}{8}$ of $\frac{5}{9}$ of $14\frac{1}{7}$ by $\frac{3}{11}$ of $\frac{2}{7}$ of $13\frac{4}{9}$. *Ans.* $7\frac{7}{44}$ and $6\frac{9}{16}$

27. Add together $2\frac{1}{13}$, $3\frac{4}{5}$, $2\frac{3}{7}$ and $5\frac{6}{11}$, and divide the sum by $2\frac{82}{91}$. *Ans.* $4\frac{937}{1210}$

28. Express $\dfrac{1\frac{3}{4}+2\frac{5}{6}}{5\frac{1}{2}+4\frac{1}{5}}$ as a simple fraction; and also multiply $\frac{5}{9}$ of $2\frac{1}{4}$ by $\frac{3}{8}$ of $\frac{7}{15}$. *Ans.* $\frac{275}{582}$ and $\frac{7}{32}$

29. A bequeathed $\frac{11}{20}$ of his estate to his elder son; the rest to his younger, who received $525 less than his brother. What was the estate? *Ans.* $5250.

30. Find the sum, difference, and product, of $3\frac{7}{8}$ and $2\frac{1}{9}$; also the quotient of their sum by the difference.
Ans. sum $5\frac{71}{72}$, diff. $1\frac{55}{72}$, prod. $8\frac{13}{72}$, quot. $3\frac{50}{127}$

31. A cargo is worth 7 times the ship: what part of the cargo is $\frac{5}{16}$ of the ship and cargo? *Ans.* $\frac{5}{14}$

32. If a railroad car runs $112\frac{3}{4}$ miles in $5\frac{1}{6}$ hours, what is the rate per hour? *Ans.* $21\frac{51}{62}$

33. Subtract $\frac{2}{3}$ of $\frac{5}{17}$ of $6\frac{4}{5}$ from $\frac{7}{8}$ of $5\frac{3}{4}$; and multiply $19\frac{5}{7}$ by $\dfrac{3}{5}$ of $\dfrac{2\frac{1}{2}}{5\frac{3}{4}}$. *Ans.* $3\frac{67}{96}$ and $5\frac{1}{7}$

34. $6\frac{1}{2}$ is what part of $10\frac{7}{11}$? and reduce to its simplest form $\frac{8}{9}-\frac{1}{7}+\frac{5}{8}-1\frac{1}{4}$. *Ans.* $\frac{11}{18}$ and $\frac{61}{504}$

35. Multiply $\dfrac{4}{5\frac{1}{3}}$, $14\frac{1}{7}$, $\dfrac{2\frac{3}{4}}{4}$, $\dfrac{5}{7\frac{1}{3}}$, $\dfrac{1\frac{1}{5}}{2\frac{3}{4}}$, and 6. *Ans.* $13\frac{1}{56}$

36. $\frac{7}{8}$ of $\frac{5}{9}$ of what number equals $9\frac{13}{18}$? *Ans.* 20.

37. A 63 gallon cask is $\frac{5}{8}$ full: $9\frac{1}{2}$ gallons being drawn off, how full will it be? *Ans.* $\frac{239}{504}$

38. If a person going $3\frac{3}{4}$ miles per hour, performs a journey in $14\frac{3}{4}$ hours, how long would he be, if he traveled $5\frac{1}{4}$ miles per hour? *Ans.* $10\frac{15}{28}$ hours.

39. A man buys $32\frac{3}{4}$ pounds of coffee at $17\frac{5}{8}$ cts. a pound: if he had got it $4\frac{2}{3}$ cts. a pound cheaper, how many more pounds would he have received? *Ans.* $11\frac{247}{311}$

IX. DECIMAL FRACTIONS.

Art. 136. A Decimal fraction derives its name from the Latin word *decem*, meaning *ten*, and is so called, because its denominator is always 1 with ciphers annexed: being 10, or the product of several 10's; thus,

$\frac{37}{100}$, $\frac{84015}{1000}$ and $\frac{692}{1000000}$ are decimal fractions.

All the rules and operations in the several cases of Common Fractions apply as well to Decimal Fractions when thus written. But, a more simple and convenient Notation has been devised for them similar to that of whole numbers.

This notation consists in writing the numerator, and placing a point, (.), so that the number of figures on the right of it shall be equal to the number of ciphers in the denominator.

The decimal fractions before given, when expressed in this way are .37 and 84.015 and .000692

The use of the point instead of the denominator, saves time and space, while no mistake is likely to occur on this account, since the denominator can be obtained as follows:

The denominator of any decimal fraction is 1 *with as many ciphers annexed as there are figures on the right of the point.*

Art. **137.** A decimal fraction, when written with a point, is simply called a *decimal.*

The places and figures on the *right* of the point are called *decimal places* and *decimal figures*, to distinguish them from the places and figures of whole numbers.

The *point*, (.), is called the *decimal* point or *separatrix;* it separates the decimal places from the places of whole numbers.

A *pure decimal* has only decimal figures; as, .02319

A *mixed decimal* has figures of whole numbers; as, 281.63

A *complex decimal* has a common fraction in its right-hand place; as, $.8\frac{1}{3}$ and $2.62\frac{1}{2}$

A whole number may be regarded as a decimal, by supposing a point to be on the right of its units' place; as, 154 = 154.

Review.—136. What is a Decimal Fraction? Why so called? Give examples. What mode of expressing them has been adopted? Illustrate it. What is the advantage of writing decimal fractions with a point? When thus written, how can the denominator be known? 137. What is a decimal fraction called, when written with a point? What are the figures on the right of the point called? What is the point called? What is a pure decimal? What is a mixed decimal? a complex decimal? Give examples. How may every whole number be regarded?

NUMERATION OF DECIMALS.

ART. **138**. Since $.6 = \frac{6}{10}$; $.06 = \frac{6}{100}$; and $.006 = \frac{6}{1000}$, any figure expresses *tenths*, *hundredths*, or *thousanths*, according as it is in the 1st, 2d, or 3d decimal place hence, these places are named respectively the *tenths'*, the *hundreths'*, the *thousandths'* place; other places are named in the same way, as seen in the

TABLE OF DECIMAL ORDERS.

Tenths. Hundredths. Thousandths. Ten-thousandths. Hundred-thousandths. Millionths. Ten-millionths. Hundred-millionths. Billionths.

1st	place	.2	read	2 tenths.
2d	..	.08	..	8 Hundreths.
3d	..	.005	..	5 Thousanths.
4th	..	.0007	..	7 Ten-thousandths.
5th	..	.00003 . . .	..	3 Hundred-thousandths.
6th	..	.000001 . . .	..	1 Milionth.
7th	..	.0000009 .	..	9 Ten-Millionths.
8th	..	.00000004 .	..	4 Hundred-millionths.
9th	..	.000000006	..	6 Billionths.

The names of the decimal orders are derived from the names of the orders of whole numbers. The table may therefore be extended to Trillionths, Quadrillionths, &c.

ART. **139**. By inspecting the table, and recollecting that 1 tenth = 10 hundredths, and 1 hundredth = 10 thousandths, it is clear that the decimal places decrease in value from left to right, like the places of whole numbers, and according to the same law, viz:

1 *in any place equals* 10 *in the next right-hand place.*

This law holds good in every mixed decimal, like 715.2309; for, in all such, the last figure of whole numbers, (5), is *units*, and the first decimal figure on the right is *tenths*, and 1 unit = 10 tenths.

REVIEW.—138. What is the name of the 1st decimal place? The 2d? Third? Why? Repeat the Table of Decimal Orders.

ART. **140**. Since decimals are subject to the same law of local value as whole numbers, like them, also, they can be read in either of two ways.

RULE FOR READING DECIMALS.

1st. *Read in succession the value of the separate figures which compose the decimal;* or, which is much more convenient,

2d. *Read the decimal as a whole number, and annex the name of the right-hand place.*

The decimal .004038, is read, 4038 millionths.

DEM.—The reason of the rule depends on the law of local value, viz: "1 *in any place equals* 10 *in the next right-hand place.*"

Commencing with the first significant figure, 4 of the 3d place equal 40 of the 4th place; 40 of the 4th place equal 400 of the 5th place, which, with the 3 already there, make 403 of the 5th place; finally, 403 of the 5th place equal 4030 of the 6th place, which, with the 8 already there, make in all 4038 of the 6th place; or, 4038 millionths, since the 6th place expresses millionths.

A mixed decimal may be read altogether as a decimal; or, the whole number may be read, then the decimal.

Thus, 71.062 may be read 71062 thousandths; or, 71 units, and 62 thousandths.

EXAMPLES TO BE READ.

1.	.9	9.	00.100	17.	41.14414
2.	.0$\frac{3}{7}$	10.	180.010	18.	411.4414
3.	.305	11.	20300.0	19.	4.114414
4.	.7200	12.	40.68031	20.	15.0046$\frac{1}{4}$
5.	.5060	13.	207.2007	21.	73002.1$\frac{4}{5}$
6.	1.008	14.	.0900001	22.	.000000$\frac{1}{9}$
7.	9.00$\frac{1}{3}$	15.	61.001001	23.	.200006
8	105.0$\frac{7}{8}$	16.	9230010.0	24.	526.000

25.	12.3333333	28.	.437800629
26.	643000.643	29.	1000000.0303
27.	4009.62007	30.	200000.000006

REVIEW.—139. How do decimal places resemble those of whole numbers? What is the law that governs both? Does this law apply to mixed decimals? Why? 140. In how many ways may decimals be read? What is the 1st? The 2d? Prove the 2d rule. How may mixed decimals be read? Give an example.

NOTATION OF DECIMALS.

ART. **141.** Write, eighty-three thousand and one billionths.

DEMONSTRATION.—The *number* of parts must be written as a whole number, (Art. 136); the right-hand figure must express parts of the given size, (See last Rule); hence, the

1st step 83001 = Numerator.
2d step .000083001 = Decimal.

RULE FOR WRITING DECIMALS.

Write the numerator; fix the point so that the right-hand figure shall be of the same name as the decimal.

REMARKS.—1. In fixing the point, it may be necessary to prefix ciphers to the numerator, as in the example just given; but in the case of an improper decimal fraction, the point will fall between two of the figures; thus, 346 tenths is written 34.6, which may also be read 34 units and 6 tenths.

2. The operations under this and the preceding rule serve to prove each other.

EXAMPLES TO BE WRITTEN.

1. Five tenths.
2. Twenty-two hundredths.
3. One hundred and four thousandths.
4. Two units and one hundredth.
5. One thousand six hundred and five ten-thousandths.
6. Eighty-seven hundred-thousandths.
7. Twenty-nine and a half ten-millionths.
8. Nineteen million and one billionths.
9. Seventy thousand and forty-two units and sixteen hundredths.
10. Two thousand units and fifty-six and a third millionths.
11. Four hundred and twenty-one tenths.
12. Six thousand hundredths.
13. Eight units and a half a hundredth.
14. Forty-eight thousand three hundred and five thousandths.
15. Thirty-three million ten millionths.
16. Four hundred thousandths
17. Four hundred-thousandths.
18. One unit and a half a billionth.
19. Sixty-six thousand and three millionths.
20. Sixty-six million and three thousandths.
21. Thirty-four and a third tenths.

REVIEW.—141. What is the rule for writing decimals? Explain it.

22. Forty-four million units and four millionths.
23. Two hundred and eighteen thousand and six billionths.
24. Ninety-six thousands.
25. Ninety-six hundreds.
26. Ninety-six tens.
27. Ninety-six units.
28. Ninety-six tenths.
29. Ninety-six hundredths.
30. Ninety-six thousandths.
31. Three hundred and fifty-eight thousand and six ten-millionths.
32. Two million millionths.
33. Four million units and four millionths.
34. Four million and four millionths.
35. Fifty-thousand and seven hundred-thousandths.
36. Three million and a half billionths.

ART. **142.** Decimal Fractions are distinguished from common fractions, by not having written denominators; and from whole numbers, by the decimal point.

The *denomination*, or *size of the parts* in any decimal, depends on the position of the *point;* so that the set of figures which expresses but one value as a whole number, may, as a *decimal*, express different values, according to the situation of the point. *Great care must be exercised, then, in placing the point correctly and distinctly.*

ART. **143.** PROPOSITION I.—*Decimal ciphers may be annexed to, or omitted from, the right of any number, and not alter its value.*

DEMONSTRATION.—The ciphers themselves are of no value; the other figures retain their places and, consequently, their values. Hence, the *value* of the number is not altered, but merely its *denomination.*

EXAMPLES.

.25 = .250
16 = 16. = 16.000
19.08300 = 19.083
200.00 = 200

REMARK.—Be careful that the ciphers annexed or omitted are *decimal* ciphers.

NOTE.—Annexing or omitting ciphers is equivalent to *multiplying or dividing both terms of the decimal fraction by* 10, 100, 1000, &c.

ART. **144.** PROPOSITION II.—*If, in any decimal, the point be moved to the* RIGHT, *the number is* MULTIPLIED *continually by* 10 *as often as a figure is passed over.*

REVIEW.—142. How are decimals distinguished from common fractions? How from whole numbers? How is the size of the parts in any decimal known? Why should great care be exercised in placing the point correctly? 143. What is Proposition 1? Prove it.

DEMONSTRATION.—For each place passed over by the point in moving to the *right*, every figure is *advanced* one step in the scale of notation, and is worth 10 times as much as before; hundreths are changed into tenths, tenths into units, units into tens, tens into hundreds, and so on; hence, the proposition is true.	EXAMPLES. .0 5 6 7 0.5 6 7 5.6 7 5 6.7 5 6 7. 5 6 7 0.

ART 145. PROPOSITION III.—*If, in any decimal, the point be moved to the* LEFT, *the number is* DIVIDED *continually by* 10 *as often as a figure is passed over.*

DEMONSTRATION.—For each place passed over by the point in moving to the *left*, every figure is *degraded* one step in the scale of notation, and is worth only $\frac{1}{10}$ as much as before; hundreds are changed into tens, tens into units, units into tenths, tenths into hundredths, and so on; hence, the proposition is true.	EXAMPLES. 2 3 4 0. 2 3 4. 2 3.4 2.3 4 .2 3 4 .0 2 3 4

NOTE.—If the point is to be moved in either direction over more places than the number can furnish, supply the deficiency by taking in ciphers, as in the last examples of these propositions.

REMARK.—These 3 propositions apply to any whole number, supposing a point to stand on the right of its units' place.

REDUCTION OF DECIMALS.

ART. 146. CASE I.—To convert a decimal into its simplest equivalent common fraction.

RULE.—*Take the decimal as it stands, for the numerator, commencing with the first significant figure; for the denominator, write* 1 *with as many ciphers annexed as there are decimal places; then reduce this fraction to its lowest terms.*

NOTE.—In reducing the fraction to its lowest terms, use no divisors but 2 and 5; since they are the only prime factors of 10, and, therefore, the only prime number that will divide the denomi-

REVIEW.—144. State Proposition 2. Prove it. 145. State Proposition 3. Prove it. How are deficient places supplied? How can these propositions apply to whole numbers? 146. What is the rule for reducing a decimal to its simplest equivalent common fraction?

nator, which is either 10, or the product of 10's. When the numerator ends in any odd number except 5, neither 2 nor 5 will divide it, (Art. 90), and the fraction will be in its lowest terms.

Reduce .039375 to its equivalent common fraction.

SOLUTION.—After obtaining the common fraction, divide both terms by 5, four times in succession; the numerator then ends in 3, and the fraction can be reduced no further. (Note.)

$$.039375 = 5)\frac{39375}{1000000} = 5)\frac{7875}{200000} = 5)\frac{1575}{40000} = 5)\frac{315}{8000} = \frac{63}{1600} \textit{ Ans.}$$

A mixed decimal like 18.0067 may be written as an improper fraction $\frac{180067}{10000}$; or, as a mixed number thus: $18\frac{67}{10000}$.

By this rule, a complex decimal is converted into a complex common fraction, which may be simplified as in Art. 132.

$$\text{Thus, } .08\tfrac{1}{3} = \frac{8\frac{1}{3}}{100} = \frac{25}{300} = \frac{5}{60} = \frac{1}{12} \textit{ Ans.}$$

The rule also serves to change a part of a pure decimal into a common fraction, thereby rendering the decimal complex, as, $6.875 = 6.87\frac{5}{10} = 6.87\frac{1}{2}$; in this way a decimal may sometimes be easily reduced to a common fraction, if the student is familiar with the aliquot parts of 100. Thus,

$$.039375 = .0393\tfrac{3}{4} = .03\tfrac{15}{16} = \frac{3\frac{15}{16}}{100} = \frac{63}{1600}, \text{ since } \frac{75}{100} = \frac{3}{4} \text{ and } \frac{93\frac{3}{4}}{100} = \frac{15}{16}.$$

REDUCE TO COMMON FRACTIONS.

1.	.25625	*Ans.* $\frac{41}{160}$	9.	$11.0\frac{5}{9}$	*Ans.* $11\frac{1}{18}$
2.	.15234375	*Ans.* $\frac{39}{256}$	10.	.390625	*Ans.* $\frac{25}{64}$
3.	2.125	*Ans.* $2\frac{1}{8}$	11.	$.1944\frac{4}{9}$	*Ans.* $\frac{7}{36}$
4.	19.01750	*Ans.* $19\frac{7}{400}$	12.	$.24\frac{4}{9}$	*Ans.* $\frac{11}{45}$
5.	$16.00\frac{1}{5}$	*Ans.* $16\frac{1}{500}$	13.	$.33\frac{1}{3}$	*Ans.* $\frac{1}{3}$
6.	$350.028\frac{4}{7}$	*Ans.* $350\frac{1}{35}$	14.	$.66\frac{2}{3}$	*Ans.* $\frac{2}{3}$
7.	$.666666\frac{2}{3}$	*Ans.* $\frac{2}{3}$	15.	.25	*Ans.* $\frac{1}{4}$
8.	.003125	*Ans.* $\frac{1}{320}$	16.	.75	*Ans.* $\frac{3}{4}$

REVIEW.—146. What divisors only need be used in reducing to the lowest terms? Why? How can it be known, by inspection, that the fraction is in its lowest terms? Why? How may a mixed decimal be changed to a common fraction? What does a complex fraction become by the application of the rule? How can a pure decimal be rendered complex? Give examples.

17.	$.16\frac{2}{3}$	*Ans.* $\frac{1}{6}$	26.	$.91\frac{2}{3}$	*Ans.* $\frac{11}{12}$
18.	$.83\frac{1}{3}$	*Ans.* $\frac{5}{6}$	27.	$.06\frac{1}{4}$	*Ans.* $\frac{1}{16}$
19.	$.12\frac{1}{2}$ or .125	*Ans.* $\frac{1}{8}$	28.	$.18\frac{3}{4}$	*Ans.* $\frac{3}{16}$
20.	$.37\frac{1}{2}$ or .375	*Ans.* $\frac{3}{8}$	29.	$.31\frac{1}{4}$	*Ans.* $\frac{5}{16}$
21.	$.62\frac{1}{2}$ or .625	*Ans.* $\frac{5}{8}$	30.	$.43\frac{3}{4}$	*Ans.* $\frac{7}{16}$
22.	$.87\frac{1}{2}$ or .875	*Ans.* $\frac{7}{8}$	31.	$.56\frac{1}{4}$	*Ans.* $\frac{9}{16}$
23.	$.08\frac{1}{3}$	*Ans.* $\frac{1}{12}$	32.	$.68\frac{3}{4}$	*Ans.* $\frac{11}{16}$
24.	$.41\frac{2}{3}$	*Ans.* $\frac{5}{12}$	33.	$.81\frac{1}{4}$	*Ans.* $\frac{13}{16}$
25.	$.58\frac{1}{3}$	*Ans.* $\frac{7}{12}$	34.	$.93\frac{3}{4}$	*Ans.* $\frac{15}{16}$

As much work is sometimes saved, by using a common fraction instead of its equivalent decimal, the pupil should be familiar with this transformation.

REMARK.—If the decimal contain a great many places, a simple approximate value can sometimes be obtained by using the first one or two figures only; for example, .3260873 is nearly $\frac{33}{100} = \frac{1}{3}$ nearly; and $1.7689 = 1\frac{75}{100}$ nearly $= 1\frac{3}{4}$ nearly.

CASE II.—TO CHANGE ANY FRACTION INTO A DECIMAL.

ART. **147.** If the fraction has 10, 100, 1000, &c., for its denominator, it can be written as a decimal, by *writing the numerator, and placing the point so that the number of figures on the right of it, shall be equal to the number of ciphers in the denominator.*

EXPRESS IN DECIMAL FORM,

$$\frac{3}{10}, \frac{8}{100}, \frac{109}{1000}, \frac{120056}{10000}, \frac{52\frac{2}{3}}{100000}, \frac{1600}{1000000},$$

Ans. .3, .08, .109, 12.0056, $.00052\frac{2}{3}$, .001600,

If the denominator of the fraction, is not 10, 100, 1000, use this

RULE.—*Divide the numerator by the denominator, annexing decimal ciphers to the former as they are needed; for each cipher annexed, make a decimal place in the quotient.*

NOTES.—1. Before annexing ciphers to the numerator, be careful to place a point on its right.

REVIEW.—147. How can a fraction whose denominator is 10, 100, 1000, &c., be written as a decimal? If the denominator is not 10, 100, 1000, &c.? What should be done before annexing the ciphers?

2. The point may be fixed in the quotient at any time, *by making the figure last written in the quotient occupy the same decimal place as the figure of the dividend last used;* in doing so, prefix ciphers to the quotient, if necessary, to make the requisite number of places.

3. The operations under this and the preceding rule serve to prove each other.

Reduce $\frac{7}{8}$ to its equivalent decimal.

OPERATION.

$$8)\underline{7.000}$$
$$.875$$
$$\tfrac{7}{8} = .875 \text{ Ans.}$$

DEMONSTRATION.—The numerator with the decimal ciphers annexed, is of the same value as before, but of a lower denomination, (Art. 143), and, when it is divided by the denominator, the quotient must be of that denomination also, and must therefore contain the same number of decimal places.

The analysis of the operation is as follows: $\frac{7}{8} = \frac{1}{8}$ of 7 units $= \frac{1}{8}$ of 7000 thousandths $=$ 875 thousandths $= .875$

REDUCE TO DECIMALS,

1. $\frac{3}{4}$	$= .75$	6. $\frac{4}{5}$	$= .8$
2. $\frac{1}{8}$	$= .125$	7. $\frac{99}{200}$	$= .495$
3. $\frac{1}{20}$	$= .05$	8. $\frac{5}{64}$	$= .078125$
4. $\frac{15}{32}$	$= .46875$	9. $\frac{13}{256}$	$= .05078125$
5. $\frac{9}{1600}$	$= .005625$	10. $\frac{1}{1024}$	$= .0009765625$

The rule converts a mixed number into a mixed decimal, and a complex into a pure decimal; thus, $9\frac{3}{8} = 9.375$, since $\frac{3}{8} = .375$; and $.26\frac{3}{25} = .2612$, since $\frac{3}{25} = .12$

11. $16\frac{1}{2}$	$= 16.5$	13. $.015\frac{1}{4}$	$= .01525$
12. $42\frac{3}{16}$	$= 42.1875$	14. $101.01\frac{3}{4}$	$= 101.0175$

15. $75119\frac{3}{80}$ . . . . $= 75119.0375$

16. $2.00\frac{1}{320}$ . . . . $= 2.00003125$

ART. **143**. Sometimes annexing ciphers does not render the numerator exactly divisible by the denominator; in that case, after the quotient has been carried out as far as desirable, the sign $(+)$ *plus* is annexed to show that there is still a remainder. Such, *having no end*, are called

REVIEW.—147. When may the point be fixed in the quotient? How? Explain the example. What effect does the rule have on a mixed number? On a complex decimal?

interminate or *infinite* decimals; thus, $\frac{2}{3}$ = .666666 +, an *interminate;* while $\frac{1}{8}$ = .125, a *terminate* decimal.

Sometimes, when the remainder omitted in an *interminate* decimal is large enough to give the next quotient figure more than 5, the last quotient figure is written 1 larger than it really is, and the sign (—) *minus* is annexed instead of (+) *plus*, to show that the quotient is written a little too large; as, $\frac{2}{3}$ = .66 +, or .67 —.

1. $\frac{7}{27}$ . . . . . . = .259259+
2. $10\frac{29}{48}$ . . . . . = 10.60417—
3. $.065\frac{2}{7}$ . . . . = .0652857+
4. $430.18\frac{1}{11}$ . . . . . . = 430.1809+

ADDITION OF DECIMALS.

ART. 149. RULE.—*Write the numbers to be added so that figures of the same denomination may be in columns; add as in whole numbers, commencing at the right, and point off the result to agree with that one of the given numbers having the most decimal places.*

PROOF.—Same as in addition of whole numbers.

NOTE.—Complex decimals, if there are any, must be made pure, (Art. 147), as far, at least, as the decimal places extend in the other numbers. If, after this, there are common fractions in the right-hand column, add them; or, neglect them, using the signs + or — as in Art. 148.

Add 23.8 and $17\frac{1}{2}$ and .0256 and $.41\frac{2}{3}$

SOLUTION.—After reducing, set figures of the same order in column; then add and carry as in whole numbers. As the right-hand figures, when added, must make the right-hand figure of the answer of the same denomination as those in the column, the point should be fixed as the rule directs.

OPERATION.

$$\begin{array}{rr} & 23.8 \\ 17\frac{1}{2} = & 17.5 \\ & .0256 \\ .41\frac{2}{3} = & .4166\frac{2}{3} \\ \hline \textit{Ans.} & 41.7422\frac{2}{3} \end{array}$$

REVIEW.—148. When the quotient can not be made exact, what is done? What are such decimals called? Why? When may the sign *minus* be used?

1. Find the sum of $1 + .9475$ *Ans.* 1.9475

2. Of $1.33\frac{1}{3}$ added to itself twice. *Ans.* 4.

3. Of 14.034, 25, $.000062\frac{1}{2}$, .0034 *Ans.* 39.0374625

4. Of 83 thousandths, 2101 hundredths, 25 tenths, and $94\frac{1}{2}$ units. *Ans.* 118.093

5. Of $.16\frac{2}{3}$, $.37\frac{1}{2}$, 5, $3.4\frac{3}{8}$, $.000\frac{7}{8}$ *Ans.* $8.980\frac{1}{24}$

6. Of 4 units, 4 tenths, 4 hundredths. *Ans.* 4.44

7. Of $.11\frac{1}{9} + .6666\frac{2}{3} + .222222\frac{2}{9}$ *Ans* 1.

8. Of $.14\frac{2}{7}$, $.018\frac{3}{5}$, 920, $.0139\frac{3}{7}$ *Ans.* 920.1754

9. Of $16.008\frac{7}{9}$, $.0074\frac{2}{3}$, $.2\frac{5}{6}$, $.00019042\frac{1}{5}$ *Ans.* $16.299768199\frac{7}{9}$

10. Of .675, 2 millionths, $64\frac{1}{8}$, and 3.49000107 *Ans.* 68.29000307

11. Of four times $4.067\frac{7}{8}$ and $.000\frac{1}{2}$ *Ans.* 16.272

12. Of 216.86301, 48.1057, .029, 1.3, 1000. *Ans.* 1266.29771

13. Add 35 units, 35 tenths, 35 hundredths, 35 thousandths. *Ans.* 38.885

14. Add ten thousand and one millionths; four hundred-thousandths; 96 hundredths; forty-seven million sixty thousand and eight billionths. *Ans.* 1.017101008

SUBTRACTION OF DECIMALS.

ART. **150.** RULE—*Write the subtrahend under the minuend, placing figures of the same denomination in columns. Subtract as in whole numbers, commencing at the right, and point the result as in addition of decimals.*

PROOF.—As in subtraction of whole numbers.

NOTES.—If either or both of the given decimals be complex, proceed as directed in the note to last rule.

REVIEW.—149. What is the rule for adding decimals? The proof? What must be done with complex decimals? If there are common fractions in the right-hand column, what should be done? Explain the example. 150. What is the rule for subtraction of decimals? The proof?

2. If the minuend has not as many decimal places as the subtrahend, annex decimal ciphers to it, or suppose them to be annexed until the deficiency is supplied.

From 6.8 subtract 2.057

SOLUTION.—Write the numbers as the rule directs; suppose ciphers to be annexed to the 8, and subtract as in whole numbers; saying 7 from 10 leaves 3, carry 1; 6 from 10 leaves 4, and so on.

$$\begin{array}{r} 6.8 \\ 2.057 \\ \textit{Ans. } 4.743 \end{array}$$

From $13.256\frac{5}{9}$ subtract $6.77\frac{1}{3}$

In this example, the complex decimals are rendered pure to the same extent, and the common fractions subtracted.

$$\begin{array}{r} 13.256\frac{5}{9} \\ 6.77\frac{1}{3} = 6.773\frac{1}{3} \\ \textit{Ans. } 6.483\frac{2}{9} \end{array}$$

EXAMPLES FOR PRACTICE.

1. Subtract 8.00717 from 19.54 *Ans.* 11.53283
2. 3 thousandths from 3000. *Ans.* 2999.997
3. 72.0001 from 72.01 *Ans.* .0099
4. Subtract $.93\frac{2}{35}$ from $1.169\frac{3}{7}$ *Ans.* $.238\frac{6}{7}$
5. How much is 19 less $8.999\frac{1}{9}$? *Ans.* $10.000\frac{8}{9}$
6. How much less is $.04\frac{1}{3}$ than .4? *Ans.* $.35\frac{2}{3}$
7. How much is $.65007 - \frac{1}{2}$? *Ans.* .15007
8. What is $2\frac{3}{4} - 1\frac{4}{5}$ in decimals? *Ans.* .95
9. Take .007601 twice from .02 *Ans.* .004798
10. Take $1.98\frac{1}{3}$ three times from 6. *Ans.* .05
11. Subtract 1 from 1.684 *Ans.* .684
12. $\frac{5}{6}$ of a millionth from $.000\frac{4}{9}$ *Ans.* $.000443\frac{11}{18}$
13. $1\frac{1}{2}$ hundredths from $49\frac{3}{8}$ tenths. *Ans.* 4.9225
14. 10000 thousandths from 10 units. *Ans.* 0
15. $24\frac{1}{2}$ tenths from 3701 thousandths. *Ans.* 1.251
16. $1\frac{7}{8}$ units from 1875 thousandths. *Ans.* 0
17. $\frac{5}{9}$ of a hundreth from $\frac{1}{18}$ of a tenth. *Ans.* 0
18. $64\frac{1}{6}$ hundredths from 100 units. *Ans.* $99.35\frac{5}{6}$

REVIEW.—150. What must be done with complex decimals? If the minuend do not contain as many decimal places as the subtrahend, what must be done? Explain the examples.

MULTIPLICATION OF DECIMALS.

ART. **151.** RULE.—*Multiply as in whole numbers, and point the product, so that it shall have as many decimal places as the multiplicand and multiplier together.*

PROOF.—As in multiplication of whole numbers.

REMARK.—If the product has not as many decimal places as required, supply the deficiency by prefixing ciphers.

Multiply 2.56 by .184

$$\frac{256}{100} \times \frac{184}{1000} = \frac{47104}{100000}$$

Hence,

$$2.56 \times .184 = .47104$$

DEMONSTRATION.—Express the decimals as common fractions; their product will be the product of their numerators divided by the product of their denominators, (Art. 130). Write the fractions in decimal form; the denominator of the product has as many ciphers as both the other denominators: and as each of these ciphers make a decimal place, (Art. 136), the product will have as many decimal places as both factors.

EXAMPLES FOR PRACTICE.

1. $1 \times .1$ . . . . $= .1$
2. $16 \times .03\frac{1}{3}$ . . . . $= .53\frac{1}{3}$
3. $.01 \times .1\frac{1}{2}$ . . . . $= .0015$
4. $.080 \times 80$ . . . . $= 6.4$
5. $37.5 \times 82\frac{1}{2}$ . . . . $= 3093.75$
6. $64.01 \times .32$ . . . . $= 20.4832$
7. $48000. \times 73.$ . . . . $= 3504000.$
8. $64.66\frac{2}{3} \times 18.$ . . . . $= 1164.$
9. $.56\frac{1}{4} \times .03\frac{1}{16}$ . . . . $= .0172\frac{17}{64}$
10. 738×120.4 . . . . 88855.2
11. $.0001 \times 1.006$ . . . . $= .0001006$
12. 34 units $\times .193$ . . . . $= 6.562$
13. 27 tenths $\times .4\frac{1}{5}$ . . . . $= 1.134$
14. $43.7004 \times .008$ . . . . $= .3496032$
15. $21.0375 \times 4.44\frac{4}{9}$ . . . . $= 93.5$

REVIEW.—151. What is the rule for Multiplication of Decimals? The proof? How are deficient places in the product to be supplied? Prove the rule.

16. 9300.701 × 251 . . . . =2334475.951
17. 430.0126 × 4000 . . . . . = 1720050.4
18. .059 × .059 × .059 . . . . = .000205379
19. 42 units × 42 tenths. . . . . = 176.4
20. $2\frac{1}{4}$ hundredths × 600 . . . . . . = 13.5
21. 7100 × $\frac{1}{8}$ of a millionth. . . = .0008875
22. 26 millions × 26 millionths. . . . . = 676.
23. 2700 hundredths × 60 tenths . . . . = 162.

24. What denomination will a figure in the tenths' place, multiplied by a figure in the hundredths' place, give? *Ans.* .1 × .01 = .001 or thousandths.

25. A figure in the units' place, by one in the hundred-thousandths' place? *Ans.* Hundred-thousandths.

26. 1 thousandth by 1 thousandth? *Ans.* Millionths.

27. 1 hundred by 1 tenth? *Ans.* Tens.

28. 1 in thousands' by 1 in tenths'? *Ans.* Hundreds.

CONTRACTED MULTIPLICATION OF DECIMALS.

ART. **152**. The methods of contracting multiplication in whole numbers, apply also to multiplication of decimals. It is only necessary to call attention to two cases.

CASE I.—TO MULTIPLY A DECIMAL BY 10, 100, 1000, &c.

RULE.—*Remove the decimal point of the multiplicand to the right, over as many places as there are ciphers in the multiplier; the result will be the product required.*

NOTES.—1. If there are not as many places to the right of the point as are required in the operation, supply the deficiency by taking in ciphers, unless the multiplicand be a complex decimal; in that case, the proper figures should be ascertained and set down, (Art. 147).

2. The rule given in Art. 54, for multiplying a whole number by 10, 100, 1000, &c., is a particular case of this, since a whole number may be regarded as a decimal, the point being on the right of the units' place, (Art. 137).

REVIEW.—152. What methods of Contracted Multiplication are used for decimals? What is the 1st Case? Give the rule. How are deficient places supplied? How does this rule apply to whole numbers?

DEMONSTRATION.—To multiply by 10, 100, 1000, &c., is the same as to multiply continually by 10; this is done by moving the point to the right, over as many places as there are ciphers in the multiplier, (Art. 144).

EXAMPLES FOR PRACTICE.

1. $56. \times 100$ . . . . = 5600.
2. $.075 \times 100$ . . . . =7.5
3. $.01\frac{3}{4} \times 1000$ . . . . = 17.5
4. 16.083×10 . . . . = 160.83
5. $10.34\frac{2}{5} \times 100000$ . . . . = 1034400.
6. $98.047\frac{1}{3} \times 1000000$ . . . . . = $98047333\frac{1}{3}$

ART. **153.** Whenever the product of two decimals is not required to contain figures below a certain order, the work may be shortened.

CASE II.—TO MULTIPLY, RESERVING A CERTAIN NUMBER OF DECIMAL PLACES IN THE PRODUCT.

RULE.—*Count off in the multiplicand, the number of decimal places to be reserved, draw a vertical line through the lowest, and write the multiplier so that its units' figure shall fall on this line. Begin at the left of the multiplier to form the partial products, always starting at that figure of the multiplicand which is as far on one side of the line, as the figure of the multiplier then in use is on the other side, carrying the tens, however, obtained by multiplying the next lower figure.*

Set the right-hand figures of these partial products in a column, add and point off the number of decimal places required.

NOTES.—1. The multiplicand should extend one figure further to the right of the line than the multiplier does to the left of it. To accomplish this, it may be necessary to annex decimal ciphers, or to convert a common fraction into a decimal.

2. If the multiplier contain a common fraction, and it becomes necessary to multiply by it, start at the same figure of the multiplicand as in the previous multiplication.

REMARK.—In carrying tens from the nearest rejected figure of the multiplicand, carry 1 ten also for any number of units over 5; thus, for 55, carry 6; for 18, carry 2.

REVIEW.—152. Demonstrate it. 153. What is Case 2? The rule? How far should the multiplicand extend to the right of the line?

Multiply 1.89361 by 3.5672, reserving 3 decimals in the product.

SOLUTION.—Count off 3 decimals in the multiplicand, draw a vertical line through the lowest (3), and proceed thus:

SHORT METHOD.	ORDINARY METHOD.
1.89361	1.89361
3.5672	3.5672
5681		378722
947	13	25527
113	113	6166
13	946	805
6.754	5680|83
	6.754|885592

Commencing with 3, the left-hand figure of the multiplier, and 3 in the multiplicand, which are both on the line, say, 3 times 3 are 9; but 3 times 6 (the next lower figure) are 18, which is nearer 2 tens than 1 ten; carrying these 2 tens to 9 we have 11; set down 1 and carry 1. Keeping to the left, the first partial product is 5681. Next, take 5 in the multiplier, and start at 9 in the multiplicand, carrying 2 for the 15 obtained by multiplying the next lower figure, 3.

The second partial product is 947, whose first figure is set under the first figure (1) of the previous product. Thus go on, using 6 and 7 in the multiplier successively, and starting at 8 and 1 in the multiplicand. As the figures of the multiplicand toward the left are then exhausted, the operation, after adding and pointing off 3 figures, is complete.

DEMONSTRATION.—The reason of the rule consists in the fact, that all the multiplications which would give rise to denominations lower than those required in the product, are omitted.

The rule will not always give the last figure correct. To insure perfect accuracy, *reserve one more figure than is required, and omit the last figure of the product.*

What is the product of $.054363\frac{7}{11}$ by 6458.19 true to 3 decimal places?

SOLUTION.—Reserve 4 decimals, one more than is required; carry out the figures of the multiplicand, (Note 1): point off 4 places in the product, 351.0906, writing the next to the last, 1, instead of 0, because the figure omitted is over 5. This gives 351.091 for the answer.

REVIEW.—153. How are deficient places supplied? What, if the multiplier contains a common fraction? In carrying tens from the rejected figures, what directions must be observed? Solve the example. What denomination in the product, is obtained by multiplying the marked figure of the multiplicand by the units' figure of the multiplier? *Ans.* The lowest.

	EXAMPLES.	*Decimals reserved.*	ANSWERS
1.	$4.7501 \times .002866$	5	.01361
2.	$804.57\frac{2}{3} \times 17.08132$	4	13743.2315
3.	75.062×460.8917	2	34595.45
	NOTE.—Take 75.062 for the multiplier.		
4.	9.012×48.75	1	439.3
5.	$4.804136 \times .010759$	6	.051688
6.	$814\frac{53}{719} \times 26\frac{35}{44}$	3	21813.475
7.	$702.61 \times 1.258\frac{7}{36}$	3	884.020
8.	$849.93\frac{3}{4} \times .0424444$	3	36.075
9.	880.695×131.72 true to units.		116005.
10.	$.025381 \times .004907$	5	.00012
11.	$64.01082 \times .03537$	6	2.264063
12.	7.24651×81.4632	3	590.324
13.	$.681472 \times .01286$	5	.00876
14.	$7.944\frac{4}{9} \times 3.69$	4	29.3150
15.	$.053497 \times .047126$	6	.002521
16.	$1380.37\frac{1}{2} \times .234\frac{5}{8}$	2	324.16

DIVISION OF DECIMALS.

ART. **154.** RULE.—*Make the decimal places of the dividend as many as those of the divisor, if they are less. Divide as in whole numbers, annexing other decimal figures to the dividend as they are needed; point the quotient so that it shall have as many decimal places as the dividend has* MORE *than the divisor.*

PROOF.—Same as in Division of whole numbers.

NOTES.—1. To extend the dividend, decimal ciphers are used; but, if the dividend is a complex decimal, make it pure.

2. If the quotient has not the number of figures required for decimal places, prefix ciphers.

3. If the dividend has the same number of decimal places as the divisor, the quotient is *units*.

REVIEW.—153. What, by multiplying each figure of the multiplier by the figure of the multiplicand as far on the other side of the line? *Ans.* The same. What multiplications are omitted? 154. What is the rule for division of decimals? The proof? How are deficient places in the dividend supplied? How in the quotient?

4. Make complex decimals pure; or, divide them like common mixed numbers; or, multiply both by the least common multiple of the denominators of the common fractions, and *then* divide.

5. If the division is not exact, it may be continued by annexing other decimal figures to the dividend; or the remainder may be written with the divisor under it, as a common fraction.

Divide .50312 by .19

```
.19).50312(2.648
      123      Ans.
       91
       152
         0
```

DEMONSTRATION.—Since the dividend is the product of the divisor and quotient, it must have as many decimal places as both of them, (Art. 151); hence, the quotient must have decimal places enough to make with those of the divisor as many as are in the dividend, which is just as many as the dividend has *more* than the divisor.

Divide 24 by 3.2

```
3.2)24.00(7.5
     1.60
        0
```

Annex a decimal cipher to the dividend, to make it have as many decimal places as the divisor; afterward, annex another, to continue the division.

Divide .07 by 21.6

```
21.6).0700(.0032407+
       520
        880
         1600
           88
```

In long division, annex the ciphers to the remainders as they occur, reckoning them still as decimal places of the dividend; here the dividend has eight decimal places, counting the ciphers annexed to the remainders.

1. Divide $.002\frac{19}{40}$ by $.06\frac{3}{5}$ *Ans.* .0375

SUGGESTION.—Convert the numbers into the pure decimals .002475 and .066; or, multiply both by 40, the least common multiple of the denominators 5 and 40, making the dividend .099, and the divisor 2.64: use the latter method generally.

EXAMPLES FOR PRACTICE.

2. $3 \div .18\frac{3}{4}$ . . . . $= 16.$
3. $4.2 \div .31\frac{1}{4}$ . . . . $= 13.44$

REVIEW.—154. If the dividend and divisor have the same number of decimal places, what is the quotient? How may complex decimals be divided? If the division is not exact, what may be done? Demonstrate the rule. Explain the examples.

4. $63 \div 4000$ . . . . $= .01575$
5. $3.15 \div 375$ . . . . $= .0084$
6. $1.008 \div 18$ . . . . $= .056$
7. $4096 \div .032$ . . . . $= 128000.$
8. $9.7 \div 97000$ . . . . $= .0001$
9. $.9 \div .00075$ . . . . $= 1200.$
10. $13 \div 78.12\frac{1}{2}$ . . . . $= .1664$
11. $12.9 \div 8.256$ . . . . $= 1.5625$
12. $81.2096 \div 1.28$ . . . . $= 63.445$
13. $12755 \div 81632$ . . . . $= .15625$
14. $2401 \div 21.4375$ . . . . $= 112.$
15. $21.13212 \div .916$ . . . . $= 23.07$
16. $36.72672 \div .5025$ . . . . . . $= 73.088$
17. $2483.25 \div 5.15625$ . . . . $= 481.6$
18. $142.0281 \div 9.2376$ . . . . . . $= 15.375$
19. $1 \div 100$ . . . . $= .01$
20. $10.1 \div 17$ . . . . $= .59412-$
21. $.001 \div 100$ . . . . $= .00001$
22. $.08\frac{1}{3} \div .12\frac{1}{2}$ . . . . $= .66\frac{2}{3} = \frac{2}{3}$
23. $.0001 \div .01$ . . . . $= .01$
24. $95.3 \div .264$ . . . . . . $= 360.984848+$
25. $1000 \div .001$ . . . . $= 1000000.$
26. Ten $\div$ 1 tenth . . . . $= 100.$
27. $.000001 \div .01$ . . . . $= .0001$
28. $.00001 \div 1000$ . . . . . . $= .00000001$
29. $16.275 \div .41664$ . . . . . . $= 39.0625$
30. 1 ten-millionth $\div$ 1 hundreth . . $= .00001$

ART. **155**. If the dividend is less than the divisor, the quotient may be expressed as a common fraction, by taking the dividend for the numerator, the divisor for the denominator, making both terms have the same number of decimal places, and then omitting the points, which is equivalent to multiplying them by the same number, (Art. 144); thus, 3.737 divided by 99.9, equals $\frac{3.737}{99.9} = \frac{3.737}{99.900} = \frac{3737}{99900} = \frac{101}{2700}$.

REVIEW.—155. If the dividend be less than the divisor, how may the quotient be written?

1. $.75 \div 2.125$	$= \frac{6}{17}$	3. $.13\frac{1}{3} \div 2.4 \quad = \frac{1}{18}$
2. $1 \div 5.5$	$= \frac{2}{11}$	4. $.041\frac{2}{3} \div .15\frac{5}{8} \quad = \frac{4}{15}$

ART. **156**. To find the denomination of the 1st quotient figure when obtained, *count off in the dividend, as many decimal places as are in the divisor, placing a dot after the last; numerate from this dot as a decimal point, either way, up to that figure of the dividend, under which the right-hand figure of the 1st product falls. This will give the denomination of the 1st quotient figure.*

Thus, in dividing .07 by 21.6, (page 115), mark the dividend, .0·700; numerate from the dot as a decimal point to the right as far as the 0, under which the right-hand figure of the first product falls: this gives *thousandths,* which is the denomination of the 1st quotient figure, 3.

CONTRACTED DIVISION OF DECIMALS.

ART. **157**. The methods of contracting division in whole numbers apply also to decimals. Only *three* cases need be noticed.

CASE I.—TO DIVIDE A DECIMAL BY 10, 100, 1000, &c.

RULE.—*Remove the decimal point of the dividend to the left, over as many places as there are ciphers in the divisor; the result will be the required quotient.*

NOTE.—If there are not as many places to the left as are required, prefix ciphers.

REMARK.—The rule in Art. 67, for dividing a whole number by 10, 100, 1000, &c., is a particular case of this, since a whole number may be considered a decimal with a point on its right. (Art. 137.)

Divide 68.075 by 10000. *Ans.* .0068075

DEMONSTRATION.—To divide by 10, 100, 1000, &c., is the same as to divide continually by 10. This is done by moving the point to the left, over as many places as there are ciphers in the divisor. (Art. 145.)

REVIEW.—155. Before reducing the common fraction, what must be done? Why? 156. How is the order of any quotient figure determined as soon as it is set down? 157. What methods of contraction are used in division of decimals? What is the 1st case? The rule? How are deficient places supplied? Why can the rule be applied to whole numbers? Prove the rule.

1. 65 ÷ 1000 . . . . = .065
2. .072$\frac{1}{9}$ ÷ 10 . . . . = .0072$\frac{1}{9}$
3. 1 ÷ 1000000 . . . . = .000001
4. 2001.2 ÷ 100 . . . . = 20.012
5. 93000 ÷ 1000 . . . . . . = 93.000 = 93.
6. 4.472 ÷ 10000 . . . . = .0004772

ART. **158.** It often happens that the quotient is not required to contain decimal figures below a certain denomination; if so, the work may be shortened.

CASE II.—TO DIVIDE, RESERVING A CERTAIN NUMBER OF DECIMALS IN THE QUOTIENT.

RULE.—*Mark that figure of the dividend, whose denomination would result from multiplying a unit of the highest denomination in the divisor, by a unit of the lowest denomination required in the quotient.*

Divide as usual, until this figure is reached; then, stop bringing down from the dividend, and at each subsequent division, drop a figure from the divisor, carrying for its tens.

Continue thus, until the divisor is reduced to a single figure, and then point off the quotient as required.

NOTE.—If the marked figure of the dividend is reached at the first multiplication, mark the figure of the divisor, whose product by the first quotient figure falls under the marked figure of the dividend; and at the next step, reject this figure with those on its right.

REMARKS.—1. If the dividend has no figure of the denomination to be marked, annex ciphers to it, or continue it further, if it is an interminate decimal, until it does.

2. In carrying tens from the rejected figures, observe the directions in Case 2 of contracted multiplication of decimals. (Art. 153, Rem.)

Divide 1.078543 by 319.562 true to 5 decimal figures.

SHORT METHOD.

```
319.562)1.078|543(.00337
           958|686
           ---
           120|
            96|
           ---
            24|
            22|
           ---
             2|
```

ORDINARY METHOD.

```
319.562)1.078|543(.00337
           958|686
           ----------
           119|8570
            95|8686
           ----------
            23 98840
            22 36934
           ----------
             1|61906
```

SOLUTION.—The highest denomination in the divisor is *hundreds*, the lowest required in the quotient is *hundred-thousandths;* and $100 \times .00001 = .001$; therefore, mark the *thousandths'* figure (8) of the dividend. As this marked figure is reached at the first multiplication mark 9 in the multiplier, since it gives the figure that falls under 8 in the dividend.

Cross the rejected figures of dividend and divisor; cross one more from the divisor at every new multiplication, carrying tens as directed: 337, with the necessary ciphers and point prefixed, is the quotient required.

DEMONSTRATION.—The reason of the rule consists in the fact, that all operations, which would involve denominations lower than those required in the quotient, are omitted.

	EXAMPLES.	*Decimals reserved.*	ANSWERS.
1.	$1 \div 2.6783$	3	.373
2.	$10.00371 \div .056248$	2	177.85
3.	$.187564 \div .00043129$	true to units.	435.
4.	$.007516362 \div 652.18$	8	.00001152
5.	$1000.86 \div 3.1415926$	7	318.5836381
6.	$61.0598314 \div 4278$	6	.014273
7.	$421.33\frac{1}{3} \div 9.104\frac{3}{7}$	4	46.2778
8.	$100 \div 3.7320508$	5	26.79492
9.	$7912.5043 \div 181.34$	1	43.6
10.	$.91\frac{4}{9} \div 216.52$	10	.0042233717
11.	$555 \div 123456789$	10	.0000044955
12.	$1 \div 111111$	6	.000009

CASE III.—TO DIVIDE BY A DECIMAL LITTLE LESS THAN 1, RESERVING DECIMALS IN THE QUOTIENT.

ART. 159. RULE.—*Multiply the dividend by what the divisor wants of being a unit; multiply this product in like manner, and continue so until the product becomes too small to affect the result as required; then add to obtain the quotient.*

Divide 3815.64 by .994, reserving 2 decimals in the quotient.

REVIEW.—158. What is Case 2? The rule? Why should no denominations of the dividend be used below the one marked? *Ans.* Because, when divided by the divisor, they give lower denominations than are required in the quotient.

SOLUTION—Multiply 3815.64 by .006, the difference between .994 and a unit; write the product, 22.894, neglecting all denominations below thousandths. Do the same to this product, and to the next. The rest of the products are too small to affect the answer as required; therefore, add and obtain the quotient, 3838.67, true to 2 decimals.

```
3815.64
  22.894
    .137
       1
3838.672
    Ans.
```

DEMONSTRATION.—The operation and demonstration are similar to those in Art. 69, for the corresponding case of whole numbers, except that no remainders occur, the product being extended in decimals as far as is necessary to obtain the correct answer.

	EXAMPLES.	*Decimals reserved.*	ANSWERS.
1.	$1000 \div .98$	2	1020.41
2.	$6215.75 \div .99\frac{1}{2}$	3	6246.985
3.	$28012 \div .993$	2	28209.47
4.	$52546.35 \div .99\frac{3}{4}$	3	52678.045
5.	$4840 \div .9875$	2	4901.27

X. CIRCULATING DECIMALS.

ART. **160.** Many common fractions, when transformed, become interminate decimals. (Art. 148.) These have some curious and useful properties worth considering.

PROPOSITION I.

The only common fractions which can be changed into terminate decimals, are those which, reduced to their lowest terms, have no factors but 2 *and* 5 *in their denominators.*

DEMONSTRATION.—The numerator must contain all the prime factors of the denominator to be divisible by it. Every cipher annexed to the numerator multiplies it by 10, introducing 2 and 5, the factors of 10, as factors of the numerator. If the denominator has no factors but 2's and 5's, enough ciphers may be annexed to the numerator to give it as many 2's and 5's for factors as the denominator, and then the quotient is exact. But if the denominator have any factor besides 2 and 5, this factor never can be introduced into the numerator by annexing ciphers, for 2 and 5 are the only factors that can be so introduced. In such cases, the exact division is not possible.

EXAMPLES.

$\frac{3}{320} = .009375$

$\frac{5}{16} = .3125$

$\frac{2}{3} = .66666+$

$\frac{8}{11} = .72727+$

Tell whether the following common fractions can be changed into terminate or interminate decimals:

$\frac{2}{9}$, $\frac{3}{8}$, $\frac{9}{32}$, $\frac{11}{12}$, $\frac{95}{152}$, $\frac{105}{112}$, $\frac{21}{30}$, $\frac{4\frac{1}{2}}{288}$, $\frac{56\frac{1}{4}}{87\frac{1}{2}}$

PROPOSITION II.

ART. **161.** *Every interminate decimal arising from the transformation of a common fraction will be found, if the division be carried far enough, to contain the same figure, or set of figures, repeated in the same order without end.*

DEMONSTRATION.—In converting $\frac{1}{7}$ into a decimal, the remainders are successively 3, 2, 6, 4, 5, 1, and as these are all the whole numbers less than 7, the next remainder must be one of these repeated; on trial it is found to be 3, to which if a 0 be added, we have the same dividend and divisor as once before, and therefore must have the same quotient figure and remainder; and this remainder, with a cipher annexed, will yield another quotient figure and remainder, like a previous one, and so on continually.

EXAMPLES.

$\frac{1}{7}=.142857142857+$

$\frac{5}{9}=.5555+$

$\frac{2}{27}=.074074+$

ART. **162.** These interminate decimals on this account have received the name of *circulating* or *recurring decimals;* also, *repeating* or *periodical decimals.*

The figure, or set of figures, which is constantly repeated, is called the *repetend*, which signifies *to be repeated.*

Such decimals are expressed by placing a dot over the first and last figures of the first repetend, and omitting those which follow; thus, $\frac{2}{3}=.6666+=.\dot{6}$ and $\frac{1}{7}=.\dot{1}4285\dot{7}=.142857142857+$

ART. **163.** A *circulate* or *circulating decimal* has one or more figures constantly repeated in the same order.

A *repetend* is the figure or set of figures repeated.

REVIEW.—158. If the marked figure of the dividend is reached after the first multiplication, what should be done? If the dividend has not the proper place to be marked, what should be done? What directions are to be observed in carrying tens from the rejected figures? Solve the example. Prove the rule. 159. What is Case 3? The rule? Illustrate and prove it.

11

A *pure circulate* has no figures but the repetend; as, $.\dot{5}$ and $.\dot{1}2\dot{4}$

A *mixed circulate* has other figures before the repetend: as, $.208\dot{3}$ and $.3\dot{1}24\dot{7}$

A *simple repetend* has one figure; as, $.\dot{4}$

A *compound repetend* has two or more figures; as, $.\dot{5}\dot{9}$

Similar repetends begin at the same place; as, $.35\dot{6}23\dot{1}$ and $.01\dot{7}\dot{8}$, which both begin at thousandths.

Dissimilar repetends begin at different places; as, $.\dot{2}0\dot{5}$ and $.31\dot{2}46\dot{8}$

Similar and *conterminous* repetends begin and end at the same places; as, $.50\dot{3}9\dot{7}$ and $.42\dot{6}1\dot{8}$

ART. **164**. Any terminate decimal may be considered a circulate, its repetend being ciphers; as, $.35 = .35\dot{0} = .3500\dot{0}0\dot{0}$. Any simple repetend may be made compound, and any compound repetend still more compound, by taking in one or more of the succeeding repetends; as, $.\dot{3} = .\dot{3}333\dot{3}$, and $.05\dot{6}\dot{2} = .05\dot{6}26\dot{2}$, and $.\dot{2}5\dot{7} = .\dot{2}5725725\dot{7}$

When a repetend is thus enlarged, be careful to take in *no part* of a repetend without taking the whole of it; thus, if we take in 2 figures in the last example, the result, $.\dot{2}572\dot{5}$, would be incorrect, for the next figure understood being 7, shows that 25725 is not repeated. A repetend may be made to begin at any lower place by carrying its dots forward, each the same distance; thus, $.\dot{5} = .55\dot{5}$, and $.29\dot{4}\dot{1} = .294\dot{1}\dot{4}$, and $5.1\dot{8}3\dot{6} = 5.183\dot{6}8\dot{3}$

Dissimilar repetends can be made similar, by carrying the dots forward till they all begin at the same place, as the one furthest from the decimal point.

Similar repetends may be made conterminous by enlarging the repetends until they all contain the same number of figures. This number will be the least common multiple of the numbers of figures in the given repetends. For, suppose one of the repetends to have 2, another, 3, another, 4, and the last, 6 figures; in enlarging the first, figures must be taken in, 2 at a time; and in the others, 3, 4, and 6 at a time. The number of figures which may be taken in 2, 3, 4, and 6 at a time, is a common multiple of these numbers, and the least common multiple is preferred for convenience.

REDUCTION OF CIRCULATES.

CASE I.—TO REDUCE A PURE CIRCULATE TO A COMMON FRACTION.

ART. **165.** RULE.—*Write the repetend for the numerator, and for the denominator take as many 9's as there are figures in the repetend.*

DEMONSTRATION.—Take the pure circulate $.\dot{4}5\dot{6} = .456456456$, &c., and remove the decimal point to the right over one repetend; the result 456.456456, &c. $= 456.\dot{4}5\dot{6}$ is 1000 times $.\dot{4}5\dot{6}$, (Art. 144); hence, the part $456 = 999$ times $.\dot{4}5\dot{6}$; and $.\dot{4}5\dot{6} = \frac{456}{999} = \frac{152}{333}$, which agrees with the rule.

NOTE.—If the repetend begins before the decimal point at some place of whole numbers, carry the dots forward until the repetend begins at the tenths' place, and then apply the rule; thus, $\dot{2}5.\dot{6} = 25.\dot{6}2\dot{5} = 25\frac{625}{999}$.

CASE II.—TO REDUCE A MIXED CIRCULATE TO A COMMON FRACTION.

ART. **166.** RULE.—*Subtract the figures which precede the repetend from the whole circulate for the numerator; for the denominator take as many 9's as there are figures in the repetend, with as many ciphers annexed as there are decimal figures before the repetend.*

Change $.821\dot{4}3\dot{7}$ to its equivalent common fraction.

OPERATION.

DEM.—The work, by the rule in Case 1, results in $\frac{821437 - 821}{999000}$ for the answer, and this is what would be got, at once, if the rule last given were applied; and so with all other mixed circulates.

$$.821\dot{4}3\dot{7} = .821\tfrac{437}{999}$$

$$= \frac{821\frac{437}{999}}{1000} = \frac{821 \times 999 + 437}{999000}$$

$$= \frac{821(1000 - 1) + 437}{999000} =$$

$$\frac{821000 - 821 + 437}{999000} = \frac{821437 - 821}{999000}$$

$$= \frac{820616}{999000} = \frac{102577}{124875} \quad \textit{Ans.}$$

REDUCE TO COMMON FRACTIONS,

1.	$.\dot{3}$	$=\frac{1}{3}$	8.	$\dot{1}.00\dot{1}$	$=1\frac{1}{999}$
2.	$.0\dot{5}$	$=\frac{1}{18}$	9.	$.13\dot{8}$	$=\frac{5}{36}$
3.	$.\dot{1}2\dot{3}$	$=\frac{41}{333}$	10.	$.208\dot{3}$	$=\frac{5}{24}$
4.	$2.\dot{6}\dot{3}$	$=2\frac{7}{11}$	11.	$\dot{8}5.714\dot{2}$	$=85\frac{5}{7}$
5.	$.3\dot{1}$	$=\frac{14}{45}$	12.	$.\dot{0}6349\dot{2}$	$=\frac{4}{63}$
6.	$.0\dot{2}1\dot{6}$	$=\frac{4}{185}$	13.	$.4\dot{4}7619\dot{0}$	$=\frac{47}{105}$
7.	$\dot{4}8.\dot{1}$	$=48\frac{4}{27}$	14.	$.0902\dot{7}$	$=\frac{13}{144}$

ART. **167**. Circulates may be added, subtracted, multiplied, or divided, by this

GENERAL RULE FOR CIRCULATES.

Reduce the circulates to common fractions, and perform on them the operation required.

REMARK.—Circulates may be carried forward far enough to avoid any sensible error in the result, and then treated as other decimals. They can be added, subtracted, multiplied, and divided, without this preparation; as will now be explained.

ADDITION OF CIRCULATES.

ART. **168**. RULE.—*Make the repetends similar and conterminous, if they be not so; add, and point off as in ordinary decimals, increasing the right-hand column by the amount if any, which would be carried to it if the circulates were continued; then make a repetend in the sum, similar and conterminous with those above.*

Add $.25\dot{6}$, $5.34\dot{7}\dot{2}$, $24.\dot{8}1\dot{5}$, and $.9098$

DEMONSTRATION.—Make the circulates similar and conterminous, as directed in Art. 164. The first column of figures which would appear, if the circulates were continued, are the same as the first figures of the repetends, 6, 7, 1, 0, whose sum, 14, gives 1 to be carried to the right hand column. Since the last six figures in each number is a repetend, the last six figures of the sum is also a repetend.

$$
\begin{array}{r}
.2566\dot{6}6666\dot{6} \\
5.3472\dot{7}2727\dot{2} \\
24.8158\dot{1}5815\dot{8} \\
.9098\dot{0}0000\dot{0} \\
\hline
31.3295\dot{5}5209\dot{7}
\end{array}
$$

REMARK.—In finding the amount to be carried to the right hand column, it may be necessary, sometimes, to use the *two* succeeding figures in each repetend.

1. Add $.45\dot{3}$, $.06\dot{8}$, $.32\dot{7}$, $.94\dot{6}$ *Ans.* $1.79\dot{6}$
2. Add $3.0\dot{4}$, $6.45\dot{6}$, $23.\dot{3}\dot{8}$, $.2\dot{4}\dot{8}$ *Ans.* $33.1\dot{3}3\dot{4}$
3. Add $.\dot{2}\dot{5}$, $.10\dot{4}$, $.\dot{6}\dot{1}$, and $.56\dot{3}\dot{5}$ *Ans.* $1.53\dot{6}$
4. Add $\dot{1}.0\dot{3}$, $.25\dot{7}$, $5.\dot{0}\dot{4}$, $28.0445\dot{2}4\dot{5}$ *Ans.* $34.\dot{3}\dot{7}$
5. Add $.\dot{6}$, $.13\dot{8}$, $.0\dot{5}$, $.097\dot{2}$, $.041\dot{6}$ *Ans.* 1.
6. Add $9.21\dot{1}0\dot{7}$, $.\dot{6}\dot{5}$, $5.00\dot{4}$, $3.5\dot{6}2\dot{2}$ *Ans.* $18.\dot{4}\dot{3}$
7. Add $.20\dot{4}\dot{5}$, $.\dot{0}\dot{9}$, and $.25$ *Ans.* $.5\dot{4}$
8. Add $5.0\dot{7}7\dot{0}$, $.\dot{2}\dot{4}$, and $7.\dot{1}2494\dot{3}$ *Ans.* $12.\dot{4}$
9. Add $3.\dot{4}88\dot{4}$, $1.6\dot{3}\dot{7}$, $130.8\dot{1}$, $.06\dot{6}$ *Ans.* $1\dot{3}6.0\dot{0}$

SUBTRACTION OF CIRCULATES.

ART. 169. RULE.—*Make the repetends similar and conterminous, if they be not so; subtract and point off as in ordinary decimals, carrying one, however, to the right-hand figure of the subtrahend, if on continuing the circulates it be found necessary; then make a repetend in the remainder, similar and conterminous with those above.*

Subtract $9.3\dot{1}5\dot{6}$ from $12.90\dot{2}\dot{1}$

DEMONSTRATION.—Prepare the numbers for subtraction. If the circulates were continued, the next figure in the subtrahend (5) would be larger than the one above it (2); therefore, carry 1 to the right-hand figure of the subtrahend.

$$\begin{array}{r} 12.90\dot{2}1212\dot{1} \\ 9.31\dot{5}6156\dot{1} \\ \hline 3.58\dot{6}5055\dot{9} \end{array}$$

REMARK.—It may be necessary to observe more than one of the succeeding figures in the circulates, to ascertain whether 1 is to be carried to the right-hand figure of the subtrahend or not.

1. Subtract $.0\dot{0}7\dot{4}$ from $.2\dot{6}$ . . . . . . $= .\dot{2}5\dot{9}$
2. Subtract $\dot{9}.0\dot{9}$ from $15.35\dot{4}6\dot{5}$ . . . $= 6.2\dot{5}$
3. Subtract $\dot{4}.5\dot{1}$ from $18.23\dot{6}7\dot{3}$ . . . $= 13.7\dot{2}$
4. Subtract $37.01\dot{2}\dot{8}$ from 100.73 . . . $= 63.\dot{7}\dot{1}$

5. Subtract $8.\dot{2}\dot{7}$ from $10.0\dot{5}6\dot{3}$ . . $= 1.7\dot{8}36290\dot{0}$

6. Subtract $\dot{1}90.47\dot{6}$ from $199.6\dot{4}2857\dot{1}$ $= 9.1\dot{6}$

MULTIPLICATION OF CIRCULATES.

ART. **170.** RULE.—*If only one of the numbers be a circulate, make it the multiplicand, and perform the work as in ordinary decimals, carrying to the right hand figure of each product, the amount that would be necessary if the multiplicand were continued further; make the repetends in the partial products similar and conterminous, and add according to the rule already given.*

If the multiplier have a repetend, reduce it to a common fraction and add in the result obtained by using this fraction.

Multiply $.375\dot{4}$ by $17.4\dot{3}$

SOLUTION.—In forming the partial products, carry to the right-hand figures of each respectively, the numbers 1, 3, 0, arising from the multiplication of the figures that do not appear. The repetend of the multiplier being equal to $\frac{1}{3}$, $\frac{1}{3}$ of the multiplicand is $125\dot{1}4\dot{8}$, whose figures are set down under those of the multiplicand from which they were obtained. Point the several products, carry them forward, until their repetends are similar and conterminous, and add for answer.

$$
\begin{array}{r}
.375\dot{4} \\
17.4\dot{3} = 17.4\tfrac{1}{3} \\
\hline
.1501\dot{7}7\dot{7} \\
2.6281\dot{1}1\dot{1} \\
3.7544\dot{4}4\dot{4} \\
125\dot{1}4\dot{8} \\
\hline
6.5452\dot{4}8\dot{1}
\end{array}
\quad \textit{Ans.}
$$

1. $4.7\dot{3}\dot{5} \times 7.349$ . . . . . . $= 34.8001\dot{1}\dot{3}$
2. $.0\dot{7}06\dot{7} \times .9\dot{4}3\dot{2}$ . . . . $= .06\dot{6}66\dot{5}$
3. $714.3\dot{2} \times 3.45\dot{6}$ . . . . . $= 2469.173\dot{8}1\dot{4}$
4. $16.\dot{2}0\dot{4} \times 32.\dot{7}\dot{5}$ . . . . . $= 530.\dot{8}1044\dot{6}$
5. $19.0\dot{7}\dot{2} \times .208\dot{3}$ . . . . $= 3.973\dot{4}\dot{8}$
6. $10.0\dot{5}1\dot{2} \times 4.2\dot{6}\dot{3}$ . . . . $= 42.85\dot{4}8803\dot{3}$
7. $3.\dot{7}54\dot{3} \times 4.7157$ . . . . . $= 17.704\dot{5}08\dot{2}$
8. $1.\dot{2}5678\dot{4} \times 6.420\dot{8}\dot{1}$ . . . $= 8.069\dot{5}8320\dot{6}$

DIVISION OF CIRCULATES.

ART. 171. RULE.—*Make the repetends similar and conterminous. Subtract from each circulate the figures preceding its repetend, and use the remainders for the dividend and divisor respectively, omitting the dots.*

NOTE.—If the divisor is not a circulate, it will be shorter to divide as in ordinary decimals, bringing from the dividend the figures of the repetend instead of ciphers, to continue the division.

Divide $2.65\dot{3}$ by 1.8

FIRST OPERATION.	SECOND OPERATION.
$1.80\dot{0}$ $2.65\dot{3}$	$1.8)2.65\dot{3}(1.4\dot{7}4\dot{0}$
180 265	85
$1620)2388(1.4\dot{7}4\dot{0}$	133
768	73
1200	133
660	
1200	

DEMONSTRATION.—The remainders 1620 and 2388, in the 1st operation, are the numerators of the common fractions to which the circulates are equivalent, (Art. 166); and as they have the same denominators (each being as many 9's as there are figures in the repetend, with as many ciphers annexed as there are decimal figures before the repetend, Art. 166), dividing these numerators will give the same quotient as if the fractions themselves were used, and therefore the dots may be omitted.

EXAMPLES FOR PRACTICE.

1. $.\dot{7}\dot{5} \div .\dot{1}$ . . . . $= 6.\dot{8}\dot{1}$
2. $51.49\dot{1} \div 17$ . . . . $= 3.02\dot{8}$
3. $681.55\dot{9}887\dot{9} \div 94$ . . . . . $= 7.25\dot{0}637\dot{1}$
4. $90.5\dot{2}0374\dot{9} \div 6.7\dot{5}\dot{4}$ . . . . . $= 13.\dot{4}0\dot{1}$
5. $11.\dot{0}6873540\dot{2} \div .\dot{2}4\dot{5}$ . . . . . $= 45.1\dot{3}$
6. $9\ 5\dot{3}3066399\dot{7} \div 6.\dot{2}1\dot{7}$ . . . . . $= 1.5\dot{3}$
7. $3.500\dot{6}9135802\dot{4} \div 7.68\dot{4}$ . . . . . $= .4\dot{5}$

Multiplication and division of circulates can be frequently performed with advantage by the general rule for circulates, (Art. 167).

ART. **172.** There is a short method of converting a common fraction into a circulating decimal, when the denominator is a prime number, which is worth considering.

By actual division, $\frac{1}{7} = .\dot{1}4285\dot{7}$, which has this property, viz:

1st *Property.* The number of figures in the repetend is either one less than the denominator, or a half, a third, or some other exact part of this one less; thus, $\frac{7}{11} = \dot{6}\dot{3}$, the number of figures in the repetend (2) being $\frac{1}{5}$ of 10.

This is true of any fraction, whose denominator is a prime number other than 5. If the number of figures in the repetend is *one less* than the denominator, two other properties are observed.

2d *Property.* Each figure in the first half of the repetend added to the corresponding figure in the last half, makes 9; thus, in $.\dot{1}4285\dot{7}$, $1+8$, $4+5$, $2+7$, each equals 9.

3d *Property.* The same repetend serves for all fractions having the same prime denominator, whatever be their numerators, by starting at different places; thus, $\frac{1}{7} = .\dot{1}4285\dot{7}$, $\frac{2}{7} = .\dot{2}8571\dot{4}$, $\frac{3}{7} = .\dot{4}2857\dot{1}$.

Convert $\frac{16}{23}$ into a circulating decimal.

SOLUTION.—First, by actual division, $\frac{1}{23} = .043478\frac{6}{23}$; instead of $\frac{6}{23}$, put 6 times the value of $\frac{1}{23}$ just given, viz: $.043478260869\frac{13}{23}$; having 12 figures without repeating, the repetend must be of 22 figures by property first: obtain the other figures according to the 2d property, by subtracting each of the first 11 from 9; thus, $\frac{1}{23} = .\dot{0}43478260869565217391\dot{3}$; to get $\frac{16}{23}$ use this same repetend, and to ascertain the starting place, multiply the first 3 or 4 figures by 16; thus, .0434 multiplied by 16, gives .6944, showing that, we must commence at the eleventh figure; doing so, the value of $\frac{16}{23} = \dot{6}95652173913043478260\dot{8}$.

CONVERT INTO CIRCULATING DECIMALS,

$\frac{3}{13}$, $\frac{5}{17}$, $\frac{11}{19}$, $\frac{8}{29}$, $\frac{7}{31}$, $\frac{14}{37}$, $\frac{24}{41}$, $\frac{49}{67}$, $\frac{2}{101}$, $\frac{10}{59}$, $\frac{62}{73}$, $\frac{55}{89}$.

XI. COMPOUND NUMBERS.

ART. **173.** A *simple number* is of one denomination; as, 95 dollars; 2000 bushels; 4 apples.

A *compound number* consists of several simple numbers of different denominations; as, 16 dollars 75 cents; 3 yards 2 feet 8 inches.

Compound numbers are often called *denominate* numbers; they are used for measures, weights and money.

LONG OR LINEAR MEASURE

ART. **174.** Is used for measuring distance, also the length, breadth, and hight of bodies, called their *linear* dimensions.

TABLE.

12 inches, (in.)	make	1 foot, (ft.)
3 ft.		1 yard, (yd.)
$5\frac{1}{2}$ yd. or $16\frac{1}{2}$ ft.		1 rod, (rd.)
40 rd.		1 furlong, (fur.)
8 fur.		1 mile, (mi.)

REMARK.—The *rod* is sometimes called *pole* or *perch*.

NOTES.—1. 4 in. = 1 hand, used in measuring the hight of horses; 9 in. = 1 span; 3 feet = 1 pace; 18 in. = 1 cubit; 3 in. = 1 palm.

2. The scale used by carpenters has the foot divided into 12 in.: each inch divided into 12 equal parts, called lines; each line into 12 equal parts, called seconds; and each second into 12 equal parts, called thirds. The inch in these scales is also divided into eighths and sixteenths, and tenths.

MARINERS' MEASURE

ART. **175.** Is a kind of Long Measure used in estimating distances at sea.

6 feet	make	1 fathom.
120 fathoms		1 cable-length.
880 fathoms, or $7\frac{1}{3}$ cable-lengths		1 mile.

NOTE.—1 nautical league = 3 equatorial miles = 3.45771 statute miles. 60 equatorial miles = 69.1542 statute miles = 1 equatorial degree; 360 equatorial degrees = 1 great circle, or circumference of the earth.

REVIEW.—173. What is a simple number? a compound number? Give examples. What are compound numbers often called? What are they used for? 174. What is long measure used for? Repeat the table. 175. What is mariners' measure? Repeat the table.

SURVEYORS' AND ENGINEERS' MEASURE

ART. **176.** Is a kind of long measure, used in laying out roads, and running the boundaries of land.

100 links (lk.)......................make 1 chain, (ch.)
80 chains (ch.).......................... 1 mile, (mi.)

NOTES.—1. The chain is called surveyors' or Gunter's chain, from its inventor, and is 4 rods, or 66 feet in length. As it consists of 100 links, each link must be 7.92 in. long; hence, to change these denominations to the ordinary linear measure, recollect, that

1 chain = 4 rd. or 66 feet.
1 link = 7.92 in.

2. Since each link is $\frac{1}{100}$ of a chain, the number of links can be written as decimal hundredths with the whole chains. as 2.56 chains = 2 chains 56 links.

REMARK.—Inches and yards are not used in the last two kinds of measurement.

CLOTH MEASURE

ART. **177.** Is a kind of long measure used for dry goods.

$2\frac{1}{4}$ in...............................make 1 nail (na.),
4 na. or 9 in.......................... 1 quarter, (qr.)
4 qr.................................... 1 yd.

NOTES.—1. 1 ell Flemish = 3 qr. or $\frac{3}{4}$ yd.; 1 ell English = 5 qr. or $1\frac{1}{4}$ yd.; 1 ell French = 6 qr. or $1\frac{1}{2}$ yd.

2. At the custom-houses, the yard only is used, being divided into tenths and hundredths; in mercantile transactions, the yard is the unit, and the fractional parts employed are quarters, eighths, sixteenths, and half-sixteenths.

ART. **178.** The standard of all our linear measure is the yard, being identical with the imperial yard of Great Britain, which was determined as follows:

By accurate experiment at London, the length of a pendulum was ascertained, which, in a vacuum, at the level of the sea, vibrated 86400 times in a mean solar

---

REVIEW.—176. What is surveyors' measure? Repeat the table. What is the chain called? Why? How long is it? How long is a link? Why? How are chains and links written together? 177. What is cloth measure? Repeat the table. What is an ell Flemish? An ell English? An ell French?

day, or once every second. This pendulum was divided into 391393 equal parts, and 360000 of these were taken to be a yard, the pendulum itself then being 39.1393 inches long.

REMARK.—Linear measurement, as being especially necessary, was used, and to a certain degree fixed, at an earlier period than the measures of volume and weight, which have therefore been made to depend upon and be verified by the former.

The standards of long measure were at first very imperfect, being derived from different parts of the human body, such as a finger-joint, finger, hand, span, cubit or fore-arm, and yard or whole arm; but as commerce increased, and convenience demanded a change, they were rendered precise and uniform.

The ancient yard of Great Britain is said to have been determined by the length of the arm of King Henry I.

SQUARE OR SURFACE MEASURE

ART. **179**. Is used in estimating the contents of land, painters' and plasterers' work, and other surfaces.

A *square* is an even surface, bounded by *four* straight lines or *sides*. Each side is perpendicular to two others.

The size or name of any square depends upon that of its side; a square inch is a square, whose side is an inch long; a square foot, one whose side is a foot long, and so on.

ART. **180**. The unit by which all surfaces are measured is a square, whose side is a linear inch, foot, yard, rod or mile; and the size of any surface will be the number of times it contains this unit.

The simplest surface is a rectangle, which is an even surface, having four straight lines for sides, each opposite pair being equal, and perpendicular to the other pair. The ceiling and sides of a room, and sheets of paper, are examples of rectangles.

If the length and breadth of a rectangle are the same, the sides are all equal, and it is a square.

The size, or *area* of a rectangle, being the number of square measuring units it contains, can be ascertained as follows:

REVIEW.—178. What is the standard unit of length in the United States? How is it determined? What measures of length were used at first? 179. What is square measure? What is a square? A sq. inch? A sq. ft.? A sq. yd.? 180. What is the unit of measure for all surfaces?

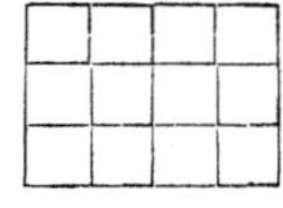

Take a rectangle 4 inches long by 3 inches wide. If upon each of the inches in the length, a square inch be conceived to stand, there will be a row of 4 square inches, extending the whole length of the rectangle, and reaching 1 inch of its width. As the rectangle contains as many such rows as there are inches in its width, its area must be equal to the number of square inches in a row (4) multiplied by the number of rows (3), = 12 square inches; hence, to find the area of a rectangle,

RULE.—*Multiply the number of linear units in the length by the number of linear units in the breadth, after expressing them in the same denomination. The product will be the area in square units of the same denomination.*

It is easy to determine the number of sq. inches in a sq. foot, of sq. feet in a sq. yd., and so on; for, since 1 sq. foot is 12 inches long by 12 inches wide, it must contain $12 \times 12 = 144$ sq. in.; and since 1 sq. yd. is 3 feet long by 3 feet wide, it must contain $3 \times 3 = 9$ sq. feet; and since 1 sq. rod is $5\frac{1}{2}$ yd. long by $5\frac{1}{2}$ yd. wide, it must contain $5\frac{1}{2} \times 5\frac{1}{2} = 30\frac{1}{4}$ sq. yd.

144 square inches (sq. in.) make 1 square foot, (sq. ft.)
9 sq. ft.............................. 1 square yard, (sq. yd.)
$30\frac{1}{4}$ sq. yd.............................. 1 square rod, (sq. rd.)

LAND MEASURE

ART. **181.** Is a kind of surface measure, used to express the contents of land.

40 perches (P.) make 1 rood, (R.)
10 sq. chains, or 4 R........................ 1 acre, (A.)
640 A...................... 1 sq. mile, (sq. mi.)

NOTE.—Since 1 chain = 4 rods, 1 sq. chain $= 4 \times 4 = 16$ sq. rods. Since links are written as decimal hundreths of a chain, chains and links can be considered as chains only; thus, 7 chains and 9 links = 7.09 chains, and the area of a square whose side is 7.09 chains, will be expressed in square chains and a decimal, as follows: $7.09 \times 7.09 = 50.2681$ square chains; there is no necessity, then, of using the denominations link or square link in practice.

REVIEW.—180. What is a rectangle? What is the rule for the area of a rectangle? Prove it. How many square inches in a square foot? Why? How many square feet in a square yard? Why? How many square yards in a square rod? Why? Repeat the table.

CUBIC OR SOLID MEASURE

ART. **182.** Is used to measure the bulk of stone, timber, masonry, and other solid work; to find the contents of cellars, and to verify measures of capacity.

A cube is a solid, bounded by six equal squares or *faces*, each opposite pair of which is perpendicular to the other four. Its length, breath and hight, then, are all equal, and each is called the *side* of the cube.

The size or name of any cube, like that of a square, depends upon its side, as cubic inch, cubic foot, cubic yard.

ART. **183**. The unit by which all solids are measured is a cube, whose side is a linear inch, foot, &c., and their size or solidity will be the number of times they contain this unit.

The simplest solid is the *rectangular* solid, which is bounded by six rectangles, called its *faces*, each opposite pair being equal, and perpendicular to the other four; a bar of soap, a candle-box, are rectangular solids. If the length, breadth, and hight are the same, the faces are squares, and the solid is a cube.

The size, or *solidity* of any rectangular solid is found as we obtain the *area* of a square (Art. 180.)

Suppose the rectangle 4 inches long by 3 inches wide, in Art. 180, to be the bottom or lower *base* of a rectangular solid, its upper face being of the same dimensions, and its hight 5 inches.

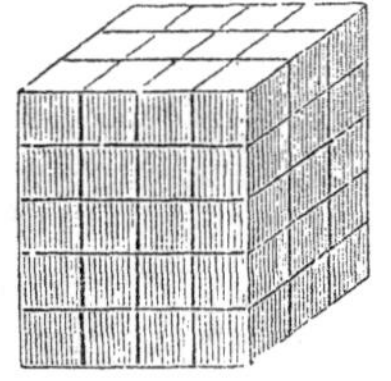

If upon each of the 12 square inches in the lower base, a cubic inch be conceived to stand, there will be a section of 12 cubic inches covering the whole bottom of the solid, and reaching 1 inch of its hight; as the solid contains as many such sections as there are inches in hight, its solidity must be equal to the number of cubic inches in a section (12) multiplied by the number of sections (5) or 60 cubic inches. But the number of cubic inches in a section is the same as the number of square

REVIEW.—181. What is land measure? Repeat the table. How many square chains in an acre? Why? Why do we not use square links? 182. What is cubic measure? What is a cube? A cubic inch? A cubic foot? A cubic yard? 183. What is the unit for all solids? What is a rectangular solid?

inches in the base, (12,) and this again is equal to the number of linear inches in the length (4), multiplied by the number in the width, (3); hence,

TO FIND THE SOLIDITY OF A RECTANGULAR SOLID,

RULE.—*Multiply the length, breadth, and hight together, after expressing them in the same denomination; the product will be the solidity in cubic units of the same denomination.*

It is easy to determine the number of cubic inches in a cubic foot, and of cubic feet in a cubic yard.

For, since 1 cubic foot is 12 inches long, 12 inches wide, and 12 inches high, its solidity will be $12 \times 12 \times 12 = 1728$ cubic inches; and, since 1 cubic yard is 3 feet long, 3 feet wide, and 3 feet high, its solidity will be $3 \times 3 \times 3 = 27$ cubic feet; hence,

1728 cubic inches (cu. in.)	make	1 cubic foot (cu. ft.)
27 cu. ft.	....	1 cubic yard (cu. yd.)

NOTES.—1. 1 tun of round timber = 40 cu. ft.; 1 tun of hewn timber = 50 cu. ft.; 1 tun of shipping = 42 cu. ft.

2. 1 cord of wood, 8 ft. long, 4 ft. wide, and 4 ft. high, contains $8 \times 4 \times 4 = 128$ cu. ft.; 1 cord foot or foot of wood, 1 ft. long, 4 ft. wide, and 4 ft. high, contains $1 \times 4 \times 4 = 16$ cu. ft.

3. 1 reduced foot, plank measure, 1 ft. long, 1 ft. wide, and 1 in. thick, contains $12 \times 12 \times 1 = 144$ cu. in.; all planks and scantling less than an inch are reckoned 1 inch thick; but, if more than 1 inch thick, allowance must be made by multiplying by that dimension.

4. 1 perch of masonry, 1 rod long, 1 ft. high and $1\frac{1}{2}$ ft. thick, contains $16\frac{1}{2} \times 1 \times 1\frac{1}{2} = \frac{33}{2} \times \frac{3}{2} = \frac{99}{4} = 24\frac{3}{4}$ cu. ft., which is usually taken 25 cubic feet in practice.

TROY OR MINT WEIGHT

ART. **184.** Is used for weighing gold, silver, jewels, in testing the strength of spirituous liquors, in philosophical experiments, and in comparing different weights.

24 grains (gr.)	make	1 pennyweight, (pwt.)
20 pwt.	....	1 ounce, (oz.)
12 oz.	....	1 pound, (lb.)

REVIEW.—183. What is the rule for its solidity? Prove it. How many cubic inches in a cubic foot? Why? How many cubic feet in a cubic yard? Why? Repeat the table. What is a cord of wood? A perch of masonry? How many cubic feet in each?

REMARKS.—1. The Troy pound is the standard of weight in the U. S. Mints; it is identical with the imperial Troy pound of Great Britain, which contains 5760 grains, 252.458 of which are equal in weight to a cubic inch of distilled water when the barometer is 30 inches, and the thermometer (Fahrenheit's) 62°.

2. The name Troy is supposed by some to be derived from Troyes, a city of France, where this weight was first introduced from the East, during the Crusades; and by others from Troy Novant, the ancient name of London. The ounce Troy is the only denomination used at the mint, all amounts of gold and silver being expressed in it and its decimal divisions.

3. The Troy pound is equal to the weight of 22.794422 cubic inches of distilled water, at the temperature of 39°.83 Fahrenheit, the barometer being 30 inches.

4. The grain was originally fixed by taking a grain of wheat from the middle of the ear, and thoroughly drying it; at first, 32 of these grains made a pennyweight, but afterward 24.

5. The pennyweight was the weight of the silver penny in use at that time, and is marked (pwt.) from (p.) for penny, and (wt.) for *weight.*

DIAMOND WEIGHT

ART. 185. Is used for weighing diamonds and other precious stones.

16 parts............make	1 carat grain	= .8	Troy grain.
4 carat grains.........	1 carat.	= 3.2	do. do.

REMARK.—The carat in this table is an absolute weight, and must be carefully distinguished from the same word used in speaking of the fineness of gold, for then it indicates the *proportion* of pure gold in a mass.

APOTHECARIES' WEIGHT

ART. 186. Is chiefly used in mixing medical prescriptions.

20 grains (gr.)..................make	1 scruple,	(℈.)
3 ℈	1 dram,	(ʒ.)
8 ʒ	1 ounce,	(℥.)
12 ℥	1 pound,	(℔.)

The pound, ounce, and grain, of this weight, are the same as those of Troy weight; the pound in each contains 12 oz. = 5760 gr.

REVIEW.—184. What is Troy Weight? Repeat the table. What is said of the Troy pound? The name Troy? What denomination only is used at the mint? What is said of the grain?

AVOIRDUPOIS OR COMMERCIAL WEIGHT

ART. **187**. Is used in commercial transactions when goods are bought or sold by the quantity. Heavy and bulky articles, as groceries, the coarser metals, drugs, &c are weighed by it.

16 drams (dr.).........make 1 ounce, (oz.)
16 oz............................ 1 pound, (lb.)
25 lb............................ 1 quarter, (qr.)
4 qr............................ 1 hundred-weight, (cwt.)
20 cwt............................ 1 tun, (T.)

NOTES.—1. A stone = 14 lb.; but a stone of fish, or butcher's meat = 8 lb., and a stone of glass = 5 lb.; a seam of glass = 24 stone = 120 lb.; 1 pack of wool = 240 lb.

2. In Great Britain, the qr. = 28 lb., the cwt. = 112 lb., the tun = 2240 lb. These values are used at the U. S. custom-houses in invoices of English goods; but generally in this country the qr. = 25 lb., the cwt. = 100 lb., and the tun = 2000 lb.

3. The lb. avoirdupois is equal to the weight of 27.7274 cu. in. of distilled water at 62° (Fah.); or 27.7015 cu. in. at 39°.83 (Fah.), the barometer at 30 in. For ordinary purposes, 1 cubic foot of water can be taken 62$\frac{1}{2}$ lb., or 1000 oz. avoirdupois.

4. The terms *gross* and *net* are used in this weight. *Gross weight* is the weight of the goods, together with the box, cask, or whatever contains them. *Net weight* is the weight of the goods alone.

REMARK.—The word *avoirdupois* is from the French *avoirs, du, pois*, signifying *goods of weight*. The lb. avoirdupois differs from the lb. Troy or apothecaries,' the former being 7000 gr., the latter, each 5760 gr. The oz. avoirdupois also differs from the oz. Troy or apothecaries.'

COMPARISON OF WEIGHTS.

ART. **188**. Since 1 lb. av. = 7000 gr. Troy, 1 oz. av. = $\frac{1}{16}$ of a lb. av., = $\frac{1}{16}$ of 7000 gr. Troy = 437$\frac{1}{2}$ gr. Troy; and 1 dr. av. = $\frac{1}{16}$ of an oz. av., = $\frac{1}{16}$ of 437$\frac{1}{2}$ gr. Troy = 27$\frac{11}{32}$ gr. Troy; and in a similar way 1 oz. Troy

REVIEW.—185. Repeat the table of diamond weight. What is said of a carat? 186. What is apothecaries' weight? Repeat the table. What denominations are identical in Troy and apothecaries' weight? 187. What is avoirdupois weight? Repeat the table. What is said of the qr., cwt., and tun, in Great Britain and the U. S. custom-houses? What elsewere? What is said of the lb. avoirdupois? What is gross weight? Net weight?

and apothecaries' = 480 gr. Troy; and 1 dr. apothecaries' = 60 gr. Troy. Hence,

1 lb. avoirdupois............	= 7000	gr. Troy or apoth.
1 oz........do..................	= $437\frac{1}{2}$	do...do........do.
1 dr........do..................	= $27\frac{11}{32}$	do...do........do.
1 lb. Troy or apoth.........	= 5760	do...do........do.
1 oz. do..........do...........	= 480	do...do........do.
1 dr. apothecaries'...........	= 60	do...do........do.

This table serves to convert denominations of one kind of weight into those of another.

WINE OR LIQUID MEASURE

Art. **189**. Is used for measuring all liquids, except ale, beer, and milk.

4 gills (gi.)..............make 1 pint, (pt.)
2 pt.............................. 1 quart, (qt.)
4 qt.............................. 1 gallon, (gal.) = 231 cu. in.

Notes.—1. $31\frac{1}{2}$ gal. = 1 barrel (bbl.); 63 gal. = 1 hogshead (hhd.); 42 gal. = 1 tierce; 84 gal. = 1 puncheon; 126 gal. = 1 pipe or butt; 2 pipes = 1 tun. These are generally mentioned as measures, but are only vessels, and are gauged and sold by the gallon, not being of uniform capacity. When the contents of cisterns, wells, &c., are expressed in hhd. or bbl., they have the values given above.

2. In England, 1 anker = 10 gal.; 1 runlet = 18 gal.

Art. **190**. The standard unit of liquid measure in the U. S. is the wine gallon, formerly used in Great Britain, containing 231 cu. in., and which contains a weight of 58372.1754 gr., = nearly $8\frac{1}{3}$ lb. av., of distilled water at 39°.83 Fahrenheit, the barometer at 30 inches.

A pint of water is generally considered 1 pound.

ALE AND BEER MEASURE

Art. **191**. Is used in measuring ale, beer, and milk, though milk is frequently sold by wine measure.

2 pints (pt.) make 1 quart, (qt.)
4 qt.................. 1 gallon, (gal.) = 282 cu. in.

Review.—187. What does avoirdupois mean? How does the lb. avoirdupois differ from the lb. Troy and apothecaries? 188. Repeat the table for comparison of weights. 189. What is wine measure?

NOTES.—1. 1 bbl. = 36 gal.; 1 hhd. = 54 gal. These are mentioned as measures, but are only vessels, and are gauged and sold by the gallon, not being of uniform capacity.

2. In England, 1 firkin = 9 gal.; 1 kilderkin = 2 firkins; 1 bbl. = 2 kilderkins.

REMARK.—The beer gallon contains 282 cu. in., or 10.179933 lb. av. of distilled water at 39°.83 Fahrenheit, the barometer at 30 inches.

DRY MEASURE

ART. **192**. Is used for measuring grain, fruit, vegetables, coal, salt, &c.

2 pints (pt.) make 1 quart, (qt.)
8 qt.................. 1 peck, (pk.)
4 pk.................. 1 bushel, (bu.) = 2150.42 cu. in.

NOTES.—1. 4 qt. or $\frac{1}{2}$ a peck = 1 dry gal. = 268.8 cu. in. nearly.

2. 1 qr. = 8 bu. = 480 lb., used in England in measuring wheat. 36 bu., and in some places 32 bu. make 1 chaldron, used in some of the United States, and formerly in Great Britain, in measuring coal; but now coal is bought and sold by weight in England, and in many parts of this country.

3. Grain is often bought and sold by weight. In England, and in many of the United States, 60 lb. of wheat, 48 lb. of barley, 56 lb. of rye or corn, and 32 lb. of oats, are each declared to make a bushel.

ART. **193**. The unit of our dry measure is the Winchester bushel, formerly used in England, and so called from the town where the standard was kept. It is 8 in. deep, and $18\frac{1}{2}$ in. diameter, and contains 2150.42 cu. in., or 77.627413 lb. av. of distilled water at 39°.83 Fahrenheit, the barometer at 30 inches.

The New York bushel is equal to the imperial bushel of Great Britain, and therefore contains 2218.192 cu. in.

COMPARISON OF MEASURES.

ART. **194**. The wine gallon contains 231 cu. in.; the beer gallon 282 cu. in.; and the dry gallon 268.8 cu. in.

REVIEW.—189. What is said of the bbl. and hhd.? 190. What is our standard of liquid measure? What does it contain? 191. What is beer measure? Repeat the table. What does the beer gallon contain? 192. What is dry measure? Repeat the table. How much is a qr. of wheat? How is grain measured? 193. What is the unit of dry measure? What does it contain? What is the New York bushel?

These were superseded in Great Britain, in 1826, by the imperial gallon, both of dry and liquid measure, which contains 277.274 cu. in., or 10 lb. av. of distilled water at 62° Fahrenheit, the barometer at 30 inches. At the same time the dry or Winchester bushel was replaced by the imperial bushel of 8 imperial gallons, containing 2218.192 cu. in.

1 wine gallon of United States	=	231	cubic inches.
1 beer gallon........................	=	282	do.
1 dry gallon..........................	=	268.8	do.
1 imperial gallon of G. Britain for dry and liquid measure	=	277.274	do.
1 dry bushel of U. S...............	=	2150.42	do.
1 imperial bushel of G. Britain	=	2218.192	do.

This table is useful in converting denominations of one measure to those of another.

APOTHECARIES' FLUID MEASURE

ART. **195.** Is used for measuring all liquids that enter into the composition of medical prescriptions.

60 minims (♏)...............	make	1 fluid drachm, (f ʒ).
8 f ʒ................................		1 fluid ounce, (f ℥).
16 f ℥................................		1 pint, (O).
8 O....................................		1 gallon, (Cong.)

NOTES.—1. Cong. is an abbreviation for *congiarium*, the Latin for gallon; O. is the initial of *octans*, the Latin for *one-eighth*, the pint being one-eighth of a gallon.

2. For ordinary purposes, 1 tea-cup = 2 wine-glasses = 8 table-spoons = 32 tea-spoons = 4 f ℥.

ART. **196.** TIME.

60 seconds (sec.)..............	make	1 minute, (min.)
60 min..................................		1 hour, (hr.)
24 hr.....................................		1 day, (da.)
7 da.......................................		1 week, (wk.)
4 wk.......................................		1 month, (mon.)
12 calendar mon......................		1 year, (yr.)
365 da....................................		1 common year.
366 da....................................		1 leap year.
100 yr.....................................		1 century, (cen.)

NOTE.—1 Solar year = 365 da. 5 hr. 48 min. 48 sec. = 365¼ da. nearly.

The denomination from which the preceding table is constructed, is the *day*, which is the interval of time be tween one mean noon and the next.

The apparent noon is the moment when the sun comes to the meridian of any place, and appears exactly half-way between rising and setting. Owing to the unequal motion of the earth around the sun, and the oblique position of its axis to its orbit, the interval between any apparent noon and the next is not uniform.

The average, however, is taken of all these intervals that occur in a year, and this average interval is called the mean solar day, which is divided as above.

A *year* is the time during which the earth makes a complete circuit about the sun, and reaches again a given point in its orbit; it contains 365 da. 5 hr. 48 min. 48 sec., or nearly $365\frac{1}{4}$ days.

The ancients were unable to find accurately the number of days in a year. They had 10, afterward 12 calendar months, corresponding to the revolutions of the moon around the earth. In the time of Julius Cæsar the year contained $365\frac{1}{4}$ days; instead of taking account of the $\frac{1}{4}$ of a day every year, the common or civil year was reckoned 365 days, and every 4th year a day was inserted, (called the *intercalary* day,) making the year then have 366 days. The extra day was introduced by repeating the 24th of February, which with the Romans was called the *sixth day before the kalends of March.* The years containing this day twice, were on this account called *bissextile,* which means *having two sixths.* By us they are generally called *leap* years.

But $365\frac{1}{4}$ days, = 365 days and 6 hours, is a little longer than the true year, which is 365 days 5 hours 48 minutes 48 seconds. The difference, 11 minutes 12 seconds, though small, produced, in a long course of years, a sensible error, which was corrected by Gregory XIII., who, in 1582, suppressed the 10 days that had been gained, by decreeing that the 5th of October should be the 15th.

To prevent difficulty in future, it has been decided to adopt the following rule.

REVIEW.—194. Repeat the table for comparison of measures. 195. What is apothecaries' fluid measure? Repeat the table. 196. Repeat the table of time. What is the unit of this table? What is a mean solar day? A year? What did the months correspond to? What was the Julian calendar? How many days were taken for a year in it? What is the true length of the solar year? What error was committed in the Julian calendar? When and by whom was it corrected? What is a leap year?

RULE FOR LEAP YEARS.

Every year that is divisible by 4 is a leap year, unless it ends with a double cipher; in which case it must be divisible by 400 to be a leap year.

Thus, 1832, 1648, 1600 and 2000 are leap years; but 1857, 1700, 1800, 1918, are not.

The *Gregorian* calendar was adopted in England in 1752. The error then being 11 days, Parliament declared the 3d of September to be the 14th, and at the same time made the year begin January 1st, instead of March 25th. Russia, and all other countries of the Greek Church, still use the Julian calendar; consequently their dates (*Old Style*) are now 12 days later than ours, (*New Style*). The error in the Gregorian calendar is small, amounting to a day in 3600 years.

ART. 197. The names of the months in their order, are January, February, March, April, May, June, July, August, September, October, November, December.

The year formerly began with March instead of January; consequently, September, October, November and December were the 7th, 8th, 9th, and 10th months, as their names indicate; being derived from the Latin numerals Septem (7), Octo (8), Novem (9), Decem (10).

ART. 198. MISCELLANEOUS TABLE.

12 things	make	1 dozen.
12 dozen or 144 things		1 gross.
12 gross or 144 dozen		1 great gross.
20 things		1 score.
56 lb.		1 firkin of butter.
100 lb.		1 quintal of fish.
196 lb.		1 bbl. of flour.
200 lb.		1 bbl. of pork.
14 lb.		1 stone.
$21\frac{1}{2}$ stone		1 pig of iron or lead.
8 pigs		1 fother.

ART. 199. The words folio, quarto, octavo, &c., used

REVIEW.—196. What is the rule for leap years in the Gregorian calendar? Where does the Julian calendar still prevail? What is the error in the Gregorian calendar? 197. Name the months in their order? What is the origin of the names, September, October, &c.?

in speaking of books, show how many leaves make a sheet of paper.

A sheet folded into 2 leaves forms a *folio*.		size.
Do.	4...do.	*quarto* or 4to.	do.
Do.	8...do.	*octavo* or 8vo.	do.
Do.	12...do.	*duodecimo* or 12mo.	do.
Do.	18...do.	18mo.	do.
Do.	36...do.	36mo.	do.

Also,

24 sheets of paper make	1 quire.
20 quires	1 ream.
2 reams	1 bundle.
5 bundles	1 bale.

CIRCULAR OR ANGULAR MEASURE

ART. **200.** Is used for latitude and longitude, and for expressing the distances between any two points on the surface of the globe, or in the heavens.

60 seconds ($''$) make	1 minute, ($'$)
$60'$	1 degree, ($^\circ$)
30°	1 sign, (s.)
360° or 12^s	1 circumference, (c.)

NOTE.—$90^\circ = 1$ quadrant or quarter of a circumference, $180^\circ =$ 1 semi-circumference. The degree being $\frac{1}{360}$ of a circumference is of different lengths in different circumferences; thus, the equator being larger than the polar circles, a degree of the former is larger than a degree of the latter.

COMPARISON OF TIME AND LONGITUDE.

ART. **201.** The *difference of longitude* of two places is the distance in degrees, minutes, and seconds, that one of them is further east or west of the established meridian than the other.

The sun appears to go entirely round the earth, (360°), from east to west, in 24 hours, crossing in succession the meridians of all places on its surface.

REVIEW.—200. What is circular or angular measure? Repeat the table. 201. What is the difference of longitude of two places? At what rate per hour does the sun appear to travel around the earth? At what rate per minute? What per second?

A place further east than another will have the sun on its meridian sooner, and, therefore, its time will be later at the rate of 1 hour for every 15° of longitude; or, (by taking $\frac{1}{60}$ of each of these quantities), at the rate of 1 minute for every 15′ of longitude; or (by taking $\frac{1}{60}$ of these), at the rate of 1 second for every 15″ of longitude Hence,

1 hour of time	= 15°	of longitude.
1 minute do	= 15′	do.
1 second do	= 15″	do.

NOTE.—*Recollect that if one place has greater east or less west longitude than another, its time must be later; and conversely, if one place has later time than another, it must have greater east or less west longitude.*

MONEY TABLES.

FEDERAL OR UNITED STATES MONEY

ART. 202. Is the currency of the United States.

10 mills (m.)	make	1 cent, (ct.)
10 ct.		1 dime, (d.)
10 d.		1 dollar, ($).
$10		1 eagle.

NOTE.—The *cent* and *mill*, which are $\frac{1}{100}$ and $\frac{1}{1000}$ of a dollar, derive their names from the Latin *centum* and *mille*, meaning *a hundred*, and *a thousand;* the *dime* which is $\frac{1}{10}$ of a dollar, is from the French word *disme*, meaning *ten.*

ART. 203. The Federal currency was authorized by act of Congress, August 8th, 1786. It has great simplicity, being on the decimal basis, and subject to the law that *one of any denomination is equal to* 10 *of the next lower;* therefore, the same Notation, Numeration, and general order of operation, can be used for Federal money as for simple numbers.

Any sum of Federal money of several denominations can be expressed as one denomination, by writing those of that denomination in units' place, those of higher denominations in places of whole numbers, and those of lower denominations in decimal places; thus,

REVIEW.—201. Why will a place that is east of another have later time? What is the table for comparing time and longitude? How do we know from the longitudes of two places, which has later time? 202. What is Federal or U. S. money? Repeat the table.

9 eagles $7 2 dimes 4 ct. and 5 m. can be written 9.7245 eagles, or $97.245, or 972.45 dimes, or 9724.5 ct. or 97245 m. It is customary to consider the dollar as the unit, and express all sums of U S. money in that denomination and its decimal divisions.

In reading U. S. money, *name the dollars and all higher denominations together as dollars, the dimes and cents as cents, and the next figure, if there is one, as mills;*

Or, *name the whole numbers as dollars, and the rest as a decimal of a dollar.*

Thus, $9.124 is read 9 dollars 12 ct. 4 mills, or 9 dollars 124 thousandths of a dollar; $175.06238 is read 175 dollars 6 ct. 2 mills, and a remainder, or, 175 dollars, 6238 hundred-thousandths of a dollar.

Art. **204**. The national coins of the United States are of gold, silver, and copper. The gold coins are the double-eagle, eagle, half-eagle, quarter-eagle, and one dollar piece. A 3 dollar piece has also been authorized.

The silver coins are the dollar, half-dollar, quarter-dollar, dime, half-dime, and 3 cent piece.

The copper coins are the cent and half-cent; the latter is now obsolete.

The mill never has been a coin; it is only a convenient name for the tenth of a cent, or thousandth of a dollar.

Pure gold and silver being too soft for coins, are mixed with baser metal, called *alloy*. By act of Congress, in 1837, our *standard* gold and silver is $\frac{9}{10}$ pure and $\frac{1}{10}$ alloy, (by weight).

The alloy in the silver coins is pure copper; in the gold coins it is copper and silver, the latter not to exceed the former in weight.

The 3 cent piece is not standard silver, being one-fourth copper. It weighs $12\frac{3}{8}$ grains. The cent weighs 168 gr.

Review.—202. Why is the cent so called? The mill? The dime? 203. How do the denominations of Federal money resemble those of simple numbers? How can sums of Federal money be read, written, added, subtracted, &c.? In writing any sum of Federal money, what single denomination is generally used? In reading any sum of Federal money, which denominations only are mentioned? What single denomination may be employed? 204. Which coins are of gold? Which of silver? Which of copper? What is the mill? What are standard gold and silver? What is the alloy for gold? What for silver?

The eagle weighs 258 gr. The half-dollar since April 1st, 1853, weighs 192 gr.

The half dollar coined before April 1st, 1853, weighs $206\frac{1}{4}$ gr.; it contains more silver than the one now coined, and is worth $53\frac{91}{128}$, instead of 50 cents.

ART. **205**. The fineness of manufactured gold is estimated in *carats*, an Arabic or Abyssinian word, signifying a small weight. In determining the purity of gold by analysis, a portion of it is taken, and, without reference to its actual weight, is called the *assay pound*, which is thus divided:

4 quarters (qr.)................make 1 assay grain, (gr.)
4 gr....................................... 1 carat, (car.)
24 car..................................... 1 assay pound

NOTES.—1. Each carat is $\frac{1}{24}$ of the mass used; if it is 18 carats gold, it is $\frac{18}{24}$ pure gold, and $\frac{6}{24}$ alloy; if it is 22 carats gold, it is $\frac{22}{24}$ pure gold, and $\frac{2}{24}$ alloy; if it is 24 carats gold, it is entirely free from alloy.

2. The quarters are written as fourths of a carat grain; thus, 19 car. $3\frac{3}{4}$ gr.

ENGLISH OR STERLING MONEY

ART. **206**. Is the currency of Great Britain.

4 farthings (qr.) make 1 penny, (d.)
12 pence.................. 1 shilling, (s.)
20 shillings............ . 1 pound, (£) = $4.84*

NOTES.—1. The Guinea, (gold), = 21s.; crown, (silver), = 5s.; half-crown = 2s. 6d.; noble = 6s. 8d.; angel = 10s.; mark = 13s. 4d.; pistole = 16s. 10d.; moidore = 27s.; 1 sovereign = 20s., (gold) = $4.84, *by act of Congress, 1842.

2. The farthing is not a coin, but stands for a quarter of a penny; thus, $5\frac{3}{4}$d. = 5 pence 3 farthings. When the penny was of silver, it was usual to mark it with a cross so deep that it could be easily broken into halves and quarters, called *half-pennies* and *fourthings*, finally, *farthings*.

REVIEW.—204. What is the weight of the eagle? The present half-dollar? What was the weight of the half dollar before 1853? How much is it now worth? 205. How is the fineness of manufactured gold estimated? Repeat the table. 206. What is sterling money? Repeat the table. How are farthings written? Which denominations are not coins? What is the origin of the name *farthing?*

3. The pound (£) is not a coin, but stands for 20s.; it is represented by the sovereign, or the bank note of £1. The pound is so called, because its equivalent, 240d. or 20s., formerly contained a pound weight of silver, the pound then being smaller than at present. A pound of standard silver is now coined into 66s.

REMARKS.—1. The symbols, £., s., d., q., are the initials of the Latin words *libra*, *solidarius*, *denarius*, *quadrans;* signifying, respectively, pound, shilling, penny, and quarter.

2. The sovereign, the standard gold coin, weighs 123.274 gr., and is 22 carats fine. The shilling, the standard silver coin, is $\frac{37}{40}$ pure silver, and $\frac{3}{40}$ copper, and weighs 87.27 gr. Pence and halfpence are made only of copper now, and each penny weighs 240 gr. = $\frac{1}{2}$ oz. Troy.

OF STATE CURRENCIES.

ART. **207**. Before our present currency was established, our accounts were kept in pounds, shillings, and pence.

In many States, the denominations shillings and pence are still retained, but not with the same values.

In New Hampshire, Massachusetts, Rhode Island, Connecticut, Virginia, Kentucky, and Tennessee,

12 d. = 1 shilling = $16\frac{2}{3}$ cents.
6 s. = \$1 = 100 cents.

In New York, North Carolina, and Ohio,

12 d. = 1 shilling = $12\frac{1}{2}$ cents.
8 s. = \$1 = 100 cents.

In New Jersey, Pennsylvania, Delaware, and Maryland,

12 d. = 1 shilling = $13\frac{1}{3}$ cents.
7 s. 6 d. or 90 d. = \$1 = 100 cents.

In South Carolina and Georgia,

12 d. = 1 shilling = $21\frac{3}{7}$ cents.
4 s. 8 d. or 56 d. = \$1 = 100 cents.

In Canada and Nova Scotia,

12 d. = 1 shilling = 20 cents.
5 s. = \$1 = 100 cents.

REVIEW.—206. What is the origin of the name *pound?* Into how many shillings is a pound of silver now coined? 207. What is the New England State currency? New York? Pennsylvania? Georgia? Canada?

FRENCH WEIGHTS AND MEASURES.

ART. **208**. The present system of weights and measures in France, adopted in 1795, is on the decimal basis.

After the unit of any measure has been determined and named, the higher denominations are made by prefixing the Greek numerals, *deca* (10), *hecto* (100), *kilo* (1000), *myria* (10000) to the name of the unit; the lower denominations are formed by prefixing the Latin numerals *deci* ($\frac{1}{10}$), *centi* ($\frac{1}{100}$), *milli* ($\frac{1}{1000}$).

FRENCH LONG MEASURE.

ART. **209**. The unit of long measure in France is the *mètre*, which is the ten-millionth part of the quadrant, extending through Paris from the equator to the pole.

1 mètre . . . . = 39.371 U. S. in.

NOTE.—1 decamètre = 10 mètres; 1 hectomètre = 100 mètres; 1 kilomètre = 1000 mètres; 1 myriamètre = 10000 mètres; 1 decimètre = $\frac{1}{10}$ mètre; 1 centimètre = $\frac{1}{100}$ mètre; 1 millimètre = $\frac{1}{1000}$ metre.

FRENCH SURFACE MEASURE.

ART. **210**. The unit of surface in France is the *are*, which is a square decamètre.

1 are . . . . = 119.6046 U. S. sq. yd.

NOTE.—1 decare = 10 ares; 1 hectare = 100 ares; 1 centiare = $\frac{1}{100}$ are.

FRENCH SOLID MEASURE.

ART. **211**. The unit of solidity in France is the *stère*, which is a cubic mètre.

1 stère . . . . = 35.31741 U. S. cu. ft.

NOTE.—1 decastère = 10 stères. 1 decistère = $\frac{1}{10}$ stère.

FRENCH WEIGHTS.

ART. **212**. The unit of weight in France is the *gramme*, which is the weight of a cubic centimètre of distilled water at the temperature of melting ice.

1 gramme . . . . = 15.434 Troy gr.

NOTE.—1 decagramme = 10 grammes; 1 hectogramme = 100 grammes; 1 kilogramme = 1000 grammes = $2\frac{1}{5}$ lb. av. nearly;

1 myriagramme = 10000 grammes. 1 decigramme = $\frac{1}{10}$ gramme; 1 centigramme = $\frac{1}{100}$ gramme; 1 milligramme = $\frac{1}{1000}$ gramme; 1 quintal = 220.55 lb. av.; 1 millier or bar = 2205.5 lb. av.

FRENCH DRY AND LIQUID MEASURE.

ART. **213**. The unit of capacity in France is the *litre,* which is a cubic decimètre.

1 litre . . . . . = 2.1135 pt. wine measure, U. S.

NOTE.—1 decalitre = 10 litres; 1 hectolitre = 100 litres; 1 kilolitre = 1000 litres; 1 myrialitre = 10000 litres. 1 decilitre = $\frac{1}{10}$ litre; 1 centilitre = $\frac{1}{100}$ litre; 1 millilitre = $\frac{1}{1000}$ litre.

ART. **214**. Some of the old weights and measures are used; as, 1 livre = $\frac{1}{2}$ a kilogramme; 1 marc = $\frac{1}{2}$ a livre; 1 once = $\frac{1}{8}$ marc; 1 gros = $\frac{1}{8}$ once; 1 grain = $\frac{1}{72}$ gros: 1 toise = 2 mètres; 1 pied or foot = $\frac{1}{3}$ mètre; 1 inch = $\frac{1}{12}$ pied or foot; 1 aune = $1\frac{1}{5}$ mètres; 1 boisseau or bushel = $12\frac{1}{2}$ litres; 1 litron = 1.074 Paris pints. When these are employed, the word *usuel* is annexed to them, signifying *customary.*

FRENCH MONEY.

ART. **215**. The unit of money is the franc, which is $\frac{9}{10}$ pure silver, and $\frac{1}{10}$ alloy, like our silver coins.

1 franc . . . . = \$.1875 or $18\frac{3}{4}$ cents.

NOTE.—1 decime = $\frac{1}{10}$ franc; 1 centime = $\frac{1}{100}$ franc.

The *livre tournois,* the former unit of French money, = $18\frac{1}{2}$ cents.

FRENCH CIRCULAR OR ANGULAR MEASURE

ART. **216**. Is the same as that of the United States and other countries.

FOREIGN WEIGHTS AND MEASURES.

ART. **217**. The pounds here mentioned are lb. avoirdupois, when not otherwise specified.

Alexandria, Egypt.—1 pik = 26.8 in.; 1 rhebebe = 4.364 bu.; 1 quillot or kisloz = 4.729 bu.; 1 rottolo forforo = .9347 lb. av.; 1 rottolo zaidino = 1.335 lb. av.; 1 rottolo zaro = 2.07 lb. av.; 1 rottolo mina = 1.67 lb. av.; 1 quintal or cantaro = 100 rottoli.

Amsterdam, Holland.—French system adopted in 1820. Old measures as follows: 1 lb. = 1.08923 lb.; 1 last = 85.25 bu.; 1 aam = 41 wine gal.; 1 stoop = $5\frac{1}{8}$ pints; 1 anker = $10\frac{1}{4}$ gal.; 1 foot = $11\frac{1}{7}$ in.; 1 ell = $27\frac{1}{12}$ in.

Antwerp, Belgium.—French system adopted in 1816. Old measures as follows: 1 lb. = $1.03\frac{1}{3}$ lb. av.; 1 schippound = 3 quintals = 310 lb. av.; 1 aam = $36\frac{1}{2}$ w. gal.; 1 viertel = 2.125 bu.; 1 stoop = 5.84 pt.

Barcelona, North of Spain.—1 lb. = .88215 lb. av.; 1 cana = .58514 yd.; 1 quartera = 1.88288 bu.; 1 carga = 32.7 w. gal.

Bombay, East Indies.—1 maund = 28 lb. av.; 1 candy = 560 lb. av.; 1 tola = 179 Troy gr.; 1 tank for pearls = 72 gr.; 1 guz = 27 in.; 1 hath = 18 in.; 1 tursoo = $1\frac{1}{8}$ in.

Bremen.—1 lb. = 1.098 lb. av.; 1 last = 80.7 bu.; 1 aam or 4 ankers = $37\frac{3}{4}$ w. gal.; 1 foot = 11.38 in.; 1 ell = .632 yd.

Batavia, East Indies.—1 pecul = 136 lb. av.; 1 catty = 1.36 lb. av.

Bencoolen, East Indies.—1 bahar = 560 lb. av.

Bahia and Rio de Janeiro, Brazil.—1 alquière of grain = 1 bu., U. S.; 1 frasco = 4.5 pt.

Cadiz, South of Spain.—1 lb. = 1.015 lb. av.; 1 vara = .9275 yd.; 1 arroba of wine = $4\frac{1}{4}$ w. gal.; 1 arroba of oil = $3\frac{3}{4}$ w. gal.; 1 moyo = 68 w. gal.; 1 botta = $127\frac{1}{2}$ w. gal.

Calcutta and Bengal Factory, East Indies.—1 maund = $74\frac{2}{3}$ lb. av.; 1 bazaar maund = $82\frac{2}{15}$ lb.; 1 tolan = 224.588 T. gr., 1 pallie = 9.08 lb. av.; 1 chittack = 45 sq. ft.; 1 biggah = 14440 sq. ft.; 1 guz = 1 yd.; 1 coss = $1\frac{3}{32}$ miles.

Canton, China.—1 tael = $1\frac{1}{3}$ oz.; 1 catty = $1\frac{1}{3}$ lb. av.; 1 pecul = $133\frac{1}{3}$ lb. av.; 1 covid or cobre = 14.625 in.; 1 li = $1897\frac{1}{2}$ ft.

Constantinople, Turkey.—1 quintal or cantaro = 124.457 lb. av.; 1 quintal of cotton = 127.2 lb. av.; 1 pik of silk = 27.9 in.; 1 pik of cotton = 27 in.; 1 kisloz = .741 bu.; 1 alma of oil = $1\frac{3}{8}$ gal.

Copenhagen, Denmark.—1 lb. = 1.1025 lb. av.; 1 anker = 10 w. gal.; 1 pot = 1.02 qt.; 1 last = 380 bu.; 1 Rhineland foot = $12\frac{1}{3}$ in.; 1 Danish ell = 2.06 ft.

Dantzic, East Prussia.—1 lb. = 1.033 lb. av.; 1 last = 620.4 w. gal.; 1 ahm of wine = $39\frac{2}{3}$ w. gal.; 1 scheffel = 1.552 bu.; 1 Dantzic foot = 11.3 in.; 1 Rhineland or Prussian foot = 12.356 in.; 1 Prussian ell = 26.256 in.; 1 last of corn = 91 bu.; 1 last of wheat, rye = 87 bu.; 1 Pru. mile = 4.8 miles

Genoa, Sardinia.— 1 lb. peso sottile, = .6989 lb. av.; 1 lb., peso grosso, = .76875 lb. av.; 1 mina = $3\frac{1}{2}$ bu.; 1 mezzarola = $39\frac{1}{4}$ w. gal; 1 barilla of oil = 17 w. gal.; 1 palmo = 9.725 in.; 1 canna = 9, 12 or 10 palmi, as it is used by manufacturers, merchants, or custom-house officers.

Hamburgh.—1 lb. = 1.068 lb. av.; 1 ahm = $38\frac{1}{4}$ w. gal.; 1 fuder = $229\frac{1}{2}$ gal.; 1 steckan of oil = $5\frac{1}{3}$ gal.; 1 last = 89.6 bu.: 1 foot = 11.289 in.; 1 Brabant ell = 27.585 in.

Havana, Cuba.—1 arroba of wine or spirits = 4.1 w. gal.; 1 fanega = 3 bu., nearly; 1 arroba of weight or 25 lb. = 25.4375 lb. av.; 1 vara = $2\frac{7}{9}$ ft.

Konigsberg.—Same as Dantzic.

La Guayra, Venezuela.—Same as Spain.

Leghorn, Tuscany.—1 lb. = .74864 lb. av., generally reckoned, = .77 lb. av.; 1 sacco of corn = 2.0739 bu.; 1 barile = 12 w. gal.; 1 braccio = 22.98 or 23 in.; 1 canna = 92 in.

Lima, Peru.—Same as Spain.

Lisbon, Portugal.—1 lb. or arratel = 1.10119 lb. av.; 1 moyo = 23.03 bu.; 1 almude = 4.37 w. gal.; 1 tonelada = $227\frac{1}{4}$ w. gal.; 1 pipe of Lisbon = 140 w. gal.; 1 pipe of port = 168 w. gal.; 1 pe or foot = 12.944 in.; 1 vara = 43.2 in.; 1 Almude of Oporto = $6\frac{5}{8}$ w. gal.

Madras, East Indies.—1 maund = 25 lb.; 1 candy = 500 lb.; 1 garce = 137 bu.; 1 Company maud = $24\frac{1}{8}$ lb.; 1 varahun = $52\frac{3}{4}$ gr.; 1 visay = 3 lb. 3 dr.; 1 baruay = $482\frac{1}{4}$ lb.; 1 gursay = $9645\frac{1}{2}$ lb.

Montevideo, Buenos Ayres.—Same as Spain.

Muscat, Arabia.—1 maund or 24 cuchas = $8\frac{3}{4}$ lb. av.

Naples, Naples.—1 cantaro grosso = $196\frac{1}{2}$ lb. av.; 1 cantaro piccolo = 106 lb.; 1 tomolo = 1.45 bu.; 1 carro = 264 w. gal.; 1 pipe wine or ludy = 132 w. gal.; 1 salma = $42\frac{3}{4}$ w. gal.; 1 canna = 6 ft. 11 in.; 1 palmo = 10.375 in.

Odessa, Russia.—Same as St. Petersburg.

Palermo, Sicily.—1 oncie = $\frac{14}{15}$ oz.; 1 salma grossa = 9.48 bu.; 1 salma generale = 7.62 bu.; 1 barile = $9\frac{3}{8}$ w. gal.; 1 caffiso of oil = $4\frac{2}{3}$ w. gal.; 1 canna = $3\frac{1}{5}$ yd.; 1 palma = $1\frac{1}{5}$ ft.

Port-au-Prince, Hayti.—Measures same as France—weights same as England, but 8 per cent. heavier.

Porto-Rico.—Same as Havana.

Rangoon, East Indies.—1 kyat or tical =.584 lb. av.; 1 Paiktha or vis = 3.65 lb. av.; 1 ten or basket = 58.4 lb. av., generally reckoned $\frac{1}{2}$ cwt.

Riga, Russia.—1 lb. = .9217 lb. av.; 1 loof = 1.9375 bu.; 1 anker = $10\frac{1}{3}$ w. gal.; 1 foot = 10.79 in., U. S.

Rotterdam, Holland.—1 last = 10.642 bu.; 1 ahm = 40 w. gal., nearly; 1 stoop = .6775 w. gal.; 1 foot = 1.02 ft., U. S. The rest like Amsterdam.

Singapore, East Indies.—1 maund of rice = 82.125 lb. av.; 1 bungkal of gold-dust = 832 gr. The rest same as Canton.

Smyrna, Turkey.—Same as Constantinople. 1 rottolo = 1.2748 lb. av.; 1 oke = 2 lb. 13 oz. 5 dr.; 1 tepper of silk = $4\frac{3}{8}$ lb., av.; 1 cheque of opium = $1\frac{5}{8}$ lb. av.; 1 chequee of goats' wool = $5\frac{5}{8}$ lb.; 1 kellow = 1.456 bu.; 1 pic = 27 in., U. S.

Stockholm, Sweden.—1 lb. or pund = .9375 lb., av.; 1 lb. of iron = $\frac{3}{4}$ lb. av.; 1 tun = $4\frac{1}{6}$ bu.; 1 ahm = $41\frac{5}{12}$ w. gal.; 1 pipe = $124\frac{1}{4}$ w. gal.; 1 foot = 11.684 in., U. S.; 1 kannor = .692 w. gal.

St. Petersburg, Russia.—1 lb. = .9026 lb. av.; 1 pood = 36.1041 lb. generally reckoned 36 lb. av.; 1 wedro = 3.14 w. gal.; 1 chetwert = 5.952 bu.; 1 sashen = 7 ft.; 1 arsheen = 28 in.; 1 foot = 1.145 ft., U. S.; 1 verst or mile = 5.3 fur.

Trebisond, Turkey.—Same as Constantinople.

Trieste, Austria.—1 lb. = 1.236 lb. av.; 1 staro = 2.34 bu.; 1 Vienna metzen = 1.723 bu.; 1 polonick = .861 bu.: 1 orna or eimer = 15 w. gal.; 1 barile = $173\frac{1}{3}$ w. gal.; 1 orna of oil = 17 w. gal.; 1 ell (for woolen goods) = 2.27 in.; 1 ell of silk = 25.2 in.

Valparaiso, Chili.—Same as Spain.

Venice, Lombardy.—1 lb., peso sottile = .66428 lb. av.; 1 lb., peso grosso = 1.05186 lb. av.; 1 staja = 2.27 bu.; 1 anfora = 137 w. gal.; 1 miro = 4.028 w. gal.; 1 braccio of wool = 26.6 in.

Vera Cruz, Mexico.—Same as Spain.

REDUCTION OF COMPOUND NUMBERS.

ART. **218**. Reduction is changing the form of a number without altering its value. It has three cases.

CASE I.—To reduce a simple number of any denomination to another denomination.

GENERAL RULE.

Multiply the given number by its unit value in the denomination required;

Or, *Divide it by the unit value of the required denomination in the one given.*

NOTE.—When there are one or more denominations between the one given and the one required, reduce the given number to each of them in succession, until the required denomination is reached.

Reduce 18 bushels to pints.

$$\begin{array}{rl} 18 & \text{bu.} \\ 4 & \\ \hline 72 & \text{pk.} \\ 8 & \\ \hline 576 & \text{qt.} \\ 2 & \\ \hline 1152 & \text{pt.} \end{array}$$

SOLUTION.—Since 1 bu. = 4 pk., 18 bu. = 18 times 4 pk. = 72 pk., and since 1 pk. = 8 qt., 72 pk. = 72 times 8 qt. = 576 qt., and since 1 qt. = 2 pt., 576 qt. = 576 times 2 pt. = 1152 pt. Or, find the unit-value of bushels in pints, thus: 1 bu. = 8 pk. = 32 qt. = 64 pt.; then multiply 18 by 64 pt. which gives 1152 pt. as before. This is called *Reduction Descending*, that is, going from a *higher* denomination (bu.) to a *lower* (pt.) and is performed by the 1st part of the rule, that is, by *multiplication*.

Reduce 236 inches to yards.

$$\begin{array}{r} 12)236 \text{ in.} \\ \hline 3)\ 19\tfrac{2}{3} \text{ ft.} \\ \hline 6\tfrac{5}{9} \text{ yd.} \end{array}$$

SOLUTION.—Since 12 inches = 1 ft., 236 inches will be as many feet as 12 in. is contained times in 236 in., which is $19\frac{2}{3}$ ft., and since 3 ft. = 1 yd., $19\frac{2}{3}$ ft. will be as many yd. as 3 ft. is contained times in $19\frac{2}{3}$ ft. which is $6\frac{5}{9}$ yd. Or, find the unit-value of yards in inches, thus; 1 yd. = 3 ft. = 36 in.; then divide 236 in. by 36 in., which gives $6\frac{5}{9}$ yd., as before. This is called *Reduction Ascending*, that is, going from a *lower* denomination (in.) to a *higher* (yd.), and is generally performed by the 2d part of the rule, that is, by *division*.

ART. **219**. Reduction Ascending is similar in principle to Reduction Descending, and can be performed by the 1st part of the rule.

Thus, in last example, instead of dividing 236 in. by 36 in. the unit value of yards, the 236 may be multiplied by $\frac{1}{36}$ yd., the unit-value of inches; for, $236 \times \frac{1}{36}$ yd. $= \frac{236}{36}$ yd. $= 6\frac{5}{9}$ yd. The operation by division is generally more convenient.

REMARK.—Reduction Descending diminishes the *size* and, therefore, increases the *number* of units given; while Reduction Ascend-

REVIEW.—218. What is reduction? How many cases? What is Case 1st? The rule? When there are several denominations between the one given and the one required, what should be done? Explain examples 1 and 2. What is Reduction Descending? How is it generally performed? What is Reduction Ascending? How is it generally performed? 219. How may it be performed?

ing increases the *size*, and, therefore, diminishes the *number* of units given. This is further evident from the fact, that the multipliers in Reduction Descending are *larger* than 1; but in Reduction Ascending *smaller* than 1.

Reduce $\frac{3}{8}$ gal. to gills.

SOLUTION.

$$\frac{3 \times \not{4} \times \not{2} \times 4}{\not{8}} = 12 \text{ gills}$$

REMARK.—The rule also applies when the number to be reduced is a common or decimal fraction. Indicate the operations and then cancel.

Reduce $5\frac{5}{7}$ gr. to ℥.

SOLUTION.—Although this is Reduction Ascending, use the first part of the rule, multiplying by the successive unit values, $\frac{1}{20}$, $\frac{1}{3}$ and $\frac{1}{8}$.

$$5\tfrac{5}{7}\text{ gr.} = \frac{\overset{2}{\not{40}}}{7} \times \frac{1}{\not{20}} \times \frac{1}{3} \times \frac{1}{\not{8}_4} = \frac{1}{84}\text{℥}.$$

Reduce 9.375 acres to perches.

```
  9.375 A
      4
 37.500 R.
     40
Ans. 1500.0 = 1500 P.
```

Reduce 2000 seconds to hours. *Ans.* $\frac{5}{9}$ hr.

$$20\not{0}\not{0} \times \frac{1}{6\not{0}} \times \frac{1}{6\not{0}} = \frac{20}{36} = \frac{5}{9}\text{ hr.}$$

Also, .6428 dr. av. to lb. av. *Ans.* .0025109375 lb.

```
16).642800 dr.
16).040175 oz.
   .0025109375 lb.
```

1. Reduce $2\frac{1}{4}$ years to seconds. *Ans.* 70956000 sec.
2. 49 hours is what part of a week? *Ans.* $\frac{7}{24}$ wk.
3. Bring 1 circumference to seconds. *Ans.* 1296000″.
4. 25″ to the decimal of a degree. *Ans.* .00694°
5. 17.0625 rd. are how many inches? *Ans.* $3378\frac{3}{8}$
6. Bring $4\frac{1}{4}$ ft. to chains. *Ans.* $\frac{17}{264}$ chain.
7. Reduce 192 sq. in. to sq. yd. *Ans.* $\frac{4}{27}$ sq. yd.

REVIEW.—219. Does the rule apply to fractions? How can the operation be shortened? In reducing common or decimal fractions, what rules must be borne in mind? *Ans.* The rules for the multiplication and division of common and decimal fractions.

8. $6\frac{2}{3}$ cu. yd. to cubic inches. *Ans.* 311040 cu. in.

9. $117.14 to mills. *Ans.* 117140 mills.

SUGGESTION.—In all reductions of U. S. money, or any system formed on the decimal basis, the multiplications or divisions are readily performed by moving the point to the right or left, since the multipliers and divisors are 10, 100, 1000, &c., (Arts. 144 and 145)

10. Reduce 6.19 cents to dollars. *Ans.* $.0619

11. 1600 mills to dollars. *Ans.* $1.60

12. $$5\frac{3}{8}$ to mills. *Ans.* 5375 mills.

SUGGESTION.—Change $\frac{3}{8}$ to a decimal, then move the point.

13. Reduce 12 lb. av. to lb. Troy. *Ans.* $14\frac{7}{12}$ lb.

SOLUTION.—The 12 lb. av. are first reduced to gr. by the table in Art. 194, and the gr. to Troy lb. by same table.

$$12 \times 7000 \times \frac{1}{5760} = \frac{175}{12} = 14\frac{7}{12}$$

14. Reduce 33 beer gal. to wine gal. *Ans.* $40\frac{2}{7}$ w. gal.

15. 36 yd. to ells Flemish. *Ans.* 48 E. Fl.

16. $\frac{1}{2}$ in. to qr. *Ans.* $\frac{1}{18}$ qr.

17. .216 gr. to oz. Troy. *Ans.* .00045 oz. Tr.

18. £.0732 to pence. *Ans.* 17.568 d.

19. $\frac{5}{7}$ lb. to tuns. *Ans.* $\frac{1}{2800}$ T.

20. 47.3084 sq. mi. to P. *Ans.* 4844380.16 P.

21. $4\frac{1}{2}$ ℈ to ℔. *Ans.* $\frac{1}{64}$ ℔.

22. $7\frac{1}{9}$ dr. av. to lb. *Ans.* $\frac{1}{36}$ lb.

23. 50 U. S. bu. to imp. bu. *Ans.* 48.47236 nearly.

24. 1200 inches to chains. *Ans.* $1\frac{17}{33}$ ch.

25. 99 yd. to furlongs. *Ans.* $\frac{9}{20}$ fur.

26. 6.3419 C. to cu. in.
Ans. 1402726.8096 cu. in.

27. 18 fathoms to miles. *Ans.* $\frac{9}{440}$ mi.

28. How many acres in a rectangle $24\frac{1}{2}$ rd. long by 16.02 rd. wide? *Ans.* 2.4530625 acres.

29. How many cubic yd. in a box $6\frac{1}{4}$ ft. long by $2\frac{1}{2}$ ft. wide and 3 ft. high? *Ans.* $1\frac{53}{72}$ cu. yd.

REVIEW.—219. How can reductions in U. S. money be performed?

30. How many perches in a rectangular field 18.22 chains long by 4.76 ch. wide? *Ans.* 1387.6352 P.

31. Reduce 256 ʒ to dr. av. *Ans.* $561\frac{129}{175}$ dr. av.

32. 1 nautical league to feet. *Ans.* 18256.7088 ft.

33. 16.02 chains to miles. *Ans.* .20025 mile.

34. 4.29 chains to feet. *Ans.* 283.14 ft.

35. $\frac{5}{9}$ of a link to rods. *Ans.* $\frac{1}{45}$ rd.

36. $\frac{2}{7}$ of a nail to ell English. *Ans.* $\frac{1}{70}$ E. E.

37. 1.644 inches to ell Flemish. *Ans.* $.060\dot{8}$ E. Fl.

38. 35.781 sq. yd. to sq. in. *Ans.* 46372.176 sq. in.

39. 256 roods to sq. chains. *Ans.* 640 sq. ch.

SUGGESTION.—First bring to acres, then to sq. chains.

40. $13\frac{1}{3}$ tuns of round timber to cords. *Ans.* $4\frac{1}{6}$ C.

41. 6.15 tuns of hewn timber to reduced feet, plank measure, 1 inch thick. *Ans.* 3690.

42. How many perches of masonry in a rectangular solid wall 40 ft. long by $7\frac{1}{2}$ ft. high and $2\frac{2}{3}$ ft. average thickness? *Ans.* $32\frac{32}{99}$ P.

43. Reduce $\frac{13}{72}$ lb. Troy to gr. *Ans.* 1040 gr.

44. 8 pwt. to lb. *Ans.* $\frac{1}{30}$ lb.

45. How many oz. Troy in the Brazilian Emperor's diamond, which weighs 1680 carats? *Ans.* $11\frac{1}{5}$ oz.

46. Reduce 75 pwt. to ʒ. *Ans.* 30 ʒ.

47. $\frac{4}{7}$ gr. to ℥. *Ans.* $\frac{1}{840}$ ℥.

48. 19 cwt. to oz. *Ans.* 30400 oz.

49. $22\frac{3}{5}$ dr. to T. *Ans.* .000044140625 T.

50. $18\frac{3}{4}$ ʒ to dr. av. *Ans.* $41\frac{1}{7}$ dr.

51. 96 oz. av. to oz. Troy. *Ans.* $87\frac{1}{2}$ oz. Tr.

52. How many wine gal. in a tank 3 ft. long by $2\frac{1}{4}$ ft. wide and $1\frac{1}{2}$ ft. deep? *Ans.* $75\frac{57}{77}$ w. gal

53. How many U. S. bushels in a bin 9.3 ft. long by $3\frac{5}{8}$ ft. wide and $2\frac{1}{4}$ ft. deep? *Ans.* 61 bu., nearly.

54. Reduce 21 bbl. of beer (36 gal.) to bbl. of wine, ($31\frac{1}{2}$ gal.) *Ans.* $29\frac{23}{77}$

55. 1 bu. to wine gal. *Ans.* 9.31 gal., nearly.

56. 1 bu. to beer gal. *Ans.* 7.625 + gal.

SUGGESTION.—Reduce to cu. in., and then to gallons.

57. 45 fʒ to O. *Ans.* $\frac{45}{128}$ O.

58. $2\frac{1}{2}$ f℥ to m. *Ans.* 1200 m.

59. If $\frac{6}{7}$ of a piece of gold is pure, how many carats fine is it? *Ans.* $20\frac{4}{7}$

60. In $18\frac{3}{4}$ carat gold what part is pure, and what part alloy? *Ans.* $\frac{25}{32}$ and $\frac{7}{32}$

CASE II.

ART. **220.** To reduce a compound number to a simple number of any denomination.

RULE.—*Commence with the highest denomination, if it be Reduction Descending; with the lowest, if it be Reduction Ascending. Reduce those of that denomination to the denomination required, and during this reduction add in at the proper times those of the other denominations.*

NOTES.—1. If the denomination required lie between the highest and lowest of those given, reduce part of the compound number by Reduction Descending, and the rest by Reduction Ascending, and add the two results.

2. The numbers added in must be of the same denomination as those to which they are added; mistakes can be avoided by marking the denomination of each number as it is obtained.

Reduce 5 lb. 2 oz. 13 pwt. 10 gr. to grains.

SOLUTION.—Since this is Reduction Descending, commence at the 5 lb., and reduce it to 60 oz.; the 2 oz. added in made 62 oz.; this is reduced to pwt., and 13 pwt. added in, making 1253 pwt.; this finally is brought to gr., and the 10 gr. added, producing 30082 gr. for the answer.

OPERATION.

```
 lb. oz. pwt.  gr.
  5   2   13   10
 12
 --                  lb. oz.
 62 oz.           =  5   2
 20
----                 lb. oz. pwt.
1253 pwt.         =  5   2   13
  24
----
 5022
2506
-----                lb. oz. pwt. gr.
30082 gr.         =  5   2   13  10
```

REVIEW.—220. What is Case 2d? The rule? If the required denomination lies between the highest and lowest given, what is necessary? What care must be exercised in the additions?

Reduce 2 cwt. 3 qr. 18 lb. to tuns.

SOLUTION.—This being Reduction Ascending, commence with 18 lb. and reduce it to .72 qr., making 3.72 qr. altogether; this is reduced to .93 of a cwt., making 2.93 cwt. in all; and this finally is reduced to .1465 T.—the result required. The successive quotients might be put in the form of common fractions, if it were desirable.

OPERATION.

25	lb. 18.00
4	qr. 3.72 = 3 qr. 18 lb.
20	cwt. 2.930 = 2 cwt. 3 qr. 18 lb
	T. .1465 = 2 cwt. 3 qr. 18 lb.

For convenience, the different simple numbers that compose the compound number, are set in a column, the lowest at the top, and the others under it in order, so that each quotient, as it is obtained, can be placed beside those of its own denomination. If any of the denominations are vacant, a cipher must be placed in the column to correspond.

Placing each quotient beside the number of its own denomination is the *adding in* required by the rule; just as the 2 oz. 13 pwt. and 10 gr., are added in in the first example.

REMARK.—The rule may be stated thus: *Reduce each of the simple numbers that compose the compound number, to the required denomination, and add the results*—but in practice it is not so convenient as the one given.

EXAMPLES FOR PRACTICE.

1. 4 cu. yd. 5 cu. ft. 256 cu. in. to cu. ft.

SOLUTION.—4 cu. yd. 5 cu. ft. = 113 cu. ft., and 256 cu. in. = 256 × $\frac{1}{1728}$ cu. ft. = $\frac{4}{27}$ cu. ft. *Ans.* 113$\frac{4}{27}$ cu. ft.

2. 17 mi. 3 fur. 38 rd. to rd. *Ans.* 5598 rd.
3. 8 rd. 1$\frac{1}{2}$ ft. to in. *Ans.* 1602 in.
4. 43 ft. 5 in. to rd. *Ans.* 2$\frac{125}{198}$
5. 9 mi. 22 rd. 10.6175 ft. to fur. *Ans.* 72.566+

SUGGESTION.—When the divisor is 5$\frac{1}{2}$, 30$\frac{1}{4}$, &c., multiply both dividend and divisor by the denominator of the fraction (2, 4, &c.), and *then* divide. The quotient will not be altered.

6. 464 yd. 2 ft. 8$\frac{2}{3}$ in. to miles. *Ans.* .26415+

REVIEW.—220. Explain Example 1st. What convenient form is adopted for Reduction Ascending? If any denomination is vacant, what is necessary? How might the rule have been stated more simply?

7. 1 mi. 3 fur. 7.2 in. to yd. *Ans.* 2420.2 yd.

8. 29 fathoms $4\frac{2}{3}$ ft. to miles. *Ans.* $.033\dot{7}\dot{8}$ mi.

9. 17 yd. 3 qr. 2 na. to na. *Ans.* 286 na.

10 5 yd. 4 in. to ells English. *Ans.* $4.0\dot{8}$ E. Eng.

11 44 E. Flem. 2 na. 2 in. to qr. *Ans.* $132.7\dot{2}$ qr.

12. 1 qr. 2 na. 1.785 in. to na. *Ans.* $6.79\dot{3}$ na.

13. 29 E. Flem. 1 qr. $2\frac{1}{2}$ na. to yd. *Ans.* 22.15625 yd.

14. 75 yd. 3 qr. $2\frac{2}{5}$ na. to feet. *Ans.* 227.7 ft.

15. 4 sq. rd. 13 sq. yd. 5 sq. ft. 98 sq. in. to sq. in.
Ans. 174482 sq. in.

16. 7 sq. ft. 120.54 sq. in. to sq. yd. *Ans.* .8708—

17. 63 sq. rd. $10\frac{4}{9}$ sq. in. to sq. ft. *Ans.* $17151\frac{533}{648}$

18. 1650 A. 3 R. 24.64 P. to sq. miles.
Ans. 2.5795375 sq. mi.

19. 301 A. 1 R. $18\frac{3}{4}$ P. to sq. ch. *Ans.* $3013.\dot{6}7\frac{3}{16}$

20. 1 sq. mi. 424 A. to perches. *Ans.* 170240 P.

21. 21 A. 35 P. to sq. chains. *Ans.* 212.1875 sq. ch.

22. 1 cu. yd. 24 cu. ft. 876 cu. in. to cu. yd. *Ans.* $1\frac{3529}{3888}$

23. What part of a cord of wood is a pile $7\frac{1}{2}$ ft. long by 5 ft. wide and $1\frac{3}{4}$ ft. high? *Ans.* $\frac{525}{1024}$

24. 83 cords 115 cu. ft. 1600 cu. in. to tuns of round timber. *Ans.* $268\frac{269}{540}$ T.

SUGGESTION.—First to cu. ft.; then to tuns of round timber.

25. 7 tuns of hewn timber 32 cu. ft. 480 cu. in to cords. *Ans.* $2\frac{2273}{2304}$ C.

26. 3 lb. 7 oz. $14.23\frac{1}{3}$ pwt. to gr. *Ans.* 20981.6 gr.

27. 19 pwt. 6 gr. to lb. *Ans.* $\frac{77}{960}$ lb.

28. 4 lb. $22\frac{1}{4}$ gr. to oz. *Ans.* $48\frac{89}{1920}$ oz.

29. 77 oz. 12 pwt. 10.464 gr. to lb. *Ans.* $6.46848\dot{3}$.

30. 68 lb. $8\frac{3}{5}$ oz. to gr. *Ans.* 395808 gr.

31. 49 carats $2\frac{7}{16}$ gr. to oz. Troy. *Ans.* $\frac{127}{384}$ oz. Tr.

32. A diamond weighing 3 oz. 16 pwt. is how many carats? *Ans.* 570 carats.

33. 3 ℔ 7 ʒ 12.24 gr. to ℥. *Ans.* 36.9005 ℥

34. 9 ℥ $1\frac{1}{2}$ ℈ to gr. *Ans.* 4350 gr.

35. 2 ʒ 2 ℈ $14\frac{1}{4}$ gr. to ℥. *Ans.* $.363\frac{1}{48}$ ℥

36. 1 ℔ $1\frac{1}{8}$ gr. to ʒ. *Ans.* $96\frac{3}{160}$ ʒ.

37. 4ʒ 2℈ 15.5 gr. to ℔. *Ans.* .051302$\frac{1}{72}$ ℔.

38 10 T. 9 cwt. 2 qr. 23 lb. to lb. *Ans.* 20973 lb.

39 3 qr. 18 lb. 15 oz. 10.08 dr. to cwt.
Ans. .9397$\frac{11}{16}$

40. 1 T. 6 cwt. 1 qr. 24 lb. 2 oz. 4$\frac{3}{5}$ dr. to lb.
Ans. 2649.14$\frac{19}{64}$

41. 1 qr. 12$\frac{1}{2}$ oz. to tuns. *Ans.* .0128$\frac{29}{32}$ T.

42. 762 lb. 8 oz. 3 dr. to cwt. *Ans.* 7.625$\frac{15}{128}$ cwt.

43. 12 lb. av. 9 oz. 10 dr. to Troy gr. *Ans.* 88210$\frac{15}{16}$ gr.

Sug.—First reduce to dr.; then to gr. by 27$\frac{11}{32}$. (Art. 188.)

44. 6℥ 3ʒ 1℈ 15.232 gr. to lb. av. *Ans.* .442176.

45. 5ʒ 2℈ 7 gr. to pwt. *Ans.* 14$\frac{11}{24}$ pwt.

46. 15 lb. Troy, 11 oz. 4 pwt. 9.085 gr. to dr. av.
Ans. 3356.71168 dr. av.

47. 6 oz. 10.48 pwt. to ʒ. *Ans.* 52.192 ʒ.

48. 37 gal. 2 qt. 1 pt. 3 gills to qt. *Ans.* 150$\frac{7}{8}$ qt.

49. 4 gal. $\frac{1}{2}$ pt. to gills. *Ans.* 130 gills.

50. 3 qt. 1 pt. 2.3 gills to gal. *Ans.* .946$\frac{7}{8}$ gal.

51. 52 gal. 1 qt. 1.052 gills to pt. *Ans.* 418.263 pt.

52. 1 gal. 3 qt. 1$\frac{1}{3}$ gills to gal. *Ans.* 1$\frac{19}{24}$ gal.

53. 47 bu. 3 pk. 2 qt. to pt. *Ans.* 3060 pt.

54. 2 pk. 6 qt. 1.8 pt. to bu. *Ans.* .71$\frac{9}{16}$ bu.

55. 3 bu. $\frac{1}{2}$ pt. to pk. *Ans.* 12$\frac{1}{32}$ pk.

56. 3 pk. 1.09$\frac{1}{3}$ pt. to bu. *Ans.* .767$\frac{1}{12}$ bu.

57. 2 pk. 7$\frac{1}{2}$ qt. to pt. *Ans.* 47 pt.

58. 8 bu. 3$\frac{3}{4}$ pk. to qt. *Ans.* 286 qt.

59. 286 bu. 3 pk. 1 qt. to imp. bu. *Ans.* 278.02—

Sug.—First to U. S. bu.; then to cu. in.; then to imp. bu.

60. 99 w. gal. 1 qt. 1 pt. to imp. gal. *Ans.* 82.7904.

61. 67 beer gal. 3 qt. 1.94 pt. to dry qt.
Ans. 285.32567—

62. 56 w. gal. 1 qt. 2$\frac{7}{8}$ gills to beer qt. *Ans.* 184.603—

63. 13 bu. 1 pk. 7 qt. 1.35 pt. to w. gal. *Ans.* 125.58—

64. 27 gal. 3 qt. 1 pt. of beer to gills. *Ans.* 1088$\frac{72}{77}$

Review.—220. Is it as convenient in this form? When the divisor is 5$\frac{1}{2}$, 30$\frac{1}{4}$, &c., what should be done?

65. 14 f℥ 5 fʒ 48 ♏ to Cong. *Ans.* .115039$\frac{1}{16}$ Cong.
66. 1 O 3 f℥ 6$\frac{1}{2}$ fʒ to ♏. *Ans.* 9510 ♏.
67. 3 fʒ 35 ♏ to f℥. *Ans.* $\frac{43}{96}$ f℥.
68. 2 O 7 f℥ 4 fʒ 58 ♏ to f℥. *Ans.* 39.62$\frac{1}{12}$ f℥.
69. 2 yr. 108 da. 18 hr. 40 min. to sec. *Ans.* 72470400 sec.
70. 6 yr. 44 da. 8 hr. 35 min. to wk. *Ans.* 319$\frac{391}{2016}$
71. 21 hr. 4 min. 54.6 sec. to da. *Ans.* .8784097$\dot{2}$ da.
72. 3 wk. 5 da. 12 hr. 26 min. 11$\frac{1}{9}$ sec. to hr. *Ans.* 636$\frac{707}{1620}$ hr.
73. 74 da. 16 hr. 45 min. 23.028 sec. to yr. *Ans.* .2046525567, nearly.
74. How many sec. in a solar year? *Ans.* 31556928.
75. 9° 13.6″ to minutes. *Ans.* 540.22$\dot{6}$′
76. 42′ 57$\frac{1}{2}$″ to degrees. *Ans.* .71597$\dot{2}$°
77. 163° 28′ 7″ to seconds. *Ans.* 588487″.
78. 7s 12° 6′ 20″ to degrees. *Ans.* 222.10$\dot{5}$°
79. 4′ 32.756″ to circum. *Ans.* .00021046, nearly.
80. $76 5 ct. 2$\frac{3}{8}$ m. to mills. *Ans.* 76052$\frac{3}{8}$
81. 9 ct. 9.1054 mills to $. *Ans.* $.0991054
82. $391 7 mills to cents. *Ans.* 39100.7 ct.
83. $84 32$\frac{4}{7}$ ct. to mills. *Ans.* 84325$\frac{5}{7}$ m.
84. 20 eagles $6 1 dime 2$\frac{1}{2}$ ct. to $. *Ans.* 206\frac{1}{8}$
85. 4 eagles 3.142 mills to dimes. *Ans.* 400.03142
86. 5 dimes 9 ct. 6 mills to eagles. *Ans.* .0596
87. £304 19s. 2$\frac{1}{2}$d. to £. *Ans.* £304$\frac{461}{480}$
88. 12s. 10$\frac{1}{4}$d. to s. *Ans.* 12$\frac{41}{48}$s.
89. £58 7s. 11d. to d. *Ans.* 14015 d.
90. £25 4$\frac{3}{4}$d. to s. *Ans.* 500$\frac{19}{48}$s.
91. Reduce 1 cwt to pwt. *Ans.* 29166$\frac{2}{3}$
92. How many ounces of gold weigh as much as a pound of lead? *Ans.* 14$\frac{7}{12}$

CASE III.—TO REDUCE A SIMPLE NUMBER OF ANY DENOMINATION TO A COMPOUND NUMBER.

ART. **221**. If the given number is a whole number, it may be reduced to a compound number by this

RULE.—*Reduce the given number to the next higher denomination, reserving the remainder; reduce the quotient to the next higher denomination, and reserve the remainder. Continue thus until the highest denomination has been reached, or until the quotient is so small as not to admit of further reduction. The last quotient with the several remainders, form the compound number required.*

NOTE.—Each remainder is of the same denomination as the dividend from which it is obtained.

REMARK.—The operations under this and the preceding rule serve to prove each other.

ART. 222. If the given simple number is a common or decimal fraction, it may be reduced to a compound number of *lower* denominations by this

RULE.—*Reduce the given fraction to the next lower denomination, reserving the whole number, if any, of that denomination. If there is a fraction in this result, reduce it to the next lower denomination, reserving the whole number, if any, of that denomination. Continue this process until no fraction occurs, or until the lowest denomination has been reached. The whole numbers reserved, with the last fraction, if any, will be the compound number required.*

NOTE.—If the given simple number be a mixed number, the whole number which it contains may be reduced to a compound number by the first of the rules given above, and the fraction by the last rule, and the two results united.

Reduce 1706 inches to a compound number.

SOLUTION.—The operation is similar to that in the 2d example, Art. 218, except that each remainder is written as a whole number of its own denomination, instead of a fraction of the next higher denomination; as 142 ft. 2 in. instead of $142\frac{2}{12}$ ft., and 47 yd. 1 ft. instead of $47\frac{1}{3}$ yd.

```
12)1706 in.
 3) 142 ft. 2 in.
     47 yd. 1 ft. 2 in.
```

REVIEW.—221. What is Case 3d? What is the rule for reducing a simple whole number to a compound number? How is the denomination of each remainder known? How are operations under this rule verified? 222. What is the rule for reducing a simple fractional number to a compound number? How is a simple mixed number reduced to a compound number?

Reduce $\frac{2}{3}$ of a wine gallon to a compound number.

SOLUTION.—Multiply $\frac{2}{3}$ gal. by 4; the result is $2\frac{2}{3}$ qt.; cut off whole number 2 as part of the compound number required. Multiply $\frac{2}{3}$ qt. by 2; the result is $1\frac{1}{3}$ pt.: cut off the 1 pt. and reduce the $\frac{1}{3}$ pt. to $1\frac{1}{3}$ gills. This being the limit of the table, the operation must stop, and the compound number required is 2 qt. 1 pt. $1\frac{1}{3}$ gills.

w. gal.	qt.	pt.	gill
$\frac{2}{3}$ =	2	1	$1\frac{1}{3}$

```
      4
qt. 2|2/3
     |2
 pt. 1|1/3
      |4
gills 1 1/3
```

Reduce 4905.06185 lb. av. to a compound number.

```
25)4905 lb.                    .06185 lb.
   4)196 qr. 5 lb.                 16
   20)49 cwt.                   .9896 oz.
       2 T. 9 cwt. 5 lb.           16
                              15.8336 dr. Ans.
```

SOLUTION.—The fraction .06185, reduced to lower denominations, is 15.8336 dr.; the whole number 4905 lb. reduced to higher denominations, is 2 T. 9 cwt. 5 lb. These results united give 2 T. 9 cwt. 5 lb. 15.8336 dr. for the answer.

Reduce 1657 yd. to a compound number.

SOLUTION.—Divide the 1657 yd. by $5\frac{1}{2}$, to reduce it to rd. To do this conveniently, multiply both dividend and divisor by 2, making the divisor 11, and the dividend 3314 half-yards, without altering the quotient. The remainder 3 is also half-yards (Art. 221, Note), and is therefore written $\frac{3}{2}$ yd. = $1\frac{1}{2}$ yd.; the $\frac{1}{2}$ yd. is then reduced to a compound number by the last rule, and joined to the result already obtained.

```
5 1/2)1657 yd.
2        2
11)3314 half yd.
  40)301 rd. 3 half yd.
      7 fur. 21 rd. 1 1/2 yd.
```

But $\frac{1}{2}$ yd. = 1 ft. 6 in.

Ans. 7 fur. 21 rd. 1 yd. 1 ft 6 in.

REDUCE TO COMPOUND NUMBERS,

	EXAMPLES.	ANSWERS.
1.	$44753\frac{1}{2}$ ft.	= 8 mi. 3 fur. 32 rd. 5 ft. 6 in.
2.	99.75 yd.	= 18 rd. 2 ft. 3 in.
3.	$\frac{5}{6}$ fur.	= 33 rd. 1 yd. 2 ft. 6 in.

EXAMPLES. ANSWERS.

4. 8760531 in. . = 138 mi. 2 fur. 5 rd. 1 ft. 9 in

5. 904.178 fath. . . . . = 904 fath. 1.068 ft.

6. $887\frac{1}{3}$ na. . . . . . . = 55 yd. 1 qr. $3\frac{1}{3}$ na.

7. 69.1525 yd. . . . . . = 69 yd. 2.44 na.

8. $601\frac{1}{12}$ qr. . . . . . = 150 yd. 1 qr. $\frac{1}{3}$ na.

9. 52.003 E. Fl. . . . . . = 52 E. Fl. .036 na.

10. 1811.0625 sq. ft. = 201 sq. yd. 2 sq. ft. 9 sq. in.

11. 300027 sq. in. = 231 sq. yd. 4 sq. ft. 75 sq. in.

12. 64.10826 P. . . . . . = 1 R. 24.10826 P.

13. $832\frac{15}{64}$ sq. yd. = 832 sq. yd. 2 sq. ft. $15\frac{3}{4}$ sq. in.

14. 1.10475 sq. mi. . . = 1 sq. mi. 67 A. 6.4 P.

15. 322.7372 A. . . . = 322 A. 2 R. 37.952 P.

16. 706.2814 sq. ch. . . = 70 A. 2 R. 20.5024 P.

17. 1567804 sq. in. = 1209 sq. yd. 6 sq. ft. 76 sq. in.

18. $87\frac{5}{8}$ sq. ft. . . . = 9 sq. yd. 6 sq. ft. 90 sq. in.

19. $931\frac{17}{32}$ A. . . . = 1 sq. mi. 291 A. 2 R. 5 P.

20. $\frac{3}{85}$ sq. mi. . . . . . = 22 A. 2 R. $14\frac{2}{17}$ P.

21. $\frac{5}{9}$ sq. rd. . . . . = 16 sq. yd. 7 sq. ft. 36 sq. in.

22. $\frac{7}{11}$ cu. yd. . . . . . = 17 cu. ft. $314\frac{2}{11}$ cu. in.

23. 1013854 cu. in. = 21 cu. yd. 19 cu. ft. 1246 cu. in.

24. .0038 yr. . . = 1 da. 9 hr. 17 min. 16.8 sec.

25. 65.387 cu. ft. = 2 cu. yd. 11 cu. ft. 668.736 cu. in.

26. 4.2045 cu. yd.
= 4 cu. yd. 5 cu. ft. 901.152 cu. in.

27. 18.9142 mi.
= 18 mi. 7 fur. 12 rd. 2 yd. 2 ft. 11.712 in.

28. $4015238\frac{7}{8}$ min. ($365\frac{1}{4}$ da. to a yr.)
Ans. 7 yr. 231 da. 14 hr. 38 min. $52\frac{1}{2}$ sec.

29. $13\frac{17}{18}$ C. . . . . = 13 C. $120\frac{8}{9}$ cu. ft.

30. $8.56\frac{4}{9}$ T. hewn timber. . . = 8 T. $28\frac{2}{9}$ cu. ft.

31. 4000 gr. Troy. . . . . . = 8 oz. 6 pwt. 16 gr.

32. $78\frac{5}{7}$ lb. Troy . . = 78 lb. 8 oz. 11 pwt. $10\frac{2}{7}$ gr.

33. 267.3 pwt. . . . = 1 lb. 1 oz. 7 pwt. 7.2 gr.

34. $\frac{10}{13}$ oz. Troy. . . . . . . = 15 pwt. $9\frac{3}{13}$ gr.

35. 45.54 oz. Tr. . = 3 lb. 9 oz. 10 pwt. 19.2 gr.

36. 692 pwt. . . . . = 2 lb. 10 oz. 12 pwt

	EXAMPLES.	ANSWERS.
37.	13.7644 ℔. . .	= 13 ℔. 9 ℥ 1 ʒ 1 ℈ 2.944 gr.
38.	$805\frac{22}{45}$ ʒ.	= 8 ℔. 4 ℥ 5 ʒ 1 ℈ $9\frac{1}{3}$ gr.
39.	$\frac{25}{36}$ ℥.	= 5 ʒ 1 ℈ $13\frac{1}{3}$ gr.
40.	90562 ℈	= 314 ℔. 5 ℥ 3 ʒ 1 ℈.
41.	$\frac{15}{22}$ ℔.	. 8 ℥ 1 ʒ 1 ℈ $7\frac{3}{11}$ gr.
42.	170053.62 gr. Apoth.	= 29 ℔. 6 ℥ 2 ʒ 13.62 gr.
43.	$467\frac{3}{5}$ ℈.	= 1 ℔ 7 ℥ 3 ʒ 2 ℈ 12 gr.
44.	.3701 ʒ.	= 1 ℈ 2.206 gr.
45.	$\frac{8}{15}$ T. . .	= 10 cwt. 2 qr. 16 lb. 10 oz. $10\frac{2}{3}$ dr.
46.	.4815 lb. av.	= 7 oz. 11.264 dr.
47.	$809\frac{11}{12}$ cwt.	= 40 T. 9 cwt. 3 qr. 16 lb. 10 oz. $10\frac{2}{3}$ dr.
48.	422031 lb.	= 211 T. 1 qr. 6 lb.
49.	$733\frac{3}{4}$ qr. . . .	= 9 T. 3 cwt. 1 qr. 18 lb. 12 oz.
50.	$\frac{17}{24}$ cwt.	= 2 qr. 20 lb. 13 oz. $5\frac{1}{3}$ dr.
51.	$8\frac{4}{35}$ gal.	= 8 gal. $3\frac{23}{35}$ gills.
52.	10072 gills.	= 314 gal. 3 qt.
53.	$95\frac{2}{9}$ qt.	= 23 gal. 3 qt. $1\frac{7}{9}$ gills.
54.	301.46 pt. . .	= 37 gal. 2 qt. 1 pt. 1.84 gills.
55.	$808\frac{1}{5}$ qt. dry measure . .	= 25 bu. 1 pk. $\frac{2}{5}$ pt.
56.	2191009.3 dr.	= 4 T. 5 cwt. 2 qr. 8 lb. 10 oz. 1.3 dr.
57.	$560052\frac{73}{84}$ oz. av.	= 17 T. 10 cwt. 3 lb. 4 oz. $13\frac{19}{21}$ dr.
58.	365.2422414 days.	*Ans.* 365 da. 5 hr. 48 min. 49.65696 sec.
59.	£$\frac{10}{33}$.	= 6s. $\frac{8}{11}$d.
60.	$\$33\frac{5}{8}$	= $33.625.
61.	$\frac{2}{3}$ wk.	= 4 da. 16 hr.
62.	.7 f℥	= 5 fʒ 36 ♏.
63.	$\frac{9}{16}$ hr.	= 33 min. 45 sec.
64.	$\frac{6}{23}$ bu.	= 1 pk. $\frac{16}{23}$ pt.
65.	.555£.	= 11s. $1\frac{1}{5}$d.
66.	$\frac{9}{14}$ gal.	= 2 qt. $1\frac{1}{7}$ pt.
67.	67.76s.	= £3. 7s. 9.12d.
68.	27.35°	= 27° 21′.
69.	32.4 O.	= 4 Cong. 6 f℥ 3 fʒ 12 ♏
70.	$\frac{2}{5}$ Cong.	= 3 O. 3 f℥ 1 fʒ 36 ♏.

	EXAMPLES.	ANSWERS.
71	.075 qt.	= .6 gill.
72.	3.07 pk.	= 3 pk. 1.12 pt.
73.	$46\frac{19}{20}$ fʒ.	= 5 f℥ 6 fʒ 57 m.
74.	260234″.	= 72° 17′ 14″.
75.	1246.05′.	= 20° 46′ 3″.
76.	$1856\frac{3}{4}$d.	= £7 14s. $8\frac{3}{4}$d.
77.	$2819\frac{3}{28}$ s.	= £140 19s. $1\frac{2}{7}$d.
78.	.8054 bu.	= 3 pk. 1 qt. 1.5456 pt.
79.	100000 m.	= 13 O 2 fʒ 40 m.
80.	.1934 sign.	= 5° 48′ 7.2″.
81.	$75\frac{3}{4}$ dimes.	= \$7.57 ct. 5 m.
82.	17.052 pk.	= 4 bu. 1 pk. .832 pt.
83.	$2\frac{10}{11}$ circum.	= 2 c. 327° 16′ $21\frac{9}{11}$″.
84.	19019.2 m.	= 2 O. 7 f℥ 4 fʒ 59.2 m.
85.	84312 mills.	= \$84.312.
86.	$\frac{29}{65}$ of a dollar.	= 44 ct. $6\frac{2}{13}$ m.
87.	$\frac{9}{19}$ of an eagle.	= \$4.736$\frac{16}{19}$
88.	2093.57 cents.	= \$20.9357.
89.	$\frac{9}{19}$ of a degree.	= 28′ $25\frac{5}{19}$″.
90.	$6407\frac{1}{4}$ pt. dry meas.	= 100 bu. 3 qt. $1\frac{1}{4}$ pt.
91.	10808107.87 sec.	= 125 da. 2 hr. 15 min. 7.87 sec.
92.	6.045964 yr.	= 6 yr. 16 da. 18 hr. 38 min. 40.704 sec.
93.	17000.12 da. ($365\frac{1}{4}$ da. to a yr.)	*Ans.* 46 yr. 198 da. 14 hr. 52 min. 48 sec.
94.	22303.5d.	*Ans.* £92 18s. $7\frac{1}{2}$d.

REMARK.—If our tables of Weights and Measures were on the decimal basis, like U. S. Money and the French tables, the same rules and methods would do for compound as for simple numbers. Common fractions would also occur less frequently, the comparatively complicated processes that arise from their use would be avoided, and the computations required in ordinary business transactions would be much shortened and simplified.

As it is, Compound numbers must be treated somewhat differently from Simple numbers; though the rules and operations are not entirely new, but rather modifications of those already explained.

ADDITION OF COMPOUND NUMBERS.

ART **223.** RULE.—*Write the numbers to be added, units of the same denomination in a column, reducing any fractions to lower denominations, until none are found in any but the right-hand column. Add the right-hand column, and reduce the result, if large enough, to the next higher denomination; write the remainder, if any, under the column added, and carry the quotient to the next column. Add the next column, reduce, set down, and carry as before, and continue so until all the columns have been added.*

PROOF.—Same as in addition of simple numbers.

NOTE.—If the right-hand column contain common or decimal fractions, add them according to the usual rules; if any of the higher denominations in the answer has a fraction, reduce it to lower denominations, and add it in.

Add 3 bu. $2\frac{1}{4}$ pk.; 1 pk. $1\frac{1}{2}$ pt.; 5 qt. 1 pt.; 2 bu. $1\frac{1}{3}$ qt; and .125 pt.

bu.	pk.	qt.	pt.	
3	2	2	0	= 3 bu. $2\frac{1}{4}$ pk.
	1	0	$1\frac{1}{2}$	= 1 pk. $1\frac{1}{2}$ pt.
		5	1	= 5 qt. 1 pt.
2	0	1	$\frac{2}{3}$	= 2 bu. $1\frac{1}{3}$ qt.
			$\frac{1}{8}$	= .125 pt.
6	0	1	$1\frac{7}{24}$	= *Ans.*

SOLUTION.—Reduce the fraction in each number to lower denominations, (Art. 222,) and place units of the same kind in columns. The right-hand column, when added, gives $3\frac{7}{24}$ pt. = 1 qt. $1\frac{7}{24}$ pt.; write the $1\frac{7}{24}$ and add the 1 qt. with the next column, making 9 qt. = 1 pk. 1 qt.; write the 1 qt. and carry the 1 pk. to the next column, making 4 pk. = 1 bu.; as there are no pk. left, set down a cipher and carry 1 bu. to the next column, making 6 bu.

Add 2 rd. 9 ft. $7\frac{1}{4}$ in.; 13 ft. 5.78 in.; 4 rd. 11 ft. 6 in.; 1 rd. $10\frac{2}{3}$ ft.; 6 rd. 14 ft. $6\frac{5}{8}$ in.

rd.	ft.	in.
2	9	7.25
	13	5.78
4	11	6
1	10	8
6	14	6.625
16	$9\frac{1}{2}$	9.655
	but $\frac{1}{2}$ ft. =	6.
16	10	3.655

SOLUTION.—The numbers are prepared, written, and added, as in the last example; the answer is 16 rd. $9\frac{1}{2}$ ft. 9.655 in. The $\frac{1}{2}$ foot is then reduced to 6 inches, (Note), and added to the 9.655 in., making 15.655 in. = 1 ft. 3.655 in. Write the 3.655 in., and carry the 1 ft., which gives 16 rd. 10 ft. 3.655 in. for the final answer.

1. Add $\frac{2}{7}$ mi.; 3 fur. $26\frac{1}{3}$ rd.; 10 mi. 14 rd. 7 ft. 6 in.; 5.24 fur.; 37 rd. 16 ft. $2\frac{1}{4}$ in.; 1 mi. 12 ft. 8.726 in.
Ans. 12 mi. 4 fur. 20 rd. 9 ft. $4.633\frac{1}{7}$ in.

2. 6.19 yd.; 2 yd. 2 ft. $9\frac{3}{4}$ in.; 1 ft. 4.54 in.; 10 yd. 2.376 ft.; $\frac{3}{4}$ yd.; $1\frac{5}{8}$ ft.; $\frac{7}{8}$ in. *Ans.* 21 yd. 2 ft. 3.517 in.

3. 3 yd. 2 qr. 3 na. $1\frac{1}{2}$ in.; 1 qr. $2\frac{2}{5}$ na.; 6 yd. 1 na. 2.175 in.; 1.63 yd.; $\frac{2}{3}$ qr.; $\frac{5}{9}$ na.
Ans. 12 yd. 1 na. 0.755 in.

4. 4 E. Fr. 4 qr. 2 na.; 7 E. Fr. 5 qr. $1\frac{1}{2}$ na.; 3 qr. 3 na. 1 in.; $\frac{7}{9}$ E. Fr.; 1.6 na.; $\frac{2}{5}$ in. *Ans.* 14 E. Fr. 3 na. $\frac{7}{8}$ in.

5. 2 E. Fl. 1 qr. $1\frac{1}{3}$ na.; 5 E. Fl. $3\frac{3}{4}$ na.; 2 qr. $2\frac{4}{9}$ na.; 1 E. Fl. 2 qr. 3.8 na.; $\frac{5}{6}$ E. Fl. *Ans.* 11 E. Fl. 1 qr. $1\frac{59}{180}$ na.

6. 15 sq. yd. 5 sq. ft. 87 sq. in.; $16\frac{1}{2}$ sq. yd.; 10 sq. yd. 7.22 sq. ft.; 4 sq. ft. 121.6 sq. in.; $\frac{11}{32}$ sq. yd.
Ans. 43 sq. yd. 7 sq. ft. 37.78 sq. in.

7. 101 A. 2 R. 18.35 P.; 66 A. 1 R. $34\frac{1}{2}$ P.; 20 A.; 12 A. 2.84 R.; 5 A. $13.33\frac{1}{3}$ P.
Ans. 205 A. 3 R. $19.78\frac{1}{3}$ P.

8. 23 cu. yd. 14 cu. ft. 1216 cu. in.; 41 cu. yd. 6 cu. ft. 642.132 cu. in.; 9 cu. yd. 25.065 cu. ft.; $\frac{7}{15}$ cu. yd.
Ans. 75 cu. yd. 4 cu. ft. 1279.252 cu. in.

9. $\frac{5}{6}$ C.; $\frac{3}{8}$ cu. ft.; 1000 cu. in;
Ans. 107 cu. ft. 1072 cu. in.

10. 2 lb. Tr. $6\frac{2}{3}$ oz.; $1\frac{3}{4}$ lb.; 12.68 pwt.; 11 oz. 13 pwt. $19\frac{1}{5}$ gr.; $\frac{3}{8}$ lb., $\frac{15}{16}$ oz.; $\frac{5}{9}$ pwt.
Ans. 5 lb. Tr. 9 oz. 9 pwt. $2.85\frac{1}{3}$ gr.

11. 8℥ 14.6 gr.; 4.18℥; $7\frac{5}{12}$ʒ; 2ʒ 2℈ 18 gr.; 1℥ 12 gr.; $\frac{1}{5}$℈. *Ans.* 1 ℔. 2℥ 4ʒ 1℈.

12. $\frac{5}{16}$ T.; 9 cwt. 1 qr. 22 lb.; 3.06 qr.; 4 T. 8.764 cwt.; 3 qr. 6 lb.; $\frac{7}{12}$ cwt. *Ans.* 5 T. 6 cwt. 2 qr. $14\frac{7}{30}$ lb.

13. .3 lb. av.; $\frac{5}{9}$ oz.; $\frac{6}{7}$ dr. *Ans.* 5 oz. $6\frac{172}{315}$ dr.

14. 6 gal. $3\frac{1}{3}$ qt.; 2 gal. 1 qt. 3.32 gills; 1 gal. 2 qt. $\frac{1}{2}$ pt.; $\frac{3}{5}$ gal.; $\frac{2}{9}$ qt.; $\frac{7}{8}$ pt. *Ans.* 11 gal. 2 qt. $.46\frac{4}{9}$ gill.

15. 4 gal. 3 gills; 10 gal. 3 qt. $1\frac{1}{2}$ pt.; 8 gal. $\frac{2}{3}$ pt.; 5.64 gal.; 2.3 qt.; 1.27 pt.; $\frac{3}{5}$ gill. *Ans.* 29 gal. 2 qt. $.22\frac{2}{3}$ gill.

REVIEW.—222. What would be the advantages if our weights and measures were on the decimal basis? 223. What is the rule for addition of compound numbers? What is the proof? If the right-hand column contains common or decimal fractions, what must be done? If any of the higher denominations of the answer has a fraction, what must be done?

16. Add 1 bu. $\frac{1}{2}$ pk.; $\frac{6}{11}$ bu.; 3 pk. 5 qt. $1\frac{1}{4}$ pt.; 9 bu. 3.28 pk.; 7 qt. 1.16 pt.; $\frac{5}{12}$ pk. *Ans.* 12 bu. 3 pk. .46$\frac{19}{33}$ pt.

17. Add $\frac{3}{8}$ bu.; $\frac{2}{5}$ pk.; $\frac{7}{9}$ qt.; $\frac{2}{3}$ pt. *Ans.* 2 pk. $\frac{28}{45}$ pt.

18. Add 6 f℥ 2 fʒ 25 m; $2\frac{1}{2}$ f℥; 7 fʒ 42 m; 1 f℥ $2\frac{2}{3}$ fʒ; 3 f℥ 6 fʒ 51 m. *Ans.* 14 f℥ 7 fʒ 38 m.

19. Add $\frac{1}{2}$ wk.; $\frac{1}{2}$ da.; $\frac{1}{2}$ hr.; $\frac{1}{2}$ min.; $\frac{1}{2}$ sec.
Ans. 4 da. 30 min. $30\frac{1}{2}$ sec.

20. Add 3.26 yr. (365 da. each); 118 da. 5 hr. 42 min. $37\frac{1}{2}$ sec.; 63.4 da.; $7\frac{4}{5}$ hr.; 1 yr. 62 da. 19 hr. $24\frac{4}{5}$ min.; $\frac{73}{192}$ da. *Ans.* 4 yr. 340 da. 1 hr. 14 min. $55\frac{1}{2}$ sec.

21. Add 27° 14′ 55.24″; 9° $18\frac{1}{4}$″; 1° $15\frac{1}{3}$′; 116° 44′ 23.8″ *Ans.* 154° 14′ 57.29″

22. Add \$84 1 ct. 5.27 m.; 67 ct. 8 m.; \$25 9 ct. $2\frac{3}{8}$ m.; $\frac{9}{16}$ of a dol.; $25\frac{3}{4}$ ct. *Ans.* \$110 60 ct. 5.645 m.

23. \$$\frac{1}{2}$; $\frac{1}{4}$ ct.; $\frac{1}{8}$ m. *Ans.* 50 ct. $2\frac{5}{8}$ m.

24. Add \$3 7 m.; \$5 20 ct.; \$100 2 ct. 6 m.; \$19 $\frac{1}{8}$ ct. *Ans.* \$127 23 ct. $4\frac{1}{4}$ m.

25. Add £21 6s. $3\frac{1}{2}$d.; £5 $17\frac{3}{4}$s.; £9.085; 16s. $7\frac{1}{4}$d.; £$\frac{5}{12}$. *Ans.* £37 10s. 8.15d.

SUBTRACTION OF COMPOUND NUMBERS.

ART. **224.** RULE.—*Prepare and write the numbers as in addition of compound numbers, placing the subtrahend below. Commence at the right, and proceed to the left, subtracting each lower number from the one above, and setting the remainder below. If a lower number is larger than the one above it, add to the upper as many units of its denomination as make one of the next higher; subtract and carry* 1 *to the next figure.*

PROOF.—Same as in subtraction of simple numbers.

NOTE.—If fractions are in the right-hand column, subtract them by the usual rules; if a fraction is in any of the higher denominations of the answer, reduce it to lower denominations, and add it in.

Subtract 1 yd. 2.45 ft. from 9 yd. 1 ft. $6\frac{1}{2}$ in.

yd.	ft.	in.
9	1	6.5
1	2	5.4
7	2	1.1

SOLUTION.—Change the $\frac{1}{2}$ in. to a decimal, making the minuend 9 yd. 1 ft. 6.5 in.; reduce the .45 ft. to inches, making the subtrahend 1 yd. 2 ft. 5.4 in. To subtract 2 ft. from the number above, add 3 ft. (= 1 yd.) to the 1 ft., making 4 ft.; set the remainder, 2 ft., below, and to compensate for the 3 ft. added above, add 1 yd. to the next lower figure, which gives 2 yd.; the remainder is then 7 yd. and the answer 7 yd. 2 ft. 1.1 in.

1. Subtract 16 rd. 8 ft. $1\frac{1}{4}$ in. from 23 rd. 1.2 ft.
Ans. 6 rd. 9 ft. 7.15 in.

2. $\frac{2}{5}$ mi. from 3 fur. 24.86 rd. *Ans.* 16.86 rd.

3. 1.35 yd. from 4 yd. 2 qr. 1 na. $1\frac{3}{4}$ in.
Ans. 3 yd. 1 qr. $\frac{2}{5}$ in.

4. $2\frac{8}{9}$ E. Fl. from $2\frac{5}{9}$ E. E. *Ans.* 1 yd. 1 qr. 1 in.

5. 2 sq. rd. 24 sq. yd. 91 sq. in. from 5 sq. rd. 16 sq. yd. $6\frac{2}{3}$ sq. ft. *Ans.* 2 sq. rd. 22 sq. yd. 8 sq. ft. 41 sq. in.

6. $.56\frac{2}{3}$ sq. yd. from 7 sq. ft. 18.27 sq. in.
Ans. 2 sq. ft. 3.87 sq. in.

7. 2 R. $19\frac{2}{9}$ P. from 11 A. *Ans.* 10 A. 1 R. $20\frac{7}{9}$ P.

8. 384 A. 1 R. 3.92 P. from 1.305 sq. mi.
Ans. 450 A. 3 R. 28.08 P.

9. $\frac{121}{192}$ cu. ft. from $\frac{11}{54}$ cu. yd. *Ans.* 4 cu. ft. 1503 cu. in.

10. 13 cu. yd. 25 cu. ft. 1204.9 cu. in. from 20 cu. yd. 4 cu. ft. 1000 cu. in. *Ans.* 6 cu. yd. 5 cu. ft. 1523.1 cu. in.

11. 9.362 oz. Troy from 1 lb. 15 pwt. 4 gr.
Ans. 3 oz. 7 pwt. 22.24 gr.

12. 5 oz. 15 pwt. from 3 lb. 22 gr.
Ans. 2 lb. 6 oz. 5 pwt. 22 gr.

13. $\frac{10}{13}$ ʒ from $\frac{6}{11}$ ℥. *Ans.* 3 ʒ 1 ℈ $15\frac{95}{143}$ gr.

14. 2 ʒ 1 ℈ 6 gr. from 4 ℥ $\frac{1}{2}$ gr. *Ans.* 3 ℥ 5 ʒ 1 ℈ $14\frac{1}{2}$ gr.

15. 56 T. 9 cwt. 1 qr. 23 lb. from 75.004 T.
Ans. 18 T. 10 cwt. 2 qr. 10 lb.

16. $\frac{13}{42}$ cwt. from 3 qr. 11 lb. 14 oz. $10\frac{3}{7}$ dr.
Ans. 2 qr. 5 lb. 15 oz. $6\frac{13}{21}$ dr.

17. $\frac{5}{24}$ lb. Troy from $\frac{6}{35}$ lb. av. *Ans.* 0.

18. 3 gal. 2 qt. $1\frac{4}{5}$ pt. from 8 gal. 1.1 qt.
Ans. 4 gal. 2 qt. $\frac{2}{5}$ pt.

19. 12 gal. 1 qt. 3 gills from 31 gal. $1\frac{1}{2}$ pt.
Ans. 18 gal. 3 qt. 3 gills.

20. .0625 bu. from 3 pk. 5 qt. 1 pt.
Ans. 3 pk. 3 qt. 1 pt.

REVIEW.—224. What is the rule for subtraction of compound numbers? What is the proof? If fractions are in the right-hand column, what must be done? What, if a fraction is in any of the higher denominations of the answer?

21. 2 pk. .84 pt. from 3 bu. $4\frac{1}{4}$ qt.
Ans. 2 bu. 2 pk. 3 qt. 1.66 pt.

22. 15 wine gal. from 15 beer gal.
Ans. $3\frac{3}{7}\frac{4}{7}$ w. gal. = 3 w. gal. 1 qt. $1\frac{7}{7}\frac{5}{7}$ gills.

SUGGESTION.—Reduce the beer gal. to w. gal. (Art. 194.)

23. 10 U. S. bu. from 10 imperial bu. of G. Britain.
Ans. 1 pk. 2 qt. 0.17+ pt.

SUGGESTION.—Reduce the imp. bu. to U. S. bu. (Art. 194).

24. 1 f℥ 4 fʒ 38 ♏ from 4 f℥ 2 fʒ.
Ans. 2 f℥ 5 fʒ 22 ♏.

25. .9 of a day from $\frac{1}{4}$ wk. *Ans.* 20 hr. 24 min.

26. 3 da. 16 hr. 47 min. 33.3 sec. from 1 wk.
Ans. 3 da. 7 hr. 12 min. 26.7 sec.

27. 275 da. 9 hr. 12 min. 59 sec. from 2.4816 yr. (allowing $365\frac{1}{4}$ days to the year.)
Ans. 1 yr. 265 da. 18 hr. 29 min. 21.16 sec.

28. 1832 yr. 8 mon. 18 da. from 1840 yr. 5 mon. 26 da., considering 1 mon. to be 30 days, as is customary in business involving time. *Ans.* 7 yr. 9 mon. 8 da.

29. What is the difference of time between Aug. 5th, 1848, and Mar. 14th, 1851? *Ans.* 2 yr. 7 mon. 9 da.

SUGGESTION.—Proceed as in last example, writing Aug. 5th in the subtrahend, as 7 mon. 5 da., and March 14th in the minuend, as 2 mon. 14 da., since these dates are respectively that long after the beginning of the year; allow 30 days for a month.

30. Find the difference of time between Sept. 22d, 1855, and July 1st, 1856. *Ans.* 9 mon. 9 da.

31. Between Dec. 31st, 1814, and April 1st, 1822.
Ans. 7 yr. 3 mon.

32. Between May 20th, 1855, and Oct. 15th, 1857.
Ans. 2 yr. 4 mon. 25 da.

33. Subtract 43° 18′ 57.18″ from a quadrant.
Ans. 46° 41′ 2.82″

34. 17° 29′ from 24° 52″. *Ans.* 6° 31′ 52″.

35. 161° 34′ 11.8″ from 180°. *Ans.* 18° 25′ 48.2″

36. $12.857 from $19. *Ans.* $6.143

37. 4 mills from $40. *Ans.* $39.996

REVIEW.—224. How is the difference of time between two dates found?

38. 1 ct. from \$1 and 1 m. *Ans.* 99 ct. 1 m.
39. 86 ct. and .6 mill from \$2.62½. *Ans.* \$1.7644
40. .08 dime from 1 ct. 8 mills. *Ans.* 1 ct.
41. $\frac{37}{40}$ ct. from \$$\frac{3}{32}$. *Ans.* 8 ct. 4½ m.
42. \$5 43 ct. 2½ m. from \$12 6 ct. 8⅓ m. *Ans.* \$6 63 ct. 5⅚ m.
43. £9 18s. 6½d. from £20. *Ans.* £10 1s. 5½d.
44. ⅝s. from $\frac{2}{45}$£. *Ans.* 3⅙d.

MULTIPLICATION OF COMPOUND NUMBERS.

ART. **225**. Since every compound number can be reduced to a simple number of either of its denominations (Art. 220), the multiplication of a compound number will only differ from the multiplication of a simple number by the reduction before and after multiplying.

GENERAL RULE.

TO MULTIPLY A COMPOUND NUMBER BY ANY SIMPLE NUMBER WHOLE OR FRACTIONAL,

Reduce the compound to a simple number of either of its denominations, and multiply as in simple numbers. The product will be a simple number of the same denomination as the multiplicand, and may be reduced to a compound number.

NOTE.—It is generally best to reduce the multiplicand to its lowest denomination.

Multiply 2 bu. 3 pk. 7 qt. 1¼ pt. by 10.8724.

```
Bu. pk. qt. pt.              10.8724
 2   3   7  1¼                = 191¼ pt.
 4                            -------
---                           27181
11 pk.                       108724
 8                          978516
---                        108724
95 qt.                   -------------
 2                     2)2079.3465 pt.
----                   ----------------
191¼ pt.               8)1039 qt. 1.3465 pt.
                       --------
                       4)129 pk. 7 qt.
                        ------
                         32 bu. 1 pk.
```

Ans. 32 bu. 1 pk. 7 qt. 1.3465 pt.

ART. **226**. If the multiplier is a whole number, the reductions may take place *during* the multiplication, instead of *before* and *after* it.

TO MULTIPLY A COMPOUND NUMBER BY A SIMPLE WHOLE NUMBER,

RULE.—*Begin at the lowest denomination; multiply each of the simple numbers that compose the compound number in succession: reduce each product to the next higher denomination, setting the remainder below the number multiplied, and carrying the quotient to the next* PRODUCT.

PROOF.—Same as in multiplication of simple numbers.

NOTES.—1. If the multiplier is a composite number, we may multiply by its factors in succession, as in Art. 53.

2. When the multiplications and reductions can not be readily performed in the mind, do the work on one side, and transfer the results.

3. If there be a fraction in the lowest denomination of the multiplicand, multiply it first; if one occurs in any of the higher denominations of the product, reduce it to lower denominations, and add it in.

Multiply 9 hr. 14 min. 8.17 sec. by 10.

da.	hr.	min.	sec.
	9	14	8.17
			10
3	20	21	21.7

SOLUTION.—Ten times 8.17 sec. =81.7 sec.=1 min. 21.7 sec. Write 21.7 sec. and carry 1 min. to the 140 min. obtained by the next multiplication. This gives 141 min.= 2 hr. 21 min. Write 21 min. and carry 2 hr. This gives 92 hr.= 3 da. 20 hr.

Multiply 12 A. 3 R. $28\frac{3}{8}$ P. by 84.

A.	R.	P.
12	3	$28\frac{3}{8}$
		7
90	1	$38\frac{5}{8}$
		12
1085	3	$23\frac{1}{2}$

SOLUTION.—Since $84=7\times12$, multiply by one of these factors, and this product by the other; the last product is the one required. The same result can be obtained by multiplying by 84 at once; performing the work on one side and transferring the results.

REVIEW.—225. What is the rule for multiplying a compound number by a simple number, whole or fractional? To which of its denominations should the multiplicand generally be reduced? 226. What is the rule for multiplying a compound number by a simple whole number? The proof?

1. Multiply 7 rd. 10 ft. 5 in. by 6. *Ans.* 45 rd. 13 ft.
2. 2 mi. 3 fur. 27 rd. by 8. *Ans.* 19 mi. 5 fur. 16 rd.
3. 16 yd. $2\frac{4}{9}$ in. by 21. *Ans.* 337 yd. 1 ft. $3\frac{1}{3}$ in.
4. 1 mi. 14 rd. $8\frac{1}{4}$ ft. by 97.
   Ans. 101 mi. 3 fur. 6 rd. 8 ft. 3 in.
5. 4 yd. 2 ft. 9.14 in. by $47\frac{3}{22}$.
   Ans. 231 yd. 2 ft. $9.73\frac{6}{11}$ in.
6. 12 E. Fl. $3\frac{1}{3}$ na. by 18. *Ans.* 221 E. Fl.
7. 6 E. E. 4 qr. 3.44 na. by 28.
   Ans. 195 E. E. 1 qr. .32 na.
8. 5 sq. yd. 8 sq. ft. 106 sq. in. by 13.
   Ans. 77 sq. yd. 5 sq. ft. 82 sq. in.
9. 41 A. 3 R. 26.1087 P. by 9.046.
   Ans. 379 A. 23.4593+ P.
10. 10 cu. yd. 3 cu. ft. 428.15 cu. in. by 67.
    Ans. 678 cu. yd. 1 cu. ft. 1038.05 cu. in.
11. 7 oz. 16 pwt. $5\frac{3}{4}$ gr. by 174.
    Ans. 113 lb. 3 oz. 5 pwt. $16\frac{1}{2}$ gr.
12. 2 ʒ 1 ℈ 13 gr. by 20. *Ans.* 6 ℥ 3 ʒ.
13. 16 cwt. 1 qr. 7.88 lb. by 11.
    Ans. 8 T. 19 cwt. 2 qr. 11.68 lb.
14. 7 lb. 6 oz. $12\frac{2}{7}$ dr. by 283.44.
    Ans. 1 T. 1 cwt. 3 lb. 15 oz. $8.98\frac{2}{7}$ dr.
15. 1 qt. $3\frac{1}{2}$ gills by 7. *Ans.* 2 gal. 2 qt. $\frac{1}{2}$ gill.
16. 5 gal. 3 qt. 1 pt. 2 gills by 35.108.
    Ans. 208 gal. 1 qt. 1 pt. 2.52 gills.
17. 26 bu. 2 pk. 7 qt. .37 pt. by 10.
    Ans. 267 bu. 7 qt. 1.7 pt.
18. 3 fʒ 48 𝔪 by 12. *Ans.* 5 f℥ 5 fʒ 36 𝔪.
19. 18 da. 9 hr. 42 min. 29.3 sec. by $16\frac{7}{11}$.
    Ans. 306 da. 4 hr. 25 min. 2 sec., nearly.
20. $1072 9 ct. 2 m. by 424. *Ans.* $454567 8 m.
21. £215 16s. $2\frac{1}{4}$d. by 75. *Ans.* £16185 14s. $\frac{3}{4}$d.
22. 10° 28′ $42\frac{1}{2}$″ by 2.754. *Ans.* 28° 51′ 27.765″

REVIEW.—226. If the multiplier is a composite number, what may be done? When the multiplying and reducing can not readily be performed in the mind, what should be done?

ART. 227. The difference of time between two places is 4 hr. 18 min. 26 sec.: what is their difference of longitude?

SOLUTION.—Every hour of time corresponds to 15° of longitude; every minute of time to 15′ of longitude; every second of time to 15″ of longitude, (Art. 201). Hence, multiplying the hours in the difference of time by 15 will give the degrees in the difference of longitude, multiplying the minutes of time by 15 will give minutes (′) of longitude, and multiplying the seconds of time by 15 will give seconds (″) of longitude; since reducing seconds to min. and min. to hr. are the same as reducing (″) to (′) and (′) to (°), the divisor in both cases being always 60, hence,

hr.	min.	sec.
4	18	26
		15
64°	36′	30″
	Diff. of Long.	

TO CHANGE DIFF. OF TIME INTO DIFF. OF LONGITUDE.

RULE.—*Multiply the difference of time by* 15, *according to the rule for Compound Multiplication, and mark the product* ° ′ ″ *instead of hr. min. and sec.*

REMARK.—The work can be shortened by cancellation, for 15 times 26 divided by $60 = \frac{\not{1}\not{5} \times 26}{\not{6}\not{0}\ 4} = 6\frac{1}{2}' = 6'\ 30''$, since $\frac{1}{2}' = 30''$. Write the 30″ and carry the 6′. Then 15 times 18 divided by 60 $= \frac{\not{1}\not{5} \times 18}{\not{6}\not{0}\ 4} = 4\frac{1}{2}° = 4°\ 30'$, since $\frac{1}{2}° = 30'$; add in the 6′ to be carried with this 30′, making 36′. Carry the 4° to the next product, 60°, making 64°; the answer is 64° 36′ 30″, as before.

1. When it is 4 o'clock P. M. at New York, it is 3 hr. 18 min. 28.4 sec. P. M. at Cincinnati: the longitude of New York is 74° 1′ 6″ W.: what is the longitude of Cincinnati? *Ans.* 84° 24′ W.

SUGGESTION.—Of two places, the one having *later* time is *east* of the other; the one having *earlier* time is *west* of the other.

2. When it is 1 P. M. at St. Louis, it is 8 hr. 14 min. $55\frac{2}{3}$ sec. P. M. at the Cape of Good Hope: the longitude of the latter place is 18° 28′ 45″ E.: what is the longitude of St. Louis? *Ans.* 90° 15′ 10″ W.

REVIEW.—227. What is the rule for converting difference of time into difference of longitude? Illustrate and prove it. Show how the operations under this rule can be shortened by cancellation.

3. A man travels from Halifax to Chicago; his watch shows 9 A. M., while the time at Chicago is 7 hr. 24 min. 24$\frac{2}{3}$ sec. A. M. The longitude of Halifax being 63° 36′ 40″ W.: what must be the longitude of Chicago?

Ans. 87° 30′ 30″ W.

4. When it is 10 A. M. at Stockholm, it is 3 hr. 24 min. 58 sec. A. M. at Wheeling: the longitude of Wheeling is 80° 42′ W.: what is the longitude of Stockholm?

Ans. 18° 3′ 30″ E.

5. Noon comes 47 min. 17 sec. sooner at Detroit than at Galveston, whose longitude is 94° 47′ 15″ W.: what is the longitude of Detroit? *Ans.* 82° 58′ W.

6. Time is 7 hr. 57 min. 26$\frac{2}{5}$ sec. later at St. Petersburgh than at New Orleans, and the longitude of the former is 30° 19′ 46″ E.: what is the longitude of the latter? *Ans.* 89° 1′ 50″ W.

7. When it is 1 P. M. at Utica, whose longitude is 75° 13′ W., it is 11 hr. 52 min. 4 sec. A. M. at Little Rock: what is the longitude of the latter?

Ans. 92° 12′ W

8. When it is 3 P. M. at Regent's Park, London, it is 9 hr. 46 min. 31.2 sec. A. M. at the University of Virginia, whose longitude is 78° 31′ 29″ W.: what is the longitude of the former? *Ans.* 9′ 17″ W.

9. When it is midnight at Madras, in India, it is 1 hr. 23 min. 16.2 sec. P. M. at Buffalo; the longitude of the former place is 80° 15′ 57″ E.: what is the longitude of the latter? *Ans.* 78° 55′ W.

10. When it is 1 A. M. at Constantinople, it is 11 hr. 13 min. 25$\frac{7}{15}$ sec. P. M. of the previous day at Paris, and the longitude of Paris is 2° 20′ 22″ E.: what is that of Constantinople? *Ans.* 28° 59′ E.

11. A ship's chronometer, set at Greenwich, points to 4 hr. 43 min. 12 sec. P. M.; the sun on the meridian: what is the ship's longitude? *Ans.* 70° 48′ W.

DIVISION OF COMPOUND NUMBERS.

ART. 228. Division of compound numbers like division of simple numbers, has two cases:

CASE I.

ART. **229.** To divide a compound number by a simple number.

Since every compound number can be reduced to a simple number of either of its denominations, (Art. 220), the division of a compound number by a simple number will only differ from the division of simple numbers by the reduction before and after dividing.

GENERAL RULE

FOR DIVIDING A COMPOUND NUMBER BY ANY SIMPLE NUMBER, WHOLE OR FRACTIONAL.

Reduce the compound number to a simple number of either of its denominations; divide as in simple numbers: the quotient will be a simple number of the same denomination as the dividend, and may be reduced to a compound number.

NOTE.—It is generally best to reduce the dividend to its lowest denomination.

Divide 17 da. 5 hr. 24 min. 19.208 sec. by 8.07

SOLUTION.—17 da. 5 hr. 24 min. 19.208 sec. = 1488259.208 sec. which divided by 8.07 gives a quotient 184418.737 + sec., and this reduces to 2 da. 3 hr. 13 min. 38.737 + sec., the quotient required.

ART. **230.** If the divisor is a whole number, the reductions may take place *during* the division, instead of *before* and *after* it.

TO DIVIDE A COMPOUND NUMBER BY A SIMPLE WHOLE NUMBER.

RULE.—*Divide that part of the dividend which is of the highest denomination first, and set the quotient below: reduce the remainder, if there is one, to the next lower denomination, add in those of that denomination in the dividend, and divide again. Continue so until the lowest denomination has been used: when, if there is a remainder, it should be expressed as a common or decimal fraction of that denomination.*

REVIEW.—228. How many cases in division of compound numbers? 229. What is the 1st case? What is the general rule for dividing a compound number by a simple number, whole or fractional? To which of its denominations is the dividend generally reduced? 230. If the divisor is a simple *whole* number, what is the rule?

PROOF.—Same as in division of simple numbers.

NOTE.—If the divisor is a composite number, we may divide by its factors in succession, as in Art. 66.

REMARK.—This rule may be used when the divisor has a common or decimal fraction, by multiplying both numbers by the denominator of this fraction, which will convert the divisor into a whole number, and yet will not alter the quotient, (Art. 75).

Divide 106 lb. 9 oz. $14\frac{2}{5}$ dr. of sugar equally among 8 men.

		lb.	oz.	dr.
	8)	106	9	$14\frac{2}{5}$
Ans.	=	13	5	$3\frac{4}{5}$

SOLUTION.—8 into 106 lb. gives a quotient 13 lb., and 2 lb. = 32 oz. to be carried to the 9 oz. making 41 oz.; 8 into 41 oz. gives a quotient 5 oz., and 1 oz. = 16 dr. to be carried to the $14\frac{2}{5}$ dr., making $30\frac{2}{5}$ dr.; 8 into $30\frac{2}{5}$ dr. gives $3\frac{4}{5}$ dr. and the operation is complete.

If $42 purchase 67 bu. 2 pk. 5 qt. $1\frac{3}{4}$ pt. of meal, how much will $1 purchase?

	bu.	pk.	qt.	pt.
6)	67	2	5	$1\frac{3}{4}$
7)	11	1	0	$1\frac{23}{24}$
	1	2	3	$1\frac{23}{168}$

SOLUTION.—Since $42 = 6 \times 7$, divide first by one of these factors, and the resulting quotient by the other; the last quotient will be the one required.

A man travels 1472 mi. 6 fur. 32 rd. in 59 days; how much a day does he average?

$$\begin{array}{rl}
 & \text{mi.}\quad \text{fur.}\quad \text{rd.}\quad \text{mi.} \\
59) & 1472 \quad 6 \quad 32(24 \\
 & 118 \\
 & 292 \\
 & 236 \\
 & 56 \text{ mi.} \\
 & 8 \\
 & 454 \text{ fur.} (7 \text{ fur.} \\
 & 413 \\
 & 41 \text{ fur.} \\
 & 40 \\
 & 1672 \text{ rd.} (28\tfrac{20}{59} \text{ rd.} \\
 & 118 \\
 & 492 \\
 & 472 \\
 & 20
\end{array}$$

SOLUTION.—As the divisor is larger than 12, and can not be separated into suitable factors, proceed as in 1st example, performing the work by long, instead of short division.

Ans. 24 mi. 7 fur. $28\frac{20}{59}$ rd.

1. Divide 16 mi. 2 fur. 29 rd. by 7.
Ans. 2 mi. 2 fur. 27 rd.

2. 37 rd. 14 ft. 11.28 in. by 18.
Ans. 2 rd. 1 ft. 8.96 in.

3. 43 E. Fl. 1 qr. $3\frac{1}{5}$ na. by 33. *Ans.* 1 E. Fl. $3\frac{47}{55}$ na.

4. 675 C. 114.66 cu. ft. by 83.
Ans. 8 C. 18.3453 + cu. ft.

5. 10 sq. rd. 29 sq. yd. 5 sq. ft. 94 sq. in. by 17.
Ans. 19 sq. yd. 4 sq. ft. $119\frac{15}{17}$ sq. in.

6. 1000 A. by 160. *Ans.* 6 A. 1 R.

7. 6 sq. mi. 35 P. by $22\frac{1}{2}$ *Ans.* 170 A. 2 R. $28\frac{2}{9}$ P.

8. 1245 cu. yd. 24 cu. ft. 1627 cu. in. by 11.303
Ans. 110 cu. yd. 6 cu. ft. 338.4 + cu. in.

9. 88 lb. 16 pwt. 17.6 gr. by 54.
Ans. 1 lb. 7 oz. 11 pwt. 10.1 + gr.

10. 3 ℥ 7 ʒ 18 gr. by 12. *Ans.* 2 ʒ 1 ℈ $16\frac{1}{2}$ gr.

11. 600 T. 7 cwt. 86 lb. by 29.06
Ans. 20 T. 13 cwt. 20 lb. 14 oz. 12+ dr.

12. 62 lb. av. 8 oz. by 96. *Ans.* 10 oz. $6\frac{2}{3}$ dr.

13. 312 gal. 2 qt. 1 pt. 3.36 gills by $72\frac{5}{8}$
Ans. 4 gal. 1 qt. 1.79+ gills.

14. 19302 bu. by 6.215
Ans. 3105 bu. 2 pk. 6 qt. 1.5+ pt.

15. 53 bu. $\frac{1}{2}$ pt. by 63. *Ans.* 3 pk. 2 qt. 1.85— pt.

16. 76 yr. 108 da. 2 hr. 38 min. 26.18 sec. by 45.
Ans. 1 yr. 254 da. 27 min. 31.25— sec.

17. 19 hr. $53\frac{2}{7}$ sec. by $7\frac{4}{5}$
Ans. 2 hr. 26 min. 16.06+ sec.

18. 152° 46′ 2″ by 9. *Ans.* 16° 58′ $26\frac{8}{9}$″

ART. **231**. Since the difference of time between two places, multiplied by 15, gives their difference of longitude, the product being marked ° ′ ″ instead of hr. min. and sec.: conversely,

TO CHANGE DIFF. OF LONGITUDE TO DIFF. OF TIME,

RULE.—*Divide the difference of longitude by* 15, *according to the rule for Compound Division, and mark the quotient, hours, minutes, and seconds, instead of* ° ′ ″

NOTE.—The division required by the rule can be shortened by canceling, in a manner similar to that explained in Art. 227.

The difference of longitude between two places is 81° 39′ 22″; what is their difference of time?

$$\begin{array}{r|lll} 15) & 81° & 39' & 22'' \\ \hline & 5\text{ hr.} & 26\text{ min.} & 37\tfrac{7}{15}\text{ sec.} \end{array}$$

SOLUTION.—15 into 81° gives 5 (marked *hr.*), and 6° to be carried. Instead of multiplying 6 by 60, adding the 39′ and then dividing, proceed thus: 15 into 6° is the same as 15 into 6 × 60′ = $\frac{6 \times \not{6}\not{0}4}{\not{1}\not{5}} = 24'$, and as 15 into 39′ gives 2′ for a quotient and 9′ remainder, the whole quotient is 26′ (marked *min.*), and remainder 9′ = 9 × 60″, which divided by 15 gives $\frac{9 \times \not{6}\not{0}4}{\not{1}\not{5}} = 36''$, which with $1\frac{7}{15}$ obtained by dividing 22″ by 15 gives $37\frac{7}{15}''$, (marked *sec.*) The ordinary mode of dividing will give the same result, and may be used if preferred.

1. What time at Columbus (long. 83° 3′ W.), when it is 4 P. M. at Baltimore, (long. 76° 37′ W.)?
Ans. 3 hr. 34 min. 16 sec. P. M.

2. What time at Copenhagen (long. 12° 34′ 57″ E.), when it is 10 P. M. at Mobile, (long. 88° 11′ W.)?
Ans. 4 hr. 43 min. $3\frac{4}{5}$ sec. A. M. the day after.

3. What time at Pittsburg (long. 79° 58′ W.), when it is 3 A. M. at Dublin, (long. 6° 20′ 30″ W.)?
Ans. 10 hr. 5 min. 30 sec. P. M. the day before.

4. When it is noon at Louisville (long. 85° 30′ W.), what time at Bangor, (long. 68° 47′ W.)?
Ans. 1 hr. 6 min. 52 sec. P. M.

5. When it is 6 P. M. at Havana (long. 82° 22′ 21″ W.), what time is it at Paramatta, (long. 151° 1′ 35″ E.)?
Ans. 9 hr. 33 min. $35\frac{11}{15}$ sec. A. M. the day after.

6. What time at Cambridge, Eng., (long. 5′ 21″ E.), when it is 9 P. M. at Cambridge, Mass., (long. 71° 7′ 21″ W.)? *Ans.* 1 hr. 44 min. $50\frac{4}{5}$ sec. A. M. the day after.

REVIEW.—230. What is the proof? If the divisor is a composite number, what may be done? How can this rule be used when the divisor has a common or decimal fraction? 231. What is the rule for converting difference of longitude into difference of time? Illustrate and prove it. How can the operations under the rule be shortened?

7. When it is 7 A. M. at Washington (long. 77° 1′ 30 W.), what time at Mexico, (long. 99° 5′ W.)?

Ans. 5 hr. 31 min. 46 sec. A. M.

CASE II.

ART. **232.** To divide one compound number by another similar compound number, the quotient being an abstract number.

RULE.—*Reduce both dividend and divisor to simple numbers of the same denomination, and then divide.*

PROOF.—Same as in division of simple numbers.

NOTE.—It is generally best to reduce the numbers to their lowest denomination; if, after reduction, one or both contain a fraction, proceed as directed in division of decimal or common fractions.

How often can a keg of 2 gal. 3 qt. $1\frac{1}{3}$ pt. be filled from a barrel of molasses containing $37\frac{1}{2}$ gal.?

SOLUTION.—$37\frac{1}{2}$ gal. $=$ 300 pt. and 2 gal. 3 qt. $1\frac{1}{3}$ pt. $= 23\frac{1}{3}$ pt.; then 300 pt. $\div 23\frac{1}{3}$ pt. $= 300 \div \frac{70}{3} = 30\not{0} \times \frac{3}{7\not{0}} = \frac{90}{7} = 12\frac{6}{7}$ times. *Ans.*

1. How many lunar months of 29 da. 12 hr. 44 min. 2.84 sec., in a solar year, (Art. 196, Note)?

Ans. 12.368 +

2. How many steps, 2 ft. 9 in. each, will a man take in going $3\frac{1}{4}$ miles? *Ans.* 6240.

3. The wheels of a locomotive are 10 ft. 5 in. in circumference, and make 8 revolutions a second; how soon will it run 100 miles? *Ans.* 1 hr. 45 min. 36 sec.

4. The new half-dollar of the U. S. contains 7 pwt. $4\frac{4}{5}$ gr. pure silver; how many dollars in 11 oz. 2 pwt. of pure silver, and if it be coined into 66 s. what is one shilling worth in U. S. currency?

Ans. $\$15.41\frac{2}{3}$, and 1 s. $= 23\frac{71}{198}$ ct.

5. How many half eagles, each weighing 5 pwt. 9 gr., and 21 car. $2\frac{2}{5}$ gr. fine, can be made of 1000 sovereigns, each weighing 5 pwt. 3.274 gr., and 22 car. fine?

Ans. $973\frac{1076}{3483}$

REVIEW.—232. What is the rule for dividing one compound number by another similar one? The proof? To which denomination is it best to reduce the numbers?

6. A comet moves $8° 17' 22\frac{1}{2}''$ in one day; in what time will it complete the circuit of the heavens, or 360°? *Ans.* 43 da. 10 hr. 16 min. 19 sec., nearly.

7. How many persons can receive each \$2 18 ct. $7\frac{1}{2}$m., out of a fund of \$59 $6\frac{1}{4}$ ct? *Ans.* 27.

8. How many half-crowns, each worth 2 s. 6 d., are in £18 7s. $10\frac{3}{4}$d.? *Ans.* $147\frac{19}{120}$

9. The Julian calendar assumed the year 365 da. 6 hr., instead of 365 da. 5 hr. 48 min. 48 sec., its true length; in how many years was a day gained? *Ans.* $128\frac{4}{7}$ yr.

10. In how many years is a day gained by the Gregorian calendar, which allows for the fraction of a day by adding in 97 days in 400 years? *Ans.* 3600 yr.

ALIQUOT PARTS.

ART. **233.** *Aliquot parts* is a useful method of finding a product, when one or both of the factors is a compound number. The following is an example of the sort of problems to which it is generally applied.

What cost 28 A. 3 R. 25 P. of land at \$16 per acre?

	\$
	16
	28
	128
	32
	\$448
2 R. = $\frac{1}{2}$	8
1 R. = $\frac{1}{2}$	4
20 P. = $\frac{1}{2}$	2
5 P. = $\frac{1}{4}$	.50
	\$462.50

SOLUTION.—Multiply \$16, the price of 1 A., by 28; the product \$448 is the price of 28 A. 3 R. is made up of 2 R. and 1 R.; the former is $\frac{1}{2}$ of an A., and the latter $\frac{1}{2}$ the former; obtain the price of 2 R. by taking $\frac{1}{2}$ the price of 1 A., or $\frac{1}{2}$ of \$16 = \$8; the price of 1 R. is $\frac{1}{2}$ of this = \$4. 25 P. is equal to 20 P. and 5 P.; the former is $\frac{1}{2}$ of 1 R., and worth $\frac{1}{2}$ of \$4 = \$2: the latter being $\frac{1}{4}$ of 20 P., is worth $\frac{1}{4}$ of \$2 = \$$\frac{1}{2}$ = 50 ct. These results added, give the value of 28 A. 3 R. 25 P., = \$462.50. The same result could be obtained by reducing 28 A. 3 R. 25 P. to acres, viz: 28.90625 A., and multiplying it by 16.

ART. **234.** This method can be applied when the multiplicand is a compound number, as in the following example:

A travels 3 mi. 5 fur. 16 rd. in 1 hr.; how far will he go in 6 da. 9 hr. 18 min. 48 sec., (12 hr. to a day)?

SOLUTION.—This example is solved like the preceding, except that the multiplications and divisions are performed on a compound instead of a simple number.

	mi.	fur.	rd.
	3	5	16
			9
da. hr.	33	0	24
6 = 8 × 9	264	4	32
15 min. = $\frac{1}{4}$		7	14
3 .. = $\frac{1}{5}$		1	$18\frac{4}{5}$
30 sec. = $\frac{1}{6}$			$9\frac{4}{5}$
15 .. = $\frac{1}{2}$			$4\frac{9}{10}$
3 .. = $\frac{1}{5}$			$\frac{49}{50}$
	298	6	$24\frac{12}{25}$

One of the most valuable applications of aliquot parts is when the product is to be U. S. money; for instance,

If $47.52 is paid for the use of money 1 yr., how much ought to be paid for using it 4 yr. 7 mon. 19 da.?

SOLUTION.—When money is paid for the use of money, 1 month is reckoned 30 days, and the aliquot parts taken accordingly. The mills in the result are usually neglected if they are under 5; but, if they are 5 or over, they are counted 1 cent. The above result would be called $220.31.

	$47.52
	4
	190.08
6 mon. = $\frac{1}{2}$	23.76
1 mon. = $\frac{1}{6}$	3.96
18 da. = $\frac{6}{10}$	2.376
1 da. = $\frac{1}{30}$	.132
	$220.308

REMARK.—When the number of which parts are taken, ends in 0, the simplest way is to take the *tenths*, instead of *halves*, *fourths*, &c. In the example above, since 1 mon. = 30 da., separate 19 da. into 18 da. and 1 da.; the former is $\frac{6}{10}$ of 30 da., and the value corresponding is found by multiplying the value for 1 mon. (3.96) by $\frac{6}{10}$ = .6, which is the same as to multiply by 6, and set the figures of the product 2.376 one place farther to the right.

ART. **235.** In all questions in aliquot parts, one of the numbers indicates a *rate*, and the other is a compound number whose *value* at this rate is to be found.

RULE FOR ALIQUOT PARTS.

Multiply the number indicating the rate by the number of that denomination for whose unit the rate is given, and separate the

REVIEW.—233. What is Aliquot parts? Explain its use. 234. What kind of a number may the multiplicand be?

numbers of the other denominations into parts whose values can be obtained directly by a simple division or multiplication of one of the preceding values. Add these different values; the result will be the entire value required.

NOTE.—Sometimes one of the values may be obtained by *adding* or *subtracting two preceding* values instead of by multiplying or dividing.

EXAMPLES FOR PRACTICE.

1. If a person travel 4 mi. 5 fur. 10 rd. 12 ft. 4 in. in 1 hr., how far will he travel in 7 hr. 37 min. 28 sec.?
Ans. 35 mi. 4 fur. 6 rd. 2 ft. $\frac{4}{225}$ in.

2. What cost 86 yd. 3 qr. 2 na. of cloth at 2.43\frac{3}{4}$ per yard? (Turn $\frac{3}{4}$ into a decimal.) *Ans.* $211.76

3. Find the cost of 231 A. 1 R. 34 P. of land at $17.28 per A. *Ans.* $3999.672

4. What is the cost of 127 yd. of carpet at 1.87\frac{1}{2}$ per yd.? *Ans.* 238.12\frac{1}{2}$

		127
		1.87$\frac{1}{2}$
		$127
50 ct.	= $\frac{1}{2}$	63.50
25 ct.	= $\frac{1}{2}$	31.75
12$\frac{1}{2}$ ct.	= $\frac{1}{2}$	15.87$\frac{1}{2}$
		238.12\frac{1}{2}$

SOLUTION.—Here the only compound number is the rate expressed in Federal money. Take it as the multiplier.

The cost of 127 yd. at $1 per yd. is $127; by taking suitable parts of this, the cost of 127 yd. is found at 50 ct., 25 ct., 12$\frac{1}{2}$ ct., a yd. respectively. The sum of all these is the cost of 127 yd. at 1.87\frac{1}{2}$ a yd. Therefore, aliquot parts can be used to find the value of any number of articles, when the value of one is known in U. S. money, by finding their cost at $1 a piece, and taking such aliquot parts of this as are necessary to make the cost at the given price.

5. Find the cost of 42 cu. yd. 24 cu. ft. of earth at $1.25 a cu. yd. *Ans.* $53.61

6. Of 7 lb. 8 oz. 16 pwt. 11 gr. of gold at $15.46 an oz. *Ans.* $1435.04

REVIEW.—234. What is one of its most valuable applications? Give an example. 235. In questions in aliquot parts what relation exists between the two quantities? What is the rule for aliquot parts? How can a value sometimes be obtained? How can the value of any number of articles be found when the price of one is given in U. S. money?

7. Find the cost of 6 T. 13 cwt. 2 qr. 21 lb. of sugar, at 4.68\frac{3}{4}$ a cwt. *Ans.* $626.77

8. Of 3 lb. 7 oz. of cheese, at 15 ct. a lb. *Ans.* 51$\frac{9}{16}$ ct.

9. 1 yd. of cloth is worth 3 qt. 1 pt. 2 gills of wine: what is 47 yd. 2 qr. 1 na. worth? *Ans.* 44 gal. 2 qt. 2$\frac{7}{8}$ gill.

10. In 1 wine gallon are 231 cu. in.: how many cu. in in 24 gal. 3 qt. 1 pt. 2$\frac{1}{2}$ gills? *Ans.* 5764$\frac{11}{64}$

11. In 1 beer gallon are 282 cu. in.: how many cu. in. in 38 gal. 1 qt. 1 pt. of beer? *Ans.* 10821$\frac{3}{4}$

12. In 1 bushel are 2150.42 cu. in.: how many cu. in. in 15 bu. 1 pk. 6 qt. 1 pt? *Ans.* 33230.7 +

13. What is the value of 29 gal. 2 qt. 1 pt. of wine, at $2.25 a gal.? *Ans.* 66.65\frac{5}{8}$

14. What is the value of 46 gal. 1 qt. 1$\frac{3}{4}$ pt. of beer, at 30 ct. a gal.? *Ans.* $13.94 +

15. What is the value of 10 bu. 3 pk. 5 qt. of corn, at 62$\frac{1}{2}$ ct. a bu.? *Ans.* $6.82 —

16. If £3 6s. silver weigh 1 lb. Tr., how much will weigh 17 lb. 11 oz. 16 pwt. 9 gr.? *Ans.* £59 7s. +

17. Built 8 rd. 14 ft. 10 in. of fence in 1 da.; how much can I build in 3 wk. 5 da. 9 hr. 46 min., if 1 da. = 10 hr., and 1 wk. = 6 da.? *Ans.* 213 rd. 6 ft. 1 in., nearly.

18. If a man is 2 hr. 25 min. 38 sec. in digging a cu. yd. of earth, how long will he be in digging 44 cu. yd. 22 cu. ft.? *Ans.* 108 hr. 46 min. 31$\frac{23}{27}$ sec.

19. A comet moves 24° 6′ 49″ in 1 hr.; how far will it go in 6 hr. 14 min. 52 sec.? *Ans.* 150° 39′ 23.3″ +

20. A pendulum beating 54000 times a day, beats how often in 4 da. 3 hr. 20 min. 5 sec.? *Ans.* 223503$\frac{1}{8}$ times.

ART. **236**. Aliquot parts can be applied to making out bills in U. S. money, when the prices of the items are given in State currencies.

What cost 33$\frac{1}{2}$ gal. of wine, at 14s. 7$\frac{1}{2}$d. a gal., New England currency?

SOLUTION.—The tables of State currencies, (Art. 207), show that in the New England States, 6s. = $1. The cost at 6s. or $1 a gal. is $33.50, from which, by multiplying and taking suitable parts, the cost at 12s., 2s., 6d., 1$\frac{1}{2}$d. is found; the sum of these is the cost at 14s. 7$\frac{1}{2}$d., as required.

	33\frac{1}{2}$ = $33.50
	2
12s. = 2	$67
2s. = $\frac{1}{6}$	11.167
6d. = $\frac{1}{4}$	2.792
1$\frac{1}{2}$d. = $\frac{1}{4}$	.698
	$81.66

What cost, in New England currency,

1. 35 yd. cloth, at 7s. 6d. a yd.? *Ans.* \$43.75
2. $19\frac{3}{4}$ yd. of muslin, at 2s. 4d. a yd.? *Ans.* \$7.68
3. 26 caps, at 8s. 6d. a piece? *Ans.* \$36.83
4. $10\frac{1}{2}$ doz. copy books, at 4s. 3d. a doz.? *Ans.* \$7.44
5. 158 lb. starch, at 9d. a lb.? *Ans.* \$19.75
6. 45 lb. 10 oz. butter, at 2s. 8d. per lb.? *Ans.* \$20.28

SUGGESTION.—45 lb. 10 oz. = $45\frac{5}{8}$ lb. = 45.625 lb.

7. 34 da. work, at 5s. 4d. a day.? *Ans.* \$30.22
8. 18 doz. and 8 eggs, at 1s. 10d. a doz.? *Ans.* \$5.70
9. 72 bu. 3 pk. corn, at 3s. 3d. a bu.? *Ans.* \$39.41
10. $86\frac{7}{8}$ yd. carpet, at 10s. 5d. a yd.? *Ans.* \$150.82

ART. **237**. What cost, in New York currency,

1. $14\frac{3}{4}$ yd. of calico, at 1s. 2d. a yd.? *Ans.* \$2.15
2. 12 bbl. potatoes, containing $2\frac{1}{2}$ bu. each, at 4s. 6d a bu.? *Ans.* $\$16.87\frac{1}{2}$
3. 2 bu. 3 pk. 6 qt. of dried peaches, at 17s. 8d. a bu.? *Ans.* \$6.49
4. 4 gal. 1 qt. 1 pt. oil, at 2s. 3d. a gal.? *Ans.* \$1.23
5. 33 dictionaries, at 5s. 6d. a-piece? *Ans.* $\$22.68\frac{3}{4}$
6. 49 boxes matches, at $4\frac{1}{2}$d. a-piece? *Ans.* \$2.30

ART. **238**. What cost, in Pennsylvania currency,

1. 3 qt. 1 pt. molasses, at 2s. a qt.? *Ans.* 93 ct.
2. 1 box candles (40 lb.), at 1s. 8d. a lb.? *Ans.* \$8.89
3. 12 lb. of coffee, at 10d. a lb.? *Ans.* \$1.33
4. 9 gal. 2 qt. 1 pt. of milk, at 6d. a qt.? *Ans.* \$2.57
5. 4 wk. 5 da. wages, at 13s. 4d. a week? *Ans.* \$8.38
6. 23 rd. $10\frac{1}{2}$ ft. of fencing, at 20s. a rd.? *Ans.* \$63.03

ART. **239**. What cost, in S. Carolina currency,

1. 13 gal. 3 qt. of oil, at 3s. 6d. a gal.? *Ans.* $\$10.31\frac{1}{4}$
2. A ham of $17\frac{1}{2}$ lb., at 11d. a lb.? *Ans.* $\$3.43\frac{3}{4}$
3. 43 lb. of butter, at 1s. $4\frac{1}{2}$d. a lb.? *Ans.* \$12.67
4. $16\frac{7}{8}$ yd. of silk, at 8s. 3d. a yd.? *Ans.* \$29.83

16

ART. **240.** What cost, in Canada currency,

1. 7 gal. 1 qt. of honey, at 6s. 10d. a gal.? *Ans.* $9.91

2. 5 bu. 1 pk. 7 qt. of dried apples, at 16s. 8d. a bu.? *Ans.* $18.23

ART. **241.** MISCELLANEOUS EXAMPLES.

1. How long is a rope winding 276 times round a tree, whose circum. is 4 yd. 2 ft. $6\frac{2}{3}$ in.? *Ans.* 1339 yd. 4 in.

2. What is the area of a field, length 67 rd. 8 ft. 5 in., breadth 39 rd. 11 ft. 2 in.? *Ans.* 16 A. 2 R. 38+ P.

3. How many bbl. ($31\frac{1}{2}$ w. gal.) in a room 22 ft. 3 in. long, 16 ft. 6 in. wide, 11 ft. 4 in. high? *Ans.* $988\frac{4}{49}$.

4. How many bu. in a bin, 8 ft. 10 in. long, 4 ft. 6 in. wide, 3 ft. 2 in. high? *Ans.* 101 bu. 5 qt., nearly.

5. How much land in a rectangular field, 83.44 ch. long, 56.27 ch. wide? *Ans.* 469 A. 2 R. 2.7008 P.

6. A bought a pipe of wine (137 gal.), lost 9 gal. 2 qt. $1\frac{1}{2}$ pt. by leakage, and sold the rest at $\$2.37\frac{1}{2}$ per gal.: how much did he receive? *Ans.* $302.37—.

7. How many quart, pint, and half-pint bottles, of each an equal number, can be filled out of a cask containing 44 gal. 2 qt. 1 pt.? *Ans.* 102.

8. A man can mow in 1 day, 2 A. 3 R. 20 P. of grass: in what time will he mow 78 A. 1 R. 36 P., allowing 10 hr. to a day? *Ans.* 27 da. 2 hr. $57\frac{9}{23}$ min.

9. What is the value of 16 lb. 7 oz. 12 pwt. 3 gr. of gold, at $15.85 an oz.? *Ans.* $3163.76.

10. If 1 cu. ft. of water weigh $62\frac{1}{2}$ lb., what is the weight of the water in a room 20 ft. long, 15 ft. 5 in. wide, 9 ft. 10 in. high? *Ans.* 94 T. 14 cwt. 3 qr. $21\frac{19}{36}$ lb.

11. If a ship sail 10 mi. 6 fur. $18\frac{1}{4}$ rd. per hour, how long will it be in going 3236 mi. 2 fur. 36.508 rd.? *Ans.* 12 da. 11 hr. 28 min. 6+ sec.

12. In a rectangular field, are 160 A. 2 R. 36 P.; one side is 74.18 ch.: what is the other? *Ans.* 21.67— ch.

13. How thin is a cu. in. of gold, beaten so as to cover a space 46 ft. 10 in. by 41 ft. 8 in.? *Ans.* $\frac{1}{281000}$ in.

14. When I arrived at Cincinnati, my watch, which kept time correctly, was 42 min. fast: from which direction had I come, and how far in that direction had I traveled? *Ans.* From the east; $564\frac{3}{8}$ mi.

NOTE.—A degree of longitude at Cincinnati is about $53\frac{3}{4}$ mi

XII. RATIO.

ART. **242.** *Ratio* is a Latin word signifying *relation* or *connection;* in Arithmetic, it means the *relation of one number to another, expressed by their quotient.*

The ratio of 2 bu. to 5 bu. is $\frac{5}{2}$; of 10 yd. to 3 yd. is $\frac{3}{10}$; showing that 5 bu. are $\frac{5}{2}$ of 2 bu., and 3 yd. $\frac{3}{10}$ of 10 yd.

Ratio exists only between quantities of the same kind, since only such can be divided, one by the other.

Since a ratio is a quotient, it is an abstract number, (Art. 60), showing how *many times* one number contains the whole or part of another.

ART. **243.** The ratio of two numbers is indicated by writing them in the order in which they are mentioned, with a colon (:) between them.

The ratio of 4 to 9 is written 4 : 9; of $2\frac{1}{3}$ to 4.65, is written $2\frac{1}{3}$: 4.65; of 2 ft. 8 in. to 1 yd. 1 ft., is written 2 ft. 8 in. : 1 yd. 1 ft.

Each number is called a term of the ratio, and both together a *couplet* or *ratio.* The first term of a ratio is the *antecedent,* which means *going before;* the 2d term is the *consequent,* which means *following.*

A *simple* ratio is a single ratio of two terms; as, $3 : 4 = \frac{4}{3}$.

A *compound* ratio is the product of two or more simple ratios; as, $4 \times 5 : 3 \times 7 = \frac{3 \times 7}{4 \times 5}$ is the product of the simple ratios $4 : 3 = \frac{3}{4}$ and $5 : 7 = \frac{7}{5}$.

The value of a ratio depends not on the *absolute,* but on the *relative* size of its terms.

TO FIND THE VALUE OF ANY RATIO,

RULE.—*Express the terms in the same denomination; take the consequent as the numerator, and the antecedent as the denominator of a common fraction; this fraction reduced to its simplest form, will be the ratio required.*

REVIEW.—242. What is the meaning of Ratio? What is ratio in Arithmetic? Give examples. When can two quantities have ratio? What kind of a number is every ratio? Why? 243. How is a ratio indicated? Give examples.

NOTE.—The French method of obtaining the ratio has been adopted here. The English method makes the antecedent the numerator, and the consequent the denominator of the fraction; the ratio of 3 in. to 7 in., by the French method, is $\frac{7}{3}$; by the English, $\frac{3}{7}$.

Since the value of a ratio is equal to the consequent divided by the antecedent, it follows that

The antecedent is equal to the consequent divided by the value of the ratio; and that

The consequent is equal to the antecedent multiplied by the value of the ratio.

Hence, if the value of a ratio is known, and one of its terms, the other can be found.

What is the ratio of 9 to 15?

SOLUTION—$9:15=\frac{15}{9}=\frac{5}{3}=1\frac{2}{3}$ *Ans.*

What is the ratio of 2 A. 3 R. 25 P. to 1 A.?

SOLUTION.—2 A. 3 R. 25 P. : 1 A. or 465 P. : 160 P., $=\frac{160}{465}=\frac{32}{93}$ *Ans.*

What is the ratio of $4\frac{3}{8}$ to $2\frac{1}{2}$? *Ans.* $\frac{4}{7}$.

SUGGESTION.—Here the rule gives a complex fraction $\frac{2\frac{1}{2}}{4\frac{3}{8}}$, which is reduced by Art. 132.

What is the ratio of 7.108 to $9.26\frac{2}{3}$?

SOL.—$7.108 : 9.26\frac{2}{3}=\frac{9.26\frac{2}{3}}{7.108}=\frac{27.8}{21.324}$ (Art. 132), $=\frac{27.800}{21.324}=\frac{27800}{21324}=\frac{6950}{5331}$ (Art. 155). *Ans.*

FIND THE RATIO

1. Of 7 to 5; of 9 to 1; of 2 to 4; of $1\frac{1}{2}$ to 6; of 36 to 50; of 112 to 16; of $2\frac{2}{5}$ to $4\frac{1}{2}$; of $7\frac{5}{6}$ to $13\frac{1}{3}$; of 91 to $6\frac{1}{8}$; of 20 to $1\frac{1}{4}$; of $8\frac{5}{16}$ to $4\frac{4}{9}$.

Ans. $\frac{5}{7}$; $\frac{1}{9}$; 2; 4; $1\frac{7}{18}$; $\frac{1}{7}$; $1\frac{7}{8}$; $1\frac{33}{47}$; $\frac{7}{104}$; $\frac{1}{16}$; $\frac{640}{1197}$.

2. Of $2\frac{2}{3} : 4\frac{2}{7}$, of 6.5 : .013, of $9\frac{3}{5}$: 17.28, of $116\frac{1}{4}$: 18.75, of $4\frac{2}{3}$: 9.8, of $\frac{7}{9}:\frac{11}{12}$, of $\frac{1}{5}:\frac{1}{4}$, of 10.08 : $3\frac{3}{8}$, of 2.176 : 14.3, of $6.37\frac{1}{2}$: 34, of $9\frac{1}{4}$: 44.4.

Ans. $1\frac{17}{28}$, $\frac{1}{500}$, $1\frac{4}{5}$, $\frac{5}{31}$, $2\frac{1}{10}$, $1\frac{5}{28}$, $1\frac{1}{4}$, $\frac{75}{224}$, $6\frac{311}{544}$, $5\frac{1}{3}$, $4\frac{4}{5}$.

REVIEW.—243. What is each of the numbers called? What are both together called? Which is the antecedent? Which, the consequent? Why so called? What is a simple ratio? A compound ratio? What does the value of a ratio depend on?

3. Of 2 ft. 6 in. to 3 yd. 1 ft. 10 in. *Ans.* $4\frac{1}{3}$

4. Of 4 mi. 6 fur. 20 rd. to 1 mi. 2 fur. 16 rd. *Ans.* $\frac{104}{385}$

5. Of 13 A. 3 R. 25 P. : 6 A. 2 R. 10 P. *Ans.* $\frac{42}{89}$

6. Of 2 C. 12 cu. ft. : 15 C. *Ans.* $7\frac{11}{67}$

7. Of 3 lb. 10 oz. 6 pwt. $10\frac{1}{2}$ gr. : 2 lb. $14\frac{3}{4}$ pwt. *Ans.* $\frac{7916}{14823}$

8. Of 2 ℥ 3 ʒ 1 ℈ : 5 ℥ 7 ʒ 14.32 gr. *Ans.* $2\frac{6429}{14500}$

9. Of 13 lb. : 9 lb. 15.2 dr. *Ans.* $\frac{223}{320}$

10. Of 14 T. 12 cwt. 1 qr. 18.44 lb. : 7 cwt. $4\frac{2}{3}$ lb. *Ans.* $\frac{26425}{1096629}$

11. Of 3 qt. $1\frac{2}{5}$ gills : 8 gal. 1 pt. *Ans.* $10\frac{30}{127}$

12. Of 10 gal. 1.54 pt. : 7 gal. 2 qt. .98 pt. *Ans.* $\frac{3049}{4077}$

13. Of 56 bu. 2 pk. 1 qt. : 35 bu. 3 pk. 6.055 qt. *Ans.* $\frac{3433}{5400}$

14. Of 5 hr. 26 min. $44\frac{4}{9}$ sec. : $3\frac{2}{5}$ da. *Ans.* $14\frac{4342}{4411}$

15. Of 2 yr. 22 da. : 7 yr. 216 da. *Ans.* $3\frac{2061}{3010}$

16. Of 42° 15′ $27\frac{1}{2}$″ : 90°. *Ans.* $2\frac{7898}{60851}$

17. If the antecedent is 7 and the ratio $1\frac{1}{2}$, what is the consequent? *Ans.* $10\frac{1}{2}$

18. If the consequent is $13.42\frac{3}{4}$ and the ratio $\frac{3}{8}$, what is the antecedent? *Ans.* $35.80\frac{2}{3}$

19. If the antecedent is $2\frac{3}{5}$ and the ratio 6.048, what is the consequent? *Ans.* 15.7248

20. If the consequent is 6 yd. 2 ft. $8\frac{1}{3}$ in. and the ratio is $3\frac{1}{3}$, what is the antecedent? *Ans.* 2 yd. $2\frac{1}{2}$ in.

21. If the antecedent is 5 bu. 1.68 pt. and the ratio is $5\frac{5}{6}$, what is the consequent? *Ans.* 29 bu. 1 pk. 2 qt. $\frac{7}{15}$ pt.

22. If the antecedent is 24.075 and the ratio is .1664, what is the consequent? *Ans.* 4.00608

23. If the consequent is $\frac{3}{7}$ and the ratio $\frac{2}{9}$, what is the antecedent? *Ans.* $1\frac{13}{14}$

24. If the consequent is 27 lb. 5 oz. 14 dr. and the ratio is $\frac{3}{5}$, what is the antecedent? *Ans.* 45 lb. 9 oz. $12\frac{2}{3}$ dr.

25. If the consequent is $7.43\frac{3}{4}$ and the ratio $2\frac{1}{3}$, what is the antecedent? *Ans.* $3.18\frac{3}{4}$

REVIEW.—243. What is the rule for finding the value of a ratio? How many methods of valuing a ratio? Explain the difference. What is the antecedent equal to? The consequent? How is the ratio of two mixed numbers found? How is the ratio of two decimals found?

26. Find the value of $\$2.56\frac{1}{4} : \$10$, of $37\frac{1}{2}$ ct. : $33\frac{1}{3}$ ct., of $\$13.66\frac{2}{3} : \$4\frac{1}{2}$, of $\$22$: 22 ct. *Ans.* $3\frac{37}{41}$, $\frac{8}{9}$, $\frac{27}{82}$, $\frac{1}{100}$

ART. **244.** *Two* ratios may be formed with the same two numbers, by taking each of them in succession as the standard of comparison; thus, the ratio *between* 5 and 7 is $\frac{7}{5}$ or $\frac{5}{7}$; the former is the ratio of 5 to 7, and the latter the ratio of 7 to 5.

One of the ratios which can be formed with two numbers will be an *improper* fraction, and the other a *proper* fraction; for the sake of distinction, call the former the *increasing* ratio, and the latter the *decreasing* ratio. If the two quantities are equal, the ratio, which ever way it is formed, will be equal to 1, and therefore neither increasing nor decreasing, but a *ratio of equality.*

TO MAKE AN INCREASING OR DECREASING RATIO,

RULE.—*Write the two numbers in the form of an improper fraction, to express an increasing ratio; but in the form of a proper fraction, to express a decreasing ratio.*

1. Make an increasing ratio with 7 and 18, with $5\frac{1}{2}$ and $4\frac{3}{4}$, with 3 yd. 2 ft. and 3 yd. 1 ft. 5 in., with $6\frac{3}{8}$ and 6.38. *Ans.* $\frac{18}{7}$, $\frac{22}{19}$, $\frac{132}{125}$, $\frac{1276}{1275}$

2. Make a decreasing ratio with $2\frac{1}{3}$ and $2\frac{1}{4}$, with 12.45 and $9\frac{2}{3}$, with 3 gal. 1 pt. and 2 gal. 2 qt., with $13\frac{1}{3}$ and $15\frac{5}{8}$ *Ans.* $\frac{27}{28}$, $\frac{580}{747}$, $\frac{4}{5}$, $\frac{64}{75}$

ART. **245.** Since every ratio is a fraction whose numerator is the consequent, and denominator the antecedent, whatever is true of a fraction is true of a ratio; hence,

1st. *Multiplying the consequent or dividing the antecedent, multiplies the ratio.* (Arts. 112 and 115.)

The ratio 10 : 4 is $\frac{4}{10}=\frac{2}{5}$; if the consequent be multiplied by 3, the ratio 10 : 12 is $\frac{12}{10}=\frac{6}{5}$, which is the former ratio multiplied by 3; if the antecedent be divided by 2, the ratio 5 : 4 is $\frac{4}{5}$, which is the former ratio multiplied by 2.

2d. *Multiplying the antecedent or dividing the consequent divides the ratio.* (Arts. 113 and 114.)

REVIEW.—244. How many ratios can be formed with two numbers? Give an example. How are they distinguished? What is a ratio of equality? What is the rule for making an increasing or decreasing ratio? 245. How is a ratio multiplied? Why? How is a ratio divided? Why?

The ratio 7 : 6 is $\frac{6}{7}$; if the consequent be divided by 3, the ratio 7 : 2 is $\frac{2}{7}$, which is the former ratio divided by 3; if the antecedent be multiplied by 2, the ratio 14 : 6 is $\frac{6}{14}=\frac{3}{7}$, which is the former ratio divided by 2.

3d. *Multiplying or dividing both terms of a ratio by a number, does not alter its value.* (Arts. 116 and 117.)

The ratio 9 : 6 is $\frac{6}{9}=\frac{2}{3}$; if both terms are multiplied by 2, the ratio 18 : 12 is $\frac{12}{18}=\frac{2}{3}$ still; if both terms be divided by 3, the ratio 6 : 4 is $\frac{4}{6}=\frac{2}{3}$, the same as at first.

XIII. PROPORTION.

ART. **246**. *Proportion* is an expression of equal ratios.

The ratios 3 : 5 and 6 : 10 are equal, each being of the value $\frac{5}{3}$. Placing a double colon (: :) between them, forms the proportion 3 : 5 : : 6 : 10, read 3 *is to* 5 *as* 6 *is to* 10, or *the ratio of* 3 *to* 5 *is equal to the ratio of* 6 *to* 10.

REMARK.—Either ratio may be written first; thus, 6 : 10 : : 3 : 5 is the same as 3 : 5 : : 6 : 10, since each expresses the equality of the same ratios.

Instead of the double colon, the sign of equality is sometimes used; as, 3 : 5 = 6 : 10, is the same as 3 : 5 : : 6 : 10.

A proportion with more than two equal ratios is called a *continued proportion*, as 3 : 5 : : 6 : 10 : : 9 : 15; but a proportion in Arithmetic generally contains only two equal ratios, and has 4 terms, since each ratio has 2 terms.

Since each ratio has an antecedent and consequent, every proportion has two antecedents and two consequents, the 1st and 3d terms being the antecedents, and the 2d and 4th the consequents.

The first and last terms of a proportion are called the *extremes;* the middle terms, the *means.* All the terms are called *proportionals*, and the last term is said to be a *fourth proportional* to the other 3 in their order.

Ratio is the relation between *two numbers* shown by their *quotient:* proportion is the relation between *two ratios* shown by their *equality.* The former has *two* terms, the latter *four.*

REVIEW.—245. How is a ratio altered in form and not in value? Why? 246. What is Proportion? Give an example. How is it written? What other way? What is a continued proportion?

Three numbers are in proportion, when the 1st has the same ratio to the 2d as the 2d has to the 3d; thus, 4, 8 and 16 are in proportion, for 4 : 8 : : 8 : 16, each ratio being 2. The second term is then called a *mean proportional* between the other two; and the last term a *third proportional* to the first and second.

ART. **247.** *Variation* is a general method of expressing proportion often used, and is either *direct* or *inverse.*

Direct variation exists between two quantities when they increase together, or decrease together.

Thus, the *distance* a ship goes at a uniform rate, *varies directly* as the time it sails; which means that the ratio of any two distances is equal to the ratio of the corresponding times taken in the same order.

Inverse variation exists between two quantities when one increases as the other decreases.

Thus, the time in which a piece of work will be done, *varies inversely* as the number of men employed; which means that the ratio of any two times is equal to the ratio of the numbers of men employed for those times, taken in reverse order.

ART. **248.** Since only *equal* ratios form a proportion, to determine the truth of a proportion,

Find the value of each ratio in the proportion; if these ratios are equal, the proportion is true; if not, it is false.

Thus, 8 : 10 : : 12 : 15 is a true proportion, the ratios $\frac{10}{8}$ and $\frac{15}{12}$ being each equal to $\frac{5}{4}$; but 9 : 5 : : 3 : 2 is not, because the ratios $\frac{5}{9}$ and $\frac{2}{3}$ are not equal, the former being $\frac{5}{9}$, the latter $\frac{6}{9}$.

WHICH ARE TRUE PROPORTIONS, AND WHICH NOT?

1. 7 : 10 : : 8 : 12, and 4 : 3 : : 24 : 18.
2. 2 ft. 1 in. : 1 yd. 4 in. : : $62\frac{1}{2}$ ct. : $1.
3. $16\frac{1}{2}$: 21 : : 2 bu. 3 pk. : 3 yd. 2 qr.
4. 3 hr. 18 min. : 7 hr. 20 min. : : 2 gal. 1 qt. : 4 gal. 2 qt.

REVIEW.—246. How many ratios in a proportion generally? How many antecedents? How many consequents? What are the extremes? The means? What are the terms called? The last term? How are Ratio and Proportion distinguished? When are three numbers in Proportion? What is the second term then called? What is the third term called? 247. What is variation? What two kinds? What is direct variation? Give an example. What is inverse variation? Give an example.

5. \$5.33⅓ : \$12 : : ⅔ cwt. : 150 lb.
6. 76 yd. 1 ft. 10 in. : 13 rd. 7⅙ ft. : : \$17.18¾ : \$16⅜
7. 16.208 : 9¾ : : 9 A. 1 R. 6½ P. : 5 A. 3 R. 37.64 P.
8. 23 T. 5 cwt. 1 qr. 15 lb. : 8 T. 2 cwt. 3 qr. 14 lb. : : 2 gal. 2 qt. : 3 qt. 1 pt.
9. £3 17s. 8¾d. : £8 10s. 3¼d. : : 16¼ bu. : 35 bu. 3 pk.
10. 22° 12′ 41½″ : 58° 7′ 6.42″ : : 1.3 hr. : 3 hr. 44 min.
11. 25 lb. Tr. : 24 lb. av. : : \$6 : \$7.
12. 5 yd. 2 ft. 9 in. : 9 E, Fl. 3 na. : : \$7.75 : \$9¼

PROPOSITION.

ART. **249**. *In every true proportion, the product of the numbers in the means is equal to the product of those in the extremes.*

DEMONSTRATION—In every true proportion, as 5 : 3 : : 10 : 6, the ratios are equal, viz: $\frac{3}{5}=\frac{6}{10}$. If both terms of the first ratio be multiplied by 10, and both terms of the second ratio by 5, their values are not altered and they are still equal, viz: $\frac{3\times10}{5\times10}=\frac{6\times5}{10\times5}$

The denominators of these fractions, having the same factors, are equal; to make the fractions equal, the numerators must also be equal; that is, $3\times10=6\times5$; but 3×10 is the product of the numbers in the means, and 6×5, the product of those in the extremes; hence, the proposition is proved.

COROLLARY 1.—*Either extreme is equal to the product of the means divided by the other extreme.*

COROLLARY 2.—*Either mean is equal to the product of the extremes divided by the other mean.*

The truth or falsity of a proportion can be determined by this proposition; the corollaries serve to find any term of a proportion, when the other three are known.

Thus, if the 1st, 2d, and 3d terms of a proportion are 6, 10, and 15, it may be written 6 : 10 : : 15 : ( ); the 4th term is represented by a parenthesis, and found by Cor. 1 to be $\frac{10\times15}{6}=25$. The unknown term, 25, must be of the same denomination as the other term, 15, of the same ratio.

REMARK.—Before applying the proposition or its corollaries, the terms of each ratio must be of the same denomination.

REVIEW.—248. How do we determine the truth or falsity of any proportion? 249. What relation exists between the extremes and means in every true proportion? Prove it.

FIND THE UNKNOWN TERM OF

1. 12 : 16 : : ( ) : 5 and $3\frac{3}{4}$: ( ) : : $1\frac{1}{2}$: 2. *Ans.* $3\frac{3}{4}$ and 5.

2. 1 bu. 2 pk. 6 qt. : 6 bu. 3 pk. : : 3.87\frac{1}{2}$: ( ). *Ans.* $15.50

3. ( ) : 4 hr. 30 min. : : $2\frac{1}{2}$ mi. : 3.375 mi. *Ans.* 3 hr. 20 min.

4. 2.16 A. : ( ) : : £13 6s. 8d. : £15. *Ans.* 2.43 A.

5. 12 yd. 3 qr. : 46 yd. 3 qr. : : ( ) : 6 T. 1 cwt. *Ans.* 1 T. 13 cwt.

6. 162.56\frac{1}{4}$: 270.93\frac{3}{4}$: : 234 men. : ( ). *Ans.* 390 men.

7. 46° 31′ 9″ : ( ) : : 3 hr. 36 min. : 2 hr. 42 min. *Ans.* 34° 53′ $21\frac{3}{4}$″

8. $16 : 45 ct. : : 1 lb. 9 oz. av. : ( ). *Ans.* $11\frac{1}{4}$ dr.

9. 5 mi. 3 fur. 30 rd. : 7 mi. 10 rd. : : ( ) : 54 horses *Ans.* 42 horses.

10. 33 bu. 1 pk. of potatoes : ( ) : : 4 bu. $1\frac{1}{2}$ pk. of apples : 2 bu. 2 pk. of apples. *Ans.* 19 bu. of potatoes.

SIMPLE PROPORTION.

ART. **250.** *Simple Proportion* is a method of solving practical questions by a ratio or proportion; it is sometimes called the *Rule of Three*, because the answer is obtained by finding one term of a proportion whose other *three* terms are known.

If 5 qt. of strawberries cost 75 ct., what are 9 qt. worth at the same rate?

SOLUTION BY ANALYSIS.—If 5 qt. cost 75 ct., 1 qt. costs $\frac{1}{5}$ of 75 ct = 15 ct., and 9 qt. cost 9 times 15 ct. = 135 ct. = $1.35.

SOLUTION BY PROPORTION.—Since the cost of each quart is the same, the whole cost *varies directly* as the number of qt.; that is, 9 qt. being $\frac{9}{5}$ of 5 qt. are worth $\frac{9}{5}$ as much, or $\frac{9}{5}$ of 75 ct. = $\frac{9\times75}{5}$ ct. = $1.35. Or, the ratio of the quantities being the same as the ratio of their costs, we have 5 qt. : 9 qt. : : 75 ct. : ( ), in which the required term is found, by Cor. 1st, Art. 249, to be $\frac{9\times75}{5}$ = 135 ct.

REVIEW.—249. What is Cor. 1st? Cor. 2d? Of what use is the proposition? Of what use are the Corollaries? Give an example.

REMARK.—Simple Proportion is sometimes applied to questions which do not admit of solution in that way. In cases of doubt, resort to a close and careful analysis, or to the following test:

A question to be solved by Simple Proportion, must contain two *kinds* of quantities, and *two* of each kind three of the quantities must be known and one required.

The two given quantities which are of different kinds, must be so related, that when one is *doubled* the other is necessarily *doubled* or *halved;* in fact, they furnish the *rate* which is applied to the remaining given quantity to obtain the answer. Hence,

TO SOLVE QUESTIONS BY SIMPLE PROPORTION,

RULE.—*With the two given quantities which are of the same kind, form an increasing or decreasing ratio, according as the answer should be greater or less than the third given quantity; multiply the third quantity by this ratio, and the result will be the answer required.*

NOTE.—Express the ratio in its simplest form, and cancel when possible.

For the benefit of those who prefer to state the question in the form of a proportion before commencing the operation, we give the following

RULE.—*Write that quantity for the 3d term which is of the same denomination as the answer; next to it, in the 2d term, put the larger or smaller of the other two quantities, according as the answer should be larger or smaller than the third term. The remaining given quantity is then put in the first term, and the fourth or required term, is found by multiplying the second and third terms together, and dividing their product by the first.*

1. If $2\frac{3}{4}$ lb. of sugar are worth $7\frac{1}{2}$ lb. of rice, how much sugar is 19.8 lb. of rice worth?

SOLUTION.—$\frac{19.8}{7\frac{1}{2}}$ of $2\frac{3}{4}$ lb. $= \frac{19.8}{7\frac{1}{2}} \times \frac{11}{4} = \frac{217.8}{30} = 7.26$ lb. *Ans.*

Or, $7\frac{1}{2} : 19.8 :: 2\frac{3}{4} : (\ )$; 4th term $= \frac{19.8 \times 2\frac{3}{4}}{7\frac{1}{2}} = 7.26$ lb.

REVIEW.—249. Of what denomination will the required term be? Before applying the proposition or corollaries, what should be done? 250. What is Simple Proportion? What is it sometimes called? Why? Give an example. What is the solution by analysis? By proportion? What kind of questions properly come under simple proportion?

2. If 15 men do a piece of work in $9\frac{3}{5}$ da., how long will 36 men be in doing the same?

SOLUTION.—Since 36 men will require *less* time than 15 men to do the same work, the answer should be *less* than $9\frac{3}{5}$ da.; make a decreasing ratio, $\frac{15}{36}$, and multiply the remaining quantity by it: $\frac{15}{36}\times 9\frac{3}{5}=\frac{15}{36}\times\frac{48}{5}=4$ da. *Ans.*

3. If I walk $10\frac{1}{2}$ mi. in 3 hr., how far will I go in 10 hr., at the same rate? *Ans.* 35 mi.

4. If the fore-wheel of a carriage is 8 ft. 2 in. in circumference, and turns round 670 times, how often will the hind-wheel, which is 11 ft. 8 in. in circumference, turn round in going the same distance? *Ans.* 469 times.

5. If a horse trot 3 mi. in 8 min. 15 sec., how far can he trot in an hour at the same rate? *Ans.* $21\frac{9}{11}$ mi.

6. What is a servant's wages for 3 wk. 5 da., at $1.75 per week? *Ans.* $6.50

7. What is a bbl. of powder containing 132 lb. worth, if 15 lb. are sold for $$5.43\frac{3}{4}$? *Ans.* $47.85

8. A body of soldiers are 42 in rank when they are 24 in file: if they were 36 in rank, how many in file would there be? *Ans.* 28.

9. If a pulse beats 28 times in 16 sec., how many times a minute is that? *Ans.* 105.

10. On 15 successive squares are 614 houses: how far from No. 1 is No. 277? *Ans.* $6\frac{471}{614}$, or nearly 7 squares.

11. If a cane 3 ft. 4 in. long, held upright, casts a shadow 2 ft. 1 in. long, how high is a tree whose shadow at the same time is 25 ft. 9 in.? *Ans.* 41 ft. $2\frac{2}{5}$ in.

12. If a horse draw 25 bu. of coal, each 80 lb., how many bu. of coke, each 96 lb., can he draw? *Ans.* $20\frac{5}{6}$

13. If a farm of 160 A. rents for $450, how much should be charged for one of 840 A.? *Ans.* $2362.50

14. If 18d. sterling, equals 25 ct. U. S. money, what is a half-crown (2s. 6d.) worth? *Ans.* $41\frac{2}{3}$ ct.

15. A grocer has a false gallon, containing 3 qt. $1\frac{1}{2}$ pt.: what is the worth of the liquor that he sells for $240, and his gain by the cheat? *Ans.* $225, and $15 gain.

16. If he uses $14\frac{3}{4}$ oz. for a pound, how much does he cheat by selling sugar for $27.52? *Ans.* $2.15

REVIEW.—250. What is the rule for solving questions by Simple Proportion? How should the ratio be expressed? What is the ordinary rule for simple proportion?

17. An equatorial degree is 365000 ft.: how many ft in 80° 24′ 37″ of the same? *Ans.* $29349751\frac{7}{18}$

18. If a pendulum beats 5000 times a day, how often does it beat in 2 hr. 20 min. 5 sec.? *Ans.* $486\frac{173}{432}$ times.

19. If I do a piece of work in 108 days, of $8\frac{1}{2}$ hr., how many days of $6\frac{3}{4}$ hr. would I be? *Ans.* 136.

20. A man borrows $1750, and keeps it 1 yr. 8 mon.: how long should he lend $1200 to compensate for the favor? *Ans.* 2 yr. 5 mon. 5 da.

21. A garrison has food to last 9 mon., giving each man 1 lb. 2 oz. a day: what should be a man's daily allowance, to make the same food last 1 yr. 8 mon.? *Ans.* $8\frac{1}{10}$ oz.

22. A garrison of 560 men have provisions to last during a siege at the rate of 1 lb. 4 oz. a day per man; if the daily allowance is reduced to 14 oz. per man, how large a reinforcement could be received? *Ans.* 240 men.

23. A shadow of a cloud moves 400 ft. in $18\frac{3}{4}$ sec.: what was the wind's velocity per hour? *Ans.* $14\frac{6}{11}$ mi.

24. If 1 lb. Troy of English standard silver be worth £3 6s., what is 1 lb. av. worth? *Ans.* £4 $2\frac{1}{2}$d.

25. If I go a journey in $12\frac{3}{4}$ days, at 40 mi. a day, how long would I be at $29\frac{3}{4}$ mi. a day? *Ans.* $17\frac{1}{7}$ da.

26. If $\frac{5}{9}$ of a ship are worth $6000, what is the whole of it worth? *Ans.* $10800.

SUGGESTION.—The two similar quantities in this question, are 5 *ninths* of the ship, and 9 *ninths* of the ship.

27. If A, worth $5840, is taxed $78.14, what is B worth, who is taxed $256.01? *Ans.* $19133.59

28. What are 4 lb. 6 oz. of butter worth, at 28 ct. a lb.? *Ans.* $$1.22\frac{1}{2}$

29. If I gain $160.29 in 2 yr. 3 mon., what would I gain in 5 yr. 6 mon., at that rate? *Ans.* $391.82

30. If I gain $92.54 on $1156.75 worth of sugar, how much must I sell to gain $67.32? *Ans.* $841.50 worth.

31. If coffee costing $255 is now worth $318.75, what did $1285.20 worth cost? *Ans.* $1028.16

32. If I gain $7.75 by trading with $100, how much ought I gain on $847.56? *Ans.* $65.6859

33. A has cloth at $3.25 a yd., and B has flour at $5.50 a bbl. If, in trading, A puts his cloth at $$3.62\frac{1}{2}$, what should B charge for his flour? *Ans.* $$6.13\frac{6}{13}$

34. What is a pile of wood, 15 ft. long, $10\frac{1}{2}$ ft. high, and 12 ft. wide, worth, at $4.25 a cord? *Ans.* $62.75

35. Find 7 mon. rent, at $475 a yr. *Ans.* $277.08

36. If a boat is rowed at the rate of 6 miles an hour, and is driven 44 feet in 9 strokes of the oar, how many strokes are made in a minute? *Ans.* 108 strokes.

37. If $\frac{2}{9}$ of a yd. of cloth cost $$3\frac{1}{5}$, what is the worth of $\frac{3}{4}$ of an E. E.? *Ans.* $$13\frac{1}{2}$

In Fahrenheit's thermometer, the freezing point of water is marked 32°, and the boiling point 212°: in the Centigrade, the freezing point is 0°, and the boiling point 100°: in Reaumer's, the freezing point is 0°, and the boiling point 80°.

38. From these data, find the value of a degree of each thermometer in the degrees of the other two. *Ans.* 1° F. $= \frac{4}{9}$° R. $= \frac{5}{9}$° C.; 1° C. $= 1\frac{4}{5}$° F. $= \frac{4}{5}$° R.; 1° R. $= 1\frac{1}{4}$° C. $= 2\frac{1}{4}$° F.

39. Convert 108° F. to degrees of the other two thermometers. (First subtract 32°.) *Ans.* $33\frac{7}{9}$° R. and $42\frac{2}{9}$° C.

40. Convert 25° R. to degrees of the other two thermometers. *Ans.* $31\frac{1}{4}$° C. and $88\frac{1}{4}$° F.

41. Convert 46° C. to degrees of the other two thermometers. *Ans.* $36\frac{4}{5}$° R. and $114\frac{4}{5}$° F.

In the working of machinery, it is ascertained that *the available power is to the weight overcome, inversely as the distances they pass over in the same time.*

42. If the whole power applied is 180 lb. and moves 4 ft., how far will it lift a weight of 960 lb.? *Ans.* 6 in.

NOTE.—The *available* power is taken $\frac{2}{3}$ of the whole power, $\frac{1}{3}$ being allowed for friction and other impediments.

43. If 512 lb. be lifted 1 ft. 3 in. by a power moving 6 ft. 8 in., what is the power? *Ans.* 144 lb.

SUGGESTION.—Find the *available* power; then add $\frac{1}{2}$ of itself.

44. A lifts a weight of 1440 lb. by a wheel and axle; for every 3 ft. of rope that passes through his hands, the weight rises $4\frac{1}{2}$ in: what power does he exert? *Ans.* 270 lb.

45. A man weighing 198 lb. let himself down 54 ft. with a uniform motion, by a wheel and axle: if the weight at the hook rises 12 ft., how much is it? *Ans.* 594 lb.

46. Two bodies free to move, attract each other with forces that *vary inversely* as their weights. If the weights are 9 lb. and 4 lb. and the smaller is attracted 10 ft., how far will the larger be attracted? *Ans.* 4 ft. $5\frac{1}{3}$ in.

47. Suppose the earth and moon to approach each other in obedience to this law, their weights being 49147 and

123 respectively. How many miles would the moon move while the earth moved 250 miles? *Ans.* 99892+ mi.

Can the following questions be solved by proportion? (See *Rem.* Art. 250.)

48. If 3 men mow 5 A. of grass in a day, how many men will mow $13\frac{1}{3}$ A. in a day?

49. If 6 men build a wall in 7 da., how long would 10 men be in doing the same?

50. If I gain 15 cents each, by selling books at $4.80 a doz., what is my gain on each at $5.40 a doz.?

COMPOUND PROPORTION.

ART. **251**. *Compound Proportion* is a method of solving questions by the use of a compound ratio, or by more than one proportion.

The questions to which Compound Proportion is applied resemble those under simple proportion; but the value of the quantity required depends not on *one* pair, but on *two or more* pairs of similar quantities: for instance,

If 3 men mow 8 A. of grass in 4 da., how long would 10 men be in mowing 36 A.?

1st step. $\frac{3}{10}$ of 4 da.
2d step. $\frac{36}{8}$ of $\frac{3}{10}$ of 4 da. $= 5\frac{2}{5}$ da. *Ans.*

SOLUTION.—If the 10 men were to mow 8 A. instead of 36 A., the question would be one of Simple Proportion, and the answer would be $\frac{3}{10}$ of 4 da., as shown in the 1st step of the operation; but this result being the time in which 10 men mow 8 A, the time in which they will mow 36 A. will be $\frac{36}{8}$ of it $= \frac{36}{8}$ of $\frac{3}{10}$ of 4 da., as shown in the 2d step, which reduces, by cancellation, to $5\frac{2}{5}$ da. Hence, a question in Compound Proportion, is only a succession of questions in Simple Proportion, each of which gives a result to be used in the next, and the last result is the answer required.

After sufficient practice, the successive steps may be dispensed with, and the answer written as a compound fraction at once.

TO SOLVE QUESTIONS BY COMPOUND PROPORTION,

RULE.—*Form an increasing or decreasing ratio of each pair of similar quantities, as if the answer depended only on those two and the odd term; multiply these ratios and the odd term together; the product will be the answer.*

NOTE.—The *odd* term is the one which is unlike all the rest.

TO TEACHERS.—Pupils should analyze each example thoroughly, and give the reasons for every step: such a course will be a valuable training to the mental powers, and is the only way to clear up an otherwise obscure and difficult subject.

If the proportional form is preferred, use a succession of simple proportions, as already explained, or the rule in "Ray's Arithmetic," 3d Book, Art. 205. A simple mode of stating questions in proportion is this

RULE OF CAUSE AND EFFECT.

Separate all the quantities contained in the question into two causes and their effects.

Write, for the first term of a proportion, all the quantities that constitute the first cause; for the second term, all that constitute the second cause; for the third, all that constitute the effect of the first cause; and for the fourth, all that constitute the effect of the second cause. The required quantity may be indicated by a bracket, and found by one of the rules in Art. 249.

NOTE.—The two causes must be exactly alike in the *number* and *kind* of their terms; and so must the two effects.

1. If 6 men, in 10 days of 9 hr. each, build 25 rd. of fence, how many hours a day must 8 men work to build 48 rd. in 12 days?

SOLUTION.—6 men 10 da. and 9 hr. constitute the 1st cause, whose effect is 25 rd.; 8 men 12 da. and ( ) hr. constitute the 2d cause, whose effect is 48 rd. Hence,

6 men.		8 men.				
10 da.	:	12 da.	: :	25 rd.	:	48 rd.
9 hr.		( ) hr.				

And the required term is $\frac{6\times10\times9\times48}{25\times8\times12}=10\frac{4}{5}$ hr. (Art. 249, Cor. 2.)

2. A boy makes 8 steps, each 1 ft. 10 in., while a man makes 5 steps, each 2 ft. 8 in.: how far will the boy go while the man is going $3\frac{3}{4}$ miles? *Ans.* $4\frac{1}{8}$ mi.

REVIEW.—251. What is Compound Proportion? What does the value of the quantity required depend on? Solve the example. What is the rule? What is the odd term? What is the rule of cause and effect? What is said of the two causes? The two effects?

3. If 18 pipes, each delivering 6 gal. per minute, fill a cistern in 2 hr. 16 min., how many pipes, each delivering 20 gal. per minute, will fill a cistern $7\frac{1}{2}$ times as large as the first, in 3 hr. 24 min.? *Ans.* 27.

4. The use of $100 for 1 year is worth $8: what is the use of $4500 for 2 yr. 8 mon. worth? *Ans.* $960.

5. If 12 men mow 25 A. of grass in 2 da. of $10\frac{1}{2}$ hr., how many hours a day must 14 men work to mow an 80 A. field in 6 days? *Ans.* $9\frac{3}{5}$ hr.

6. If 4 horses draw a railroad car 9 miles an hour, how many miles an hour can a steam engine of 150 horse-power drive a train of 12 such cars, the locomotive and tender being counted 3 cars? *Ans.* $22\frac{1}{2}$ mi. per hr.

7. How many half eagles, each weighing 5 pwt. 9 gr. and made of gold $\frac{9}{10}$ pure, are equivalent to 1000 English sovereigns each weighing 5 pwt. 3.274 gr., and made of gold $\frac{11}{12}$ pure? *Ans.* $973\frac{1076}{3483}$

8. If the use of $3750 for 8 mon. is worth $68.75, what sum is that whose use for 2 yr. 4 mon. is worth $250? *Ans.* $3896.10 +.

9. If the use of $1500 for 3 yr. 8 mon. 25 da. is worth $336.25, what is the use of $100 for 1 yr. worth? *Ans.* $6.

10. If 240 panes of glass 18 in. long, 10 in. wide, glaze a house, how many panes 16 in. long by 12 in. wide will glaze a row of 6 such houses? *Ans.* 1350.

11. If it require 800 reams of paper to publish 5000 volumes of a duodecimo book containing 320 pages, how many reams will be needed to publish 24000 copies of a book, octavo size, of 550 pages? *Ans.* 9900 reams.

12. A man has a bin 7 ft. long by $2\frac{1}{4}$ ft. wide, and 2 ft. deep, which contains 28 bu. of corn: how deep must he make another, which is to be 18 ft. long by $1\frac{7}{8}$ ft. wide, in order to contain 120 bu.? *Ans.* $4\frac{4}{9}$ ft.

13. If it require 4500 bricks, 8 in. long by 4 in. wide, to pave a court-yard 40 ft. long by 25 ft. wide, how many tiles 10 in. square, will be needed to pave a hall 75 ft. long by 16 ft. wide? *Ans.* 1728.

14. If 150000 bricks are used for a house whose walls average $1\frac{1}{2}$ ft. thick, 30 ft. high, and 216 ft. long, how many will build one with walls 2 ft. thick, 24 ft. high, and 324 ft. long? *Ans.* 240000.

15. A garrison of 1800 men has provisions to last $4\frac{1}{2}$ mon. at the rate of 1 lb. 4 oz. a day to each; how long

will 5 times as much last 3500 men, at the rate of 12 oz. per day to each man? *Ans.* 1 yr. $7\frac{3}{7}$ mon.

16. What sum of money is that whose use for 3 yr., at the rate of $\$4\frac{1}{2}$ for every hundred, is worth as much as the use of \$540 for 1 yr. 8 mon., at the rate of \$7 for every hundred? *Ans.* $\$466.66\frac{2}{3}$

XIV. PERCENTAGE.

ART. **252**. PERCENTAGE is derived from the Latin phrase *per centum*, meaning *on the hundred*. It is understood to embrace all those operations in which reference is made to 100 as a unit of comparison.

Its applications are, Gain and Loss, Commission, Interest, Banking, Stocks, Insurance, Duties and Taxes.

If I have \$750 in business, and gain \$180, what is my gain on every \$100?

SOLUTION.—This is a question in Simple Proportion, and is solved as follows: \$750 : \$100 : : \$180 : ( ); the 4th term is found by Cor. 1st, Art. 249, to be

$$\frac{100 \times 180}{750} = \$24. \textit{ Ans.}$$

Hence, I gain \$24 on every \$100, or 24 *per cent.*, since *per cent.* means *on the hundred.*

The \$24, in the solution above, is called the *rate per cent.* of the \$180 to the \$750.

Though *rate per cent.* strictly means *the number on a* 100, yet as "\$24 on a \$100" conveys the same idea as "24 hundredths," and the latter is simpler and more general, it is the prevailing practice to consider the *rate per cent.* that one number is of another, as the *number of hundredths* it is of that other, and any *per cent.* of a number is so many *hundredths* of it.

Thus, 7 *per cent.* of any thing is 7 *hundredths* of it.

$2\frac{1}{2}$ do. . . . . . . is $2\frac{1}{2}$ do. do.

100 do. . . . . . . is 100 do. do. or the whole.

REVIEW.—252. What is the meaning of Percentage? What does it embrace? What are its principal applications? Solve the example. What does rate per cent. mean strictly?

2. If I have 160 sheep, and sell 35 per cent. of them, how many sheep do I sell?

SOLUTION.—The whole flock (160) being 100 per cent., the proportion is, 160 sheep : ( ) sheep : : 100 per cent. : 35 per cent.; the required term is $\frac{160 \times 35}{100} = 56$ sheep. *Ans.*

3. In a battle, 78 men are killed, which is $13\frac{1}{3}$ per cent. of the whole force; how many were engaged?

SOLUTION.—The whole force is 100 per cent.; the slain (78) are $13\frac{1}{3}$ per cent.; hence, ( ) men : 78 men : : 100 per cent. : $13\frac{1}{3}$ per cent.; the required term is

$$\frac{78 \times 100}{13\frac{1}{3}} = \frac{3 \times 78 \times 100}{40} = 585 \text{ men. } \textit{Ans.}$$

Hence, there are *three* cases of percentage, according as the first of the two numbers compared, the second, or their rate per cent., is to be found. All these cases can be solved by the

GENERAL RULE FOR PERCENTAGE.

Any two numbers are to each other, as the rates per cent. they represent of the same quantity.

NOTE.—If one of the numbers is the standard of comparison for the other, it will be 100 per cent.; if they are both referred to a third quantity, *it* will be 100 per cent., and their rates per cent. will depend upon the nature of their relations to it.

Although this rule covers the whole subject, it is too general for practical purposes; therefore a special rule will be given for each case.

CASE I.

TO FIND ANY GIVEN PER CENT. OF A GIVEN NUMBER,

ART. 253. RULE.—*Multiply the given number by the given rate per cent., and divide the product by* 100.

NOTES.—1. It often saves work to indicate and cancel.

2. The dividing by 100 may generally be done by *pointing off in the product two more decimal places than are in the multiplicand and multiplier.*

Find 9 per cent. of $182.50

SOLUTION.—9 per cent. of \$182.50 = $\frac{9}{100}$ of \$182.50 = 9 times \$182.50, divided by 100, as the rule directs.

$$\begin{array}{r} \$182.50 \\ 9 \\ \hline \$16.4250 \end{array}$$

REMARK.—We may divide by 100 first, and then multiply by the rate per cent.; or, multiply by the rate per cent. written as decimal hundredths.

1. Find 6 per cent. of $\$82.37\frac{1}{2}$ *Ans.* $\$4.94\frac{1}{4}$
2. $14\frac{1}{2}$ pr. ct. of 6 yd. 2 ft. 8 in. *Ans.* 2 ft. 11.96 in.
3. 42 per cent. of $1250. *Ans.* $525.

REMARK.—Business men use instead of the words "per cent," the character %: thus 42 per cent. is written 42 %.

4. Find $62\frac{1}{2}$ % of 1664 men. *Ans.* 1040 men.
5. 35 % of $\frac{2}{7}$. $\frac{2}{7} \times \frac{35}{100} = \frac{70}{700} = \frac{1}{10}$ *Ans.*
6. 180 % of $4\frac{2}{5}$ *Ans.* $7\frac{23}{25}$
7. 98 % of 14 cwt. 2 qr. 20 lb. = 14 cwt. 1 qr. $15\frac{3}{5}$ lb.
8. $9\frac{3}{8}$ % of 48 mi. 6 fur. 16 rd. = 4 mi. 4 fur. 24 rd.
9. $33\frac{1}{3}$ % of 127 gal. 3 qt. 1 pt. = 42 gal. 2 qt. 1 pt.

SOLUTION.—Since $\frac{33\frac{1}{3}}{100} = \frac{1}{3}$, take $\frac{1}{3}$ of the given number.

10. Find $11\frac{1}{9}$ % of $3283.47. *Ans.* $364.83
11. 40 % of 6 hr. 28 min. 15 sec. *Ans.* 2 hr. 35 min. 18 sec.
12. 675 % of 3 lb. 10 oz. 16 pwt. 22 gr. *Ans.* 26 lb. 4 oz. 4 pwt. $4\frac{1}{2}$ gr.
13. 104 % of 75 A. 1 R. 35 P. = 78 A. 1 R. 38 P.
14. $15\frac{5}{8}$ % of a book of 576 pages. *Ans.* 90 pages.
15. $23\frac{1}{3}$ % of 45 ct. *Ans.* $10\frac{1}{2}$ ct.
16. How much is $18\frac{3}{4}$ % of a cargo that weighs 416 T. 15 cwt. 20 lb.? *Ans.* 78 T. 2 cwt. 3 qr. 10 lb.
17. Find $337\frac{1}{2}$ % of $1\frac{1}{3}$ *Ans.* $4\frac{1}{2}$
18. $56\frac{1}{4}$ % of 144 cattle. *Ans.* 81 cattle.
19. $16\frac{2}{3}$ % of 1932 hogs. *Ans.* 322 hogs.
20. $87\frac{1}{2}$ % of 1634 C. 72 cu. ft. *Ans.* 1430 C. 31 cu. ft.
21. 1000 % of $\$5.43\frac{3}{4}$ *Ans.* $\$54.37\frac{1}{2}$
22. $6\frac{2}{3}$ % of $\frac{10}{11}$ *Ans.* $\frac{2}{33}$
23. $2\frac{1}{2}$ % of $\frac{4}{5}$ *Ans.* $\frac{1}{50}$

REVIEW.—252. What is the rate per cent. that one number is of another considered? What is any per cent. of a number? Give examples. Solve example 2. Also, example 3. How many cases of percentage? What are they? What is the general rule for percentage? If one of the given quantities is the standard of comparison for the other, what rate per cent. will it represent?

24. What part is 25 % of a farm? *Ans.* $\frac{25}{100}=\frac{1}{4}$

25. What part of a quantity is $6\frac{1}{4}$ % of it? also $8\frac{1}{3}$ %; 10 %; $12\frac{1}{2}$ %; $16\frac{2}{3}$ %; 20 %; $33\frac{1}{3}$ %; 50 %; $66\frac{2}{3}$ %; and 75 %? *Ans.* $\frac{1}{16}$, $\frac{1}{12}$, $\frac{1}{10}$, $\frac{1}{8}$, $\frac{1}{6}$, $\frac{1}{5}$, $\frac{1}{3}$, $\frac{1}{2}$, $\frac{2}{3}$, $\frac{3}{4}$ of it.

26. What part of a quantity is $18\frac{3}{4}$ % of it; $31\frac{1}{4}$ %; $37\frac{1}{2}$ %; $43\frac{3}{4}$ %; $56\frac{1}{4}$ %; $62\frac{1}{2}$ %; $68\frac{3}{4}$ %; $81\frac{1}{4}$ %; $83\frac{1}{3}$ %; $87\frac{1}{2}$ %; and $93\frac{3}{4}$ %?
Ans. $\frac{3}{16}$, $\frac{5}{16}$, $\frac{3}{8}$, $\frac{7}{16}$, $\frac{9}{16}$, $\frac{5}{8}$, $\frac{11}{16}$, $\frac{13}{16}$, $\frac{5}{6}$, $\frac{7}{8}$, $\frac{15}{16}$ of it.

27. How much is 100 % of a quantity; 125 % of it; 250 %; 675 %; 1000 %; $9437\frac{1}{2}$ %?
Ans. 1 time, $1\frac{1}{4}$ times, $2\frac{1}{2}$, $6\frac{3}{4}$, 10, $94\frac{3}{8}$ times it.

28. A man owning $\frac{3}{8}$ of a ship, sold 40 % of his share: what part of the ship did he sell, and what part did he still own? *Ans.* $\frac{3}{20}$ sold; $\frac{9}{40}$ left.

29. A owed B a sum of money; at one time he paid him 40 % of it; afterward he paid him 25 % of what he owed; and finally he paid him 20 % of what he then owed: how much does he still owe? *Ans.* $\frac{9}{25}$ of it.

30. What is $78\frac{1}{8}$ % of 12 T. 6 cwt. 8 lb.?
Ans. 9 T. 12 cwt. 1 qr.

31. Out of a cask containing 47 gal. 2 qt. 1 pt., leaked $6\frac{2}{3}$ %: how much was that? *Ans.* 3 gal. $1\frac{2}{5}$ pt.

32. A has an income of $1200 a year; he pays 13 % of it for board; $10\frac{2}{5}$ % for clothing; $6\frac{3}{4}$ % for books; $\frac{7}{12}$ % for newspapers; $12\frac{7}{8}$ % for other expenses: how much does he pay for each item, and how much does he save at the end of the year?

Ans. $156, board; $124.80, clothing; $81, books; $7, newspapers; $154.50, other expenses; $676.70, saved.

33. Find $\frac{5}{6}$ % of $1200. *Ans.* $10.

34. $\frac{1}{2}$ % of $47.75. *Ans.* $23\frac{7}{8}$ ct.

35. 10 % of 20 % of $13.50. *Ans.* 27 ct.

36. 40 % of 15 % of 75 % of 133.33\frac{1}{3}$. *Ans.* $6.

37. 8% of $62\frac{1}{2}$% of 150% of $462.50. *Ans.* 34.68\frac{3}{4}$

38. A man contracts to supply dressed stone for a court-house for $119449, if the rough stone costs him

REVIEW.—253. What is Case 1? The rule? How is the division by 100 performed? How may the operation be performed?

16 ct. a cu. ft.; but if he can get it for 15 ct. a cu. ft., he will deduct 3 % from his bill; how many cu. ft. would be needed, and what does he charge for dressing a cu. ft.?
Ans. 358347 cu. ft., and $17\frac{1}{3}$ ct. a cu. ft.

39. 48 % of brandy is alcohol; how much alcohol does a man swallow in 40 years if he drinks a gill of brandy 3 times a day? *Ans.* 657 gal. 1 qt. 1 pt. 2.4 gills.

CASE II.

ART. **254**. Two numbers being given, to find the rate per cent. one is of the other.

RULE.—*Multiply the number which is to be the rate per cent. by* 100, *and divide the product by the other number.*

Or, when the numbers are small, *Take such a part of* 100 *as the number which is to be the rate per cent. is of the other.*

PROOF.—With the rate per cent. thus obtained, and the number which is the standard of comparison, proceed by the last rule; if the result is the same as the other number, the work is right.

18 is how many per cent. of 276?

SOLUTION.—18 is how many per cent. of 276, means 18 is how many *hundredths* of 276. Now 1 hundredth of 276 is $\frac{276}{100}$, and as often as 18 contains this, so many hundredths will it be of 276; but $18 \div \frac{276}{100} = 18 \times \frac{100}{276}$, (Art. 131), $= 1800 \div 276$, as the rule directs.

$$\begin{array}{l} 276)1800(6\frac{12}{23} \\ \quad\;\; 1656 \\ \quad\;\; \overline{144} \\ 12\overline{)\frac{144}{276}} = \frac{12}{23} \end{array}$$

Ans. $6\frac{12}{23}$ per cent.

6 is how many per cent. of 9?

SOLUTION.—6 is $\frac{6}{9}$ or $\frac{2}{3}$ of 9, and to find how many hundredths of 9 it is, convert $\frac{2}{3}$ into hundredths. To do this, say $\frac{2}{3} = \frac{2}{3}$ of 1 unit $= \frac{2}{3}$ of 100 hundredths $= 66\frac{2}{3}$ hundredths $= 66\frac{2}{3}$ %; or, multiply both terms of $\frac{2}{3}$ by a number that will make the denominator 100; here, the multiplier is $33\frac{1}{3}$, which changes $\frac{2}{3}$ to $\frac{66\frac{2}{3}}{100} = 66\frac{2}{3}$ % as before.

1. $14.40 is how many % of $54? *Ans.* $26\frac{2}{3}$ %.

SUGGESTION.—Reduce both to ct.; and generally reduce compound numbers to the same denomination before applying the rule.

2. 9 is how many % of 6? *Ans.* 150 %.

REVIEW.—254. What is Case 2d? The rule? When the numbers are small what rule can be used? Explain example 1. Example 2.

3. 15 ct. is how many % of \$2? *Ans.* $7\frac{1}{2}$

4. 2 yd. 2 ft. 3 in. is how many % of 4 rd.? *Ans.* $12\frac{1}{2}$

5. 3 gal. 3 qt. is what % of $31\frac{1}{2}$ gal.? *Ans.* $11\frac{19}{21}$

6. $\frac{2}{3}$ is how many % of $\frac{4}{5}$? *Ans.* $83\frac{1}{3}$

Solution.—$\frac{2}{3}=\frac{10}{15}$ and $\frac{4}{5}=\frac{12}{15}$; the first is $\frac{10}{12}$ or $\frac{5}{6}$ of the last, but $\frac{5}{6}=\frac{5}{6}$ of 100 % $=83\frac{1}{3}$ %.

7. $\frac{5}{8}$ is how many % of $\frac{1}{4}$? *Ans.* 250.

8. $\frac{1}{2}$ of $\frac{3}{5}$ of $\frac{4}{7}$ is what % of $1\frac{7}{10}$? *Ans.* $10\frac{10}{119}$

9. $2\frac{1}{2}$ is how many % of $3\frac{3}{4}$? *Ans.* $66\frac{2}{3}$

10. $\frac{8}{9}$ is how many % of $\frac{11}{12}$? *Ans.* $96\frac{32}{33}$

11. $\dfrac{2\frac{1}{2}}{3}$ is how many % of $\dfrac{3\frac{3}{4}}{10}$? *Ans.* $222\frac{2}{9}$

12. \$5.12 is what % of \$640? *Ans.* $\frac{4}{5}$

13. $14\frac{3}{8}$ is how many % of $2\frac{7}{9}$? *Ans.* $517\frac{1}{2}$

14. \$3.20 is what % of \$2000? *Ans.* $\frac{4}{25}$

15. \$45 is what % of \$12? *Ans.* 375

16. 750 men is what % of 12000? *Ans.* $6\frac{1}{4}$

17. $8\frac{3}{4}$ is how many % of $\frac{5}{6}$? *Ans.* 1050.

18. \$7.29 is what % of \$216? *Ans.* $3\frac{3}{8}$

19. $\frac{3}{7}$ is how many % of $\frac{2}{9}$? *Ans.* $192\frac{6}{7}$

20. 3 qt. $1\frac{1}{2}$ pt. is what % of 5 gal. $2\frac{1}{2}$ qt.? *Ans.* $16\frac{2}{3}$

21. 16 bu. $3\frac{1}{2}$ pk is what % of 7.125 bu.? *Ans.* $236\frac{16}{19}$

22. A's money is 50 % more than B's; then B's money is how many % less than A's? *Ans.* $33\frac{1}{3}$

Solution.—Call B's money 100 %; A's money is then 150 %: their difference is 50 %, which, compared with A's money, 150 %, is $\frac{50}{150}=\frac{1}{3}=33\frac{1}{3}$ %.

23. If A's money is 10 %; $12\frac{1}{2}$ %; 25 %; $31\frac{1}{4}$ %; $43\frac{3}{4}$ %; $62\frac{1}{2}$ %; 75 %; 100 %; 150 %; 200 %; 225 %; 375 %; 1000 % more than B's; then B's is how many % less than A's?

Ans. $9\frac{1}{11}$; $11\frac{1}{9}$; 20; $23\frac{17}{21}$; $30\frac{10}{23}$; $38\frac{6}{13}$; $42\frac{6}{7}$; 50; 60; $66\frac{2}{3}$; $69\frac{3}{13}$; $78\frac{18}{19}$: $90\frac{10}{11}$ %.

24. If A has 5 %; 15 %; $17\frac{1}{2}$ %; $22\frac{1}{2}$ %; 25 %; 30 %; $33\frac{1}{3}$ %; 45 %; 50 %; $68\frac{3}{4}$ %; 75 %; 84 %; 98 %; and $99\frac{7}{8}$ % *less* money than B; then B's money is how many % more than A's?

Ans. $5\frac{5}{19}$; $17\frac{11}{17}$; $21\frac{7}{33}$; $29\frac{1}{31}$; $33\frac{1}{3}$; $42\frac{6}{7}$; 50; $81\frac{9}{11}$; 100; 220; 300; 525; 4900; 79900 %.

25. What % of a number is 8 % of 35 % of it? *Ans.* $2\frac{4}{5}$

SOLUTION.—$\frac{8}{100} \times \frac{35}{100} = \frac{7}{250} = \frac{7}{250}$ of 100 % $= 2\frac{4}{5}$ %.

26. What % of a number is $2\frac{1}{2}$ % of $2\frac{1}{2}$ % of it? *Ans.* $\frac{1}{16}$

27. What % of a number is 40 % of $62\frac{1}{2}$ % of it? *Ans.* 25.

28 12 % of $75 is what % of $108? *Ans.* $8\frac{1}{3}$

29. If $\frac{5}{12}$ of a ship be sold, what % of it is sold?

SUGGESTION.—$\frac{5}{12}$ of 100 % $= 41\frac{2}{3}$ %.

30. U. S. standard gold and silver is 9 parts pure to 1 part alloy: what % of alloy is that? *Ans.* 10.

31. English standard gold is 22 carats fine: how many % of pure metal in a sovereign? (Art. 205.) *Ans.* $91\frac{2}{3}$

32. One pound English standard silver contains 18 pwt. of alloy: how many % alloy is it? *Ans.* $7\frac{1}{2}$

33. How many % of a township 6 miles square does a man own, who has 9000 acres? *Ans.* $39\frac{1}{16}$

34. How many % of his time does a man lose, who sleeps 7 hr. out of every 24? *Ans.* $29\frac{1}{6}$

35. How many % of a quantity is 40 % of 25 % of it? also, 16 % of $37\frac{1}{2}$ % of it? also, $4\frac{1}{6}$ % of 120 % of it? also, 2 % of 80 % of $66\frac{2}{3}$ % of it? also, $\frac{3}{5}$ % of 36 % of 75 % of it? also, $6\frac{7}{8}$ % of $22\frac{1}{2}$ % of 96 % of it?
Ans. 10, 6, 5, $1\frac{1}{15}$, $\frac{81}{500}$, $1\frac{97}{200}$

36. 30 % of the whole of an article is how many % of $\frac{2}{3}$ of it? *Ans.* 45.

37. 25 % of $\frac{2}{5}$ of an article is how many % of $\frac{3}{4}$ of it? *Ans.* $13\frac{1}{3}$

CASE III.

ART. **255.** To find a number when a certain per cent of it is given.

RULE.—*Divide the given number by the rate per cent.; the quotient will be* 1 *per cent., and* 100 *times this will be* 100 *per cent., or the number required.*

PROOF.—Find the given per cent. of the answer; this must be the same as the given number.

A man gave a beggar 25 ct., which was $2\frac{1}{2}$ % of his money: how much had he?

SOLUTION.—Since 25 ct. is $2\frac{1}{2}$ %, 1 % is $\frac{25}{2\frac{1}{2}}=10$ ct., and 100 % = 100 times 10 ct. = 1000 ct. = \$10. *Ans.*

REMARK.—The work may often be simplified. Thus, 25 ct. = $2\frac{1}{2}$ %, $=\frac{2\frac{1}{2}}{100}$, $=\frac{5}{200}$, $=\frac{1}{40}$ of the number required; hence, $\frac{40}{40}$, or the whole of it = 40 times 25 ct. = \$10. Or, since 25 ct. are $2\frac{1}{2}$ %, 100 % = 40 times $2\frac{1}{2}$ % = 40 times 25 ct. = \$10, which amounts to multiplying the given number by the quotient of 100 divided by the rate per cent.

1. $6\frac{3}{4}$ is 9 % of what number? *Ans.* 75.
2. \$3.80 is 5 % of what sum? *Ans.* \$76.
3. $\frac{2}{11}$ is 80 % of what number? *Ans.* $\frac{5}{22}$
4. 16 is $1\frac{1}{2}$ % of what number? *Ans.* $1066\frac{2}{3}$
5. 2 is 45 % of what number? *Ans.* $4\frac{4}{9}$
6. \$72 is 160 % of what? *Ans.* \$45.
7. $31\frac{1}{4}$ ct. is $15\frac{5}{8}$ % of what? *Ans.* \$2.
8. $42\frac{3}{4}$ is $17\frac{1}{10}$ % of what? *Ans.* 250.
9. \$10.75 is $3\frac{1}{3}$ % of what? *Ans.* \$322.50
10. \$24 is $166\frac{2}{3}$ % of what? *Ans.* \$14.40
11. 162 men are $4\frac{4}{5}$ % of how many? *Ans.* 3375.
12. $\$2.68\frac{3}{4}$ is $\frac{1}{2}$ % of what? *Ans.* \$537.50
13. \$19.20 is $\frac{6}{10}$ % of what? *Ans.* \$3200.
14. \$12675 is 240 % of what? *Ans.* \$5281.25
15. \$189.80 is 104 % of what? *Ans.* \$182.50
16. 16 gal. 1 pt. is $6\frac{1}{7}$ % of what? *Ans.* 262 gal. 2 qt.
17. 1 oz. $15\frac{1}{2}$ dr. is $\frac{7}{8}$ % of what? *Ans.* 14 lb. 1 oz.
18. $\frac{2}{3}$ of $1\frac{4}{5}$ is 15 % of what? *Ans.* 8.
19. 10 mi. 7 fur. 36 rd. is 75 % of what? *Ans.* 14 mi. 5 fur. 8 rd.

20. 36 men of a ship's crew die, which is $42\frac{6}{7}$ % of the whole: what was her crew? *Ans.* 84 men.

21. A stock-farmer sells 144 sheep, which is $12\frac{4}{5}$ % of his flock: how many had he? *Ans.* 1125.

22. A merchant sells 35 % of his stock for \$6000: what is it all worth at that rate? *Ans.* \$17142.86

23. I shot 12 pigeons, which was $2\frac{2}{3}$ % of the flock: how many escaped? *Ans.* 438.

REVIEW.—255. What is Case 3? The rule? The proof? How may the work be simplified? Give an example.

24. A, owing B, hands him a $10 bill, and says, there is $6\frac{1}{4}$ % of your money: what was the debt? *Ans.* $160.

25. $25 is $62\frac{1}{2}$ % of A's money, and $41\frac{2}{3}$ % of B's: what did each have? *Ans.* A $40, B $60.

26. A found $5, which was $13\frac{1}{3}$ % of what he had before: how much had he then? *Ans.* $42.50

27. B lost a $3 bill, which was $31\frac{1}{4}$ % of what he had left: how much had he at first? *Ans.* $12.60

28. I drew 48 % of my funds in bank, to pay a note of $150: how much had I left? *Ans.* $162.50

29. A farmer gave his daughter at her marriage 65 A. 2 R. 26 P. of land, which was 3 % of his farm: how much land did he own? *Ans.* 2188 A. 3 R.

30. A pays $13 a month for board, which is 20 % of his salary: what is his salary? *Ans.* $780 a year.

31. Bought 8000 bu. of wheat, which was $57\frac{1}{7}$ % of my whole stock: how much had I before? *Ans.* 6000 bu.

32. Paid 40 ct. for putting in 25 bu. of coal, which was $11\frac{3}{7}$ % of its cost: what did it cost a bu.? *Ans.* 14 ct.

33. 81 men are 5 % of 60 % of what? *Ans.* 2700 men.

34. 45 is 20 % of $18\frac{3}{4}$ % of what? *Ans.* 1200.

35. If 32 % of 75 % of 800 % of a number is 1539, what is it? *Ans.* $801\frac{9}{16}$

36. A, owning 60 % of a ship, sells $7\frac{1}{2}$ % of his share for $2500: what is the ship worth? *Ans.* $$55555.55\frac{5}{9}$

37. A father, having a basket of apples, took out $33\frac{1}{3}$% of them for his children; of these, he gave $37\frac{1}{2}$ % to his son, who gave 20 % of his share to his sister, who thus got 2 apples: how many were in the basket? *Ans.* 80.

CASE IV.

ART. **256**. A number being given, which is a given per cent. *more* or *less* than another, to find that other.

RULE.—*Represent the number required by* 100 *per cent; allow for the given rate per cent. of increase or decrease; the resulting rate per cent. will represent the number given: and from this,* 100 *per cent. or the number required, can be found by Case* III.

PROOF.—Find the given per cent. of the answer, and add it to, or subtract it from, the answer, as may be necessary; the result must be the same as the given number.

I pay \$377 rent for my house, which is 16 % more than I paid last year: what was the rent then?

SOLUTION.—Call the rent last year 100 per cent.; the rent this year being 16 % more, is 116 %; divide \$337 by 116, to get 1 %, and multiply the quotient, \$3.25, by 100, to get 100 %, or the rent last year, \$325.

```
100
 16   $          $
---
116)377.00  (3.25
    290        100
    580     -------
     00     $325.00
```

I am 28 yr. old, which is $14\frac{2}{7}$ per cent. less than my brother's age: how old is he?

SOLUTION.—Call my brother's age 100 %; my age being $14\frac{2}{7}$ % less, is $85\frac{5}{7}$ %; 28 yr. divided by $85\frac{5}{7} = \frac{196}{600}$, is 1 %; and 100 times this, or $\frac{196}{6} = 32\frac{2}{3}$, is 100 %, or my brother's age.

1. 136 is 20 % less than what? *Ans.* 170.
2. \$4.80 is $33\frac{1}{3}$ % more than what? *Ans.* \$3.60
3. $\frac{5}{6}$ is 50 % more than what? *Ans.* $\frac{5}{9}$
4. $\frac{8}{15}$ is 28 % less than what? *Ans.* $\frac{20}{27}$
5. 96 da. is 100 % more than what? *Ans.* 48 da.
6. 2576 bu. is 60 % less than what? *Ans.* 6440 bu.
7. $87\frac{1}{2}$ ct. is $87\frac{1}{2}$ % less than what? *Ans.* \$7.
8. $1\frac{3}{5}$ is 500 % more than what? *Ans.* $\frac{4}{15}$
9. 3 bu. 2 pk. 7 qt. is 8 % more than what? *Ans.* 3 bu. 1 pk. $6\frac{5}{27}$ qt.
10. 42 mi. 1 fur. 20 rd. is 55 % less than what? *Ans.* 93 mi. 6 fur.
11. 2 lb. 9 oz. $4\frac{5}{6}$ dr. is 50 % less than what? *Ans.* 5 lb. 2 oz. $9\frac{2}{3}$ dr.
12. $\frac{7}{12}$ is $99\frac{5}{8}$ % less than what? *Ans.* $155\frac{5}{9}$
13. \$920.93$\frac{3}{4}$ is $337\frac{1}{2}$ % more than what? *Ans.* \$210.50
14. \$4358.06$\frac{1}{4}$ is $233\frac{1}{3}$ % more than what? *Ans.* \$1307.41$\frac{7}{8}$
15. In $64\frac{1}{2}$ gal. of alcohol, the water is $7\frac{1}{2}$ % of the spirit; how many gal. of each? *Ans.* 60 gal. sp., $4\frac{1}{2}$ gal. w.
16. A coat cost \$32; the trimmings cost 70 % less, and the making 50 % less, than the cloth: what did each cost? *Ans.* Cloth \$17.77$\frac{7}{9}$, trim. \$5.33$\frac{1}{3}$, mak. \$8.88$\frac{8}{9}$

SUGGESTION.—Cloth = 100 %; trimming = 30 %; making = 50 % their sum 180 % = \$32. Find 1 % and then the rest.

17. If a bu. of wheat makes $39\frac{1}{5}$lb. of flour, and the cost of grinding is 4 %; how many bbl. of flour can a farmer get for 80 bu. of wheat? *Ans.* $15\frac{5}{13}$ bbl.

18. How many eagles, each containing 9 pwt. 16.2 gr. of pure gold, can I get for 455.6538 oz. pure gold at the mint, allowing $1\frac{1}{2}$ % for expense of coinage? *Ans.* 928.

19. 2047 is 10 % of 110 % less than what number? *Ans.* 2300.

20. $4246\frac{1}{2}$ is 6 % of 50 % of $466\frac{2}{3}$ % more than what number? *Ans.* 3725.

21. A drew out of bank 40 % of 50 % of 60 % of 70 % of his money, and had left $1557.20; how much had he at first? *Ans.* $1700.

22. I gave away $42\frac{6}{7}$ % of my money, and had left $2; what had I at first? *Ans.* $3.50

23. A man dying, left $33\frac{1}{3}$ % of his property to his wife, 60 % of the remainder to his son, 75 % of the remainder to his daughter, and the balance, $500, to a servant; what was the whole property, and each share?

Ans. Property, $7500; wife had $2500; son $3000; daughter $1500.

24. In a company of 87, the children are $37\frac{1}{2}$ % of the women, who are $44\frac{4}{9}$ % of the men; how many of each? *Ans.* 54 men, 24 women, 9 children.

25. In a school, 5 % of the pupils are always absent, and the attendance is 570; how many on the roll, and how many absent? *Ans.* 600; and 30.

XV. APPLICATIONS OF PERCENTAGE.

ART. 257. The importance of thoroughly understanding Percentage, can not be over estimated.

In marking goods, in calculating gain or loss, the value of investments, interest, commission, &c., its applications are various, important, and of daily occurrence; and no one can be a ready and complete accountant until he is familiar with its principles and methods.

Although the special rules are the simplest and most satisfactory, especially in the first two cases, the general

REVIEW.—256. What is Case 4? The rule? The proof? Solve example.

rule (Art. 252), is easy to recollect, and serves very well for the first three cases, and even for the fourth case, after slight preparation.

Particular care should be exercised in ascertaining the quantity which is the standard of comparison on which the rate per cent. takes effect; and bear in mind that *this quantity always represents* 100 *per cent.*

REMARKS.—1. The words *per cent.* have no reference to money. All they mean is, *on*, or *by the hundred*, and a *rate per cent.* is a number of *hundredths* of the quantity, whatever it may be. It is true, the phrase is used more in speaking of money than of any other article, but it is perfectly proper to say, "Ten per cent. of the labor;" "Twenty per cent. of the cloth;" "Seven per cent. of the time;" meaning so many hundredths of the labor, cloth, and time, respectively.

2. The advantage of using *rates per cent.*, or *ratios in hundredths*, instead of ordinary ratios, is that rates *per cent.*, like all other fractions having a common denominator, are more easily compared than ordinary ratios. Thus, it is not easy to see which of the two ratios $\frac{3}{7}$ and $\frac{4}{9}$ is the greater, but as soon as they are expressed in hundredths, viz: $42\frac{6}{7}$ % and $44\frac{4}{9}$ %, the difficulty vanishes.

XVI. GAIN AND LOSS.

ART. 258. The increase or decrease which any variable quantity undergoes, is called its *gain or loss.*

The *rate of gain or loss* is the rate per cent. the gain or loss is of the quantity on which the gain or loss accrues. The quantity on which gain or loss accrues, is the standard of comparison in questions of gain or loss, and is therefore 100 per cent.

GENERAL RULE.

Represent the quantity on which gain or loss accrues by 100 *per cent., and proceed by such rule of Percentage as the nature of the question requires.*

NOTE.—In gain or loss of money, the sum of money invested, or *cost*, is the standard of comparison, and is therefore 100 per cent.

REVIEW.—257. What single rule comprehends all the cases? What care should be exercised in all questions of percentage? What must the standard of comparison always represent?

ART. **259.** There are 4 cases of Gain or Loss, solved like the 4 corresponding cases of Percentage.

CASE I.—*Given, the quantity on which gain or loss accrues, and the rate of gain or loss, to find the gain or loss.*

Having invested \$4800, my rate of gain is $13\frac{7}{8}$ %: what is my gain?

ANALYSIS.—The \$4800 being the quantity on which the gain accrues, is 100 %; the gain being $13\frac{7}{8}$ % = $13\frac{7}{8}$ hundreths of it, is found as in Case I of Percentage, by multiplying it by $13\frac{7}{8}$, and dividing the product by 100.

$$\begin{array}{rl} \$4800 & \\ 13\frac{7}{8} & \\ \hline 14400 & \\ 4800 & \\ 4200 & \\ \hline \$666.00 & \text{Gain.} \\ \$4800 & \\ \hline \$5466 & \text{Sum after gain.} \end{array}$$

REMARKS.—When the gain or loss is known, the amount *after* gain or loss, is obtained by an addition or a subtraction.

1. If my rate of gain is 25 %, how should I mark goods for sale that cost me \$8; \$7.50; \$6.25; \$4.75; \$3.87½; \$2.62½; \$1.93¾; 62½ ct.; 15 ct. a yard?
Ans. \$10; \$9.37½; \$7.81¼; \$5.93¾; \$4.84⅜; \$3.28⅛; \$2.42$\frac{3}{16}$; 78⅛ ct.; 18¾ ct. a yard.

2. If I must lose 20 % on damaged goods, how should I mark those that cost me 12½ ct.; 25 ct.; 43¾ ct.; 75 ct.; \$1.10; \$2.40; \$3.50; \$4.37½; \$5.81¼; \$6.56¼; \$7.68¾; \$8.10 a yard?
Ans. 10 ct.; 20 ct.; 35 ct.; 60 ct.; 88 ct.; \$1.92; \$2.80; \$3.50; \$4.65; \$5.25; \$6.15; \$6.48, a yard.

3. The population of a town was 1760 last year, and has increased 26¼ %: what is it now? *Ans.* 2222.

4. A city containing 42540 inhabitants, lost 11⅔ % of them by cholera: how many died, and how many are left? *Ans.* 4963 died; and 37577 left.

5. A lb. Tr. contains 5760 gr., and a lb. av. $21\frac{19}{36}$ % more: how many gr. does it contain? *Ans.* 7000 gr.

REVIEW.—257. What is said of the words per cent.? What advantage in using rates per cent. instead of ordinary ratios? Give an example. 258. What is gain or loss? What is the rate of gain or loss? In all questions of gain or loss, which quantity is 100 %? What is the general rule? In gain or loss of money which quantity is 100 %? Why? 259. How many cases of gain or loss? What is Case 1? Analyze the example. How can the quantity *after* gain or loss be found?

6. A goes 42 mi. 3 fur. 18 rd. a day; B, 15 % faster: how far does B go per day? *Ans.* 48 mi. 6 fur. 14.7 rd.

7. U. S. standard gold is $\frac{9}{10}$ pure, and English standard gold is $1\frac{23}{27}$ % purer: how pure is it? *Ans.* $\frac{11}{12}$ pure.

8. U. S. standard silver is $\frac{9}{10}$ pure, and English standard silver is $2\frac{7}{9}$ % purer: how pure is it? *Ans.* $\frac{37}{40}$ pure.

9. The cost of publishing a book is 50 ct. a copy; if the expense of sale be 10 % of this, and the profit 25 %: what does it sell for by the copy? *Ans.* $67\frac{1}{2}$ ct.

10. A began business with $5000: the 1st year he gained $14\frac{3}{4}$ %, which he added to his capital; the 2d year he gained 8 %, which he added to his capital; the 3d year he lost 12 %, and quit: how much better off was he than when he started? *Ans.* $452.92

11. A bought a farm of government land at $1.25 an acre; it cost him 160 % to fence it, 160 % to break it up, 80 % for seed, 100 % to plant it, 100 % to harvest it, 112 % for threshing, 100 % for transportation: each acre produced 35 bu. of wheat, which he sold at 70 ct. a bushel: how much did he gain on every acre above all expenses the first year? *Ans.* $13.10

12. What must I sell a horse for, that cost me $150, to gain 35 %? *Ans.* $202.50

13. A gave $4850 for his house, and offers it for 20 % less: what is his price? *Ans.* $3880.

14. Bought hams at 8 ct. a lb.; the wastage is 10 %: how must I sell them to gain 30 %? *Ans.* $11\frac{5}{9}$ ct. a lb.

SUGGESTION.—Since a lb. wastes 10 %, or $\frac{1}{10}$, I get only $\frac{9}{10}$ lb for 8 ct., and a lb. at that rate costs $8\frac{8}{9}$ ct.; to which 30 % must be added, to get the selling price.

15. I bought a cask of brandy containing 46 gal. at $2.50 per gal.; if 6 gal. leak out, how must I sell the rest, so as to gain 25 %? *Ans.* 3.59\frac{3}{8}$ per gal.

16. I started in business with $10000, and gained 20 % the first year, and added it to what I had; the 2d year I gained 20 %, and added it to my capital; the 3d year I gained 20 %. What had I then? *Ans.* $17280.

ART. **260.** CASE II.—*Given, the quantity on which gain or loss accrues, and the gain or loss, to find the rate of gain or loss.*

REMARK.—If the quantity on which gain or loss accrues, and the quantity *after* gain or loss, are given, their difference is the gain or loss, and the question would come under this case.

The population of Cincinnati in 1840 was 45000, and in 1850 was 165000; how many per cent. had it increased in the interval?

ANALYSIS.—Here the gain is 120000, which, compared with 45000, the quantity on which the gain accrues, is $\frac{120000}{45000} = \frac{8}{3} = \frac{8}{3}$ of 100 % $= 266\frac{2}{3}$ %. *Ans.*

1. If I double my money, how many per cent. do I gain? *Ans.* 100.

2. If I lose half my goods, how many per cent. do I lose? *Ans.* 50.

3. If I buy at $1 and sell at $9, how many per cent. do I gain? *Ans.* 800.

4. If I buy at $1 and sell at $4, how many per cent. do I gain? *Ans.* 300.

5. If I buy at $4 and sell at $1, how many per cent. do I lose? *Ans.* 75.

6. If I sell $\frac{5}{9}$ of an article for what the whole cost me, how many per cent. do I gain? *Ans.* 80.

7. If I sell $\frac{5}{8}$ of an article for what $\frac{3}{7}$ of it cost me, how many per cent. do I lose? *Ans.* $31\frac{3}{7}$

8. If a person sell 14 oz. of candles for a pound, how many per cent. does he gain? *Ans.* $14\frac{2}{7}$

9. How many % larger is the earth's equatorial diameter (15850 mi.) than its polar diameter (15798 mi.)? *Ans.* $\frac{1}{3}$ % nearly.

10. If I sell an article for $\frac{1}{3}$ of its cost, how many per cent. do I lose? *Ans.* $66\frac{2}{3}$

11. The U. S. half dollars coined since 1853, contain 8 pwt. of standard silver; those coined before, contain 8 pwt. $14\frac{1}{4}$ gr. of standard silver; how many per cent. more valuable are the latter than the former? *Ans.* $7\frac{27}{64}$

12. A log 1 ft. 6 in. thick, is sawn into 13 boards each $1\frac{1}{4}$ in. thick: what % is wasted? *Ans.* $9\frac{13}{18}$

13. The U. S. wine gallon contains 231 cu. in., and the beer gallon 282 cu. in.: how many % larger is the latter

REVIEW.—260. What is Case 2? The rule? If the quantity on which gain or loss accrues, and the quantity after gain or loss, are given, how do we proceed? Analyze the example.

than the former? and how many % smaller is the former than the latter? *Ans.* $22\frac{6}{77}$ % larger; $18\frac{4}{47}$ % smaller.

14. The U. S. dry gallon or half peck, contains 268.8 cu. in.: how many % larger is it than the wine gallon? and how many % smaller than the beer gallon?

Ans. $16\frac{4}{11}$ % larger; $4\frac{32}{47}$ % smaller.

15. The imperial gallon of Great Britain contain 277.274 cu. in.: how many % is it larger than our wine and dry gallons? and how many % smaller than our beer gallon?

Ans. $20\frac{37}{1155}$ % larger; $3\frac{205}{1344}$ % larger; $1\frac{953}{1410}$ % less.

16. U. S. gold ($\frac{9}{10}$ pure) is how many % less pure than English gold, ($\frac{11}{12}$ pure)? *Ans.* $1\frac{9}{11}$

17. Gold 22 carats is how many per cent. purer than gold 18 carats fine? and how many % pure is each?

Ans. $22\frac{2}{9}$ % purer; first, $91\frac{2}{3}$ % pure; 2d, 75 % pure.

18. If I pay for a lb. of sugar, and get a lb. Troy, what % do I lose, and what % does the grocer gain by the cheat? *Ans.* $17\frac{5}{7}$ % loss; $21\frac{19}{36}$ % gain.

19. A, having failed, pays B $1750 instead of $2500, which he owed him: what % does B lose? *Ans.* 30.

20. Sugar bought at $6\frac{1}{4}$ ct. a lb., is sold at $7\frac{1}{8}$ ct. a lb.: what is the rate of gain? *Ans.* 14 %.

21. An article has lost 20 % by wastage, and is sold for 40 % above cost: what is the gain per cent.? *Ans.* 12.

SOLUTION.—20 % of the cost is wasted; 80 % of the cost remains available, on which a gain of 40 % is realized; 40 % of 80 % = 32 %, which, added to 80 %, makes 112 %; deduct the cost, 100 %; the net gain is 12 %.

22. If my retail gain is $33\frac{1}{3}$ %, and I sell at wholesale for 10 % less than at retail, what is my gain % at wholesale? *Ans.* 20.

23. Bought a lot of glass; lost 15 % by breakage: at what % above cost must I sell the remainder to clear 20 % on the whole? *Ans.* $41\frac{3}{17}$

24. If a bu. of corn is worth 35 ct. and makes $2\frac{1}{2}$ gal. of whisky, which sells at 24 ct. a gal., what is the profit of the distiller? *Ans.* $71\frac{3}{7}$ %.

ART. **261.** CASE III.—*Given, the gain or loss, and the rate of gain or loss, to find the quantity on which gain or loss accrues.*

By selling a lot for $34\frac{3}{8}$ % more than I gave, my gain is \$423.50: what did it cost me?

Analysis.—Since $34\frac{3}{8}$ % = \$423.50, 1 % = \$423.50 ÷ $34\frac{3}{8}$, = \$12.32; and 100 %, or the whole cost = 100 times \$12.32 = \$1232; as in Case 3 of Percentage.

Remark.—After the quantity on which gain or loss accrues is known, the gain or loss may be added to it or subtracted from it, to get the quantity after gain or loss accrues.

1. How large sales must I make in a year at a profit of 8 % to clear \$2000? *Ans.* \$25000.

2. I lost \$50 by selling sugar at $22\frac{1}{2}$ % below cost: what was the cost? *Ans.* \$$222.22\frac{2}{9}$

3. If I sell tea at $13\frac{1}{3}$ % gain, I make 10 ct. a lb.. how much a pound did I give? *Ans.* 75 ct.

4. I lost a $2\frac{1}{2}$ dollar gold coin, which was $7\frac{1}{7}$ % of all I had: how much had I? *Ans.* \$35.

5. A and B each lost \$5, which was $2\frac{7}{9}$ % of A's and $3\frac{1}{3}$ % of B's money: which had the most money, and how much? *Ans.* A had \$30 more than B.

6. I gained this year \$2400, which is 120 % of my gain last year, and that is $44\frac{4}{9}$ % of my gain the year before: what were my gains the two previous years? *Ans.* \$2000 last year.; \$4500 year before.

7. The dogs killed 40 of my sheep, which was $4\frac{1}{6}$ % of my flock: how many had I left? *Ans.* 920.

Art. **262**. Case iv.—*Given, the quantity after gain or loss has accrued, and the rate of gain or loss, to find the quantity on which gain or loss accrues.*

Sold goods for \$25.80, by which I gained $7\frac{1}{2}$ %: what was the cost?

Solution.—The cost is 100 %; the \$25.80 being $7\frac{1}{2}$ % more, is $107\frac{1}{2}$ %; then 1 % = \$25.80 ÷ $107\frac{1}{2}$ = 24 ct., and 100 % or the cost = 100 times 24 ct. = \$24; as in Case 4 of Percentage.

Remark.—After the quantity *on which* gain or loss accrues is found, the difference between it and the quantity *after* gain or loss accrues, will be the gain or loss.

Review.—261. What is Case 3d? Analyze the example. How can the quantity after loss or gain be found? 262. What is Case 4th? Analyze the example. How is the gain or loss found?

1. Sold cloth at \$3.85 a yd.; my gain was 10 %: how much a yard did I pay? *Ans.* \$3.50

2. Gold pens, sold at \$5 a piece, yield a profit of $33\frac{1}{3}$ %: what did they cost a piece? *Ans.* \$3.75

3. Sold out for \$952.82 and lost 12 %: what was the cost? and what would I have got if I had sold out at a gain of 12 %? *Ans.* \$1082.75 and \$1212.68

4. Sold my horse at 40 % gain; with the proceeds I bought another, and sold him for \$238, losing 20 %: what did each horse cost me?

Ans. \$212.50 for 1st, \$297.50 for 2d.

5. Sold flour at an advance of $13\frac{1}{3}$ %; invested the proceeds in flour again, and sold this lot at a profit of 24 %, realizing \$3952.50: how much did each lot cost me? *Ans.* 1st lot, \$2812.50; 2d lot, \$3187.50

6. An invoice of goods purchased in New York, cost me 8 % for transportation, and I sold them at a gain of $16\frac{2}{3}$ % on their total cost on delivery, realizing \$1260: what were they invoiced at? *Ans.* \$1000.

7. The population of a village increased 50 % each year on the previous one, for four successive years, and at the end of the 5th was 405: what was it at the end of each previous year? *Ans.* 80, 120, 180, 270.

8. For 6 years my property increased each year on the previous, 100 %, and became worth \$100000: what was it worth at first? *Ans.* \$1562.50

9. A lost at play 50 % of his money the 1st game, 50 % of the remainder the 2d, and so on for 10 successive games, when he was reduced to his last dollar: what had he at first? *Ans.* \$1024.

XVII. COMMISSION AND BROKERAGE.

Art. **263.** One who buys or sells property, makes investments, collects debts, or transacts other business for the benefit, and at the advice of another, is a *commission merchant*, *agent*, or *factor*.

When the commission merchant lives in a different country or part of the country from his employer, he is frequently called *correspondent* or *consignee;* goods sent

Review.—263. What is a commission merchant, agent, or factor?

to him to be sold are called a *consignment*, and the person sending them a *consignor*.

The charge made by a commission merchant for transacting another's business, is called his *commission*, and is estimated at a certain rate per cent. of the sum invested or realized for the other's benefit. This *rate per cent.* is the *rate of commission*.

BROKERAGE is a charge of the same nature as commission, but generally smaller.

ART. **264**. The *amount of the sale, purchase, or collection*, is the standard of comparison, and is, therefore, 100 %.

The *net proceeds* of a sale or collection is the sum left after deducting the commission and other charges.

GENERAL RULE FOR COMMISSION.

Represent by 100 *per cent. the quantity on which the commission or brokerage is charged, (which is always the amount of the sale, purchase, or collection,) and then proceed, by such rule of Percentage as the nature of the question requires.*

Commission has 4 cases, solved like the corresponding cases of Percentage.

CASE I.

ART. **265**. Given, the amount of the sale, purchase, or collection, and the rate of commission, to find the commission.

A commission merchant makes sales during a year amounting to \$168475.37$\frac{1}{2}$, on which his charge was 2$\frac{1}{2}$ %: how much did his commission come to?

ANALYSIS.—The amount of sales, \$168475.37$\frac{1}{2}$, is 100 %; the commission, being 2$\frac{1}{2}$ %, is found by multiplying by 2$\frac{1}{2}$, and dividing by 100, as in Case I of Percentage; this gives \$4211.88.

REMARK.—If the commission (\$4211.88) be subtracted from the amount of sales (\$168475.37$\frac{1}{2}$), the remainder (\$164263.49$\frac{1}{2}$) will be the amount paid to the consignors.

1. I collect for A, \$268.40, and have 5 % commission: how much do I pay over? *Ans.* \$254.98

REVIEW.—263. What is a consignor? A consignment? A consignee? How is a commission merchant paid? What is the rate of commission? What is brokerage?

2. I sell for B, 650 bbl. of flour, at \$7.50 a bbl., 28 bbl. of whisky, 35 gal. each, at $22\frac{1}{2}$ ct. a gal.: what is my commission at $2\frac{1}{4}$ %? *Ans.* \$114.65

3. Received on commission 25 hhd. sugar (36547 lb.); of which I sold 10 hhd. (16875 lb.) at 6 ct. a lb., and 6 hhd. (8246 lb.) at 5 ct. a lb., and the rest at $5\frac{1}{2}$ ct. a lb.: what is my commission at 3 %? *Ans.* \$61.60

4. A lawyer charged 8 % for collecting a note of \$648.75: what is his fee, and the net proceeds? *Ans.* \$51.90, and \$596.85

5. A lawyer, having a debt of \$1346.50 to collect, compromises by taking 80 %, and charges 5 % for his fee: what is his fee, and the net proceeds? *Ans.* \$53.86, and \$1023.34

6. Bought for C, a carriage for \$950, a pair of horses for \$575, and harness for \$120; paid charges for keeping, packing, shipping, &c., \$18.25; freight, \$36.50: what was my commission, at $3\frac{1}{3}$ %, and the whole bill? *Ans.* \$54.83, and \$1754.58

NOTE.—Commission is charged *only* on the amount of the purchase.

7. Sold 500000 lb. of pork, at $5\frac{1}{2}$ ct. a lb.: what is my commission, at $12\frac{1}{2}$ %? *Ans.* \$3437.50

8. An insurance agent has 10 % of all sums received for his company: what does he make in a year, if he receives for the company, \$28302.75? *Ans.* \$2830.27$\frac{1}{2}$

9. An insurance agent has 5 % of all sums received for his company, and 5 % of what remains at the end of the year after payment of losses: what will he make, if he receives for his company, \$47363.87$\frac{1}{2}$, and pays losses, \$31344.50? *Ans.* \$3169.16

10. An architect charges $3\frac{1}{2}$ % for designing and superintending a building, which cost \$27814.60: what is his fee? *Ans.* \$973.51

11. A factor has $2\frac{3}{4}$ % commission, and $3\frac{1}{2}$ % for guaranteeing payment: if the sales are \$6231.25, what does he get? *Ans.* \$389.45

REVIEW.—264. What quantity is the standard of comparison? What does it represent then? What are the net proceeds of a sale or collection? What is the general rule for commission and brokerage? How many cases of commission? 265. What is Case 1? Analyze the example. What are the net proceeds?

12. An architect charges $1\frac{1}{4}$ % for plans and specifications, and $2\frac{7}{8}$ % for superintending: what does he make, if the building costs $14902.50? *Ans.* $614.73

13. A broker sells for me 10 hhd. sugar (9256 lb.), at 5 ct. a lb.; what is his brokerage, at $\frac{3}{4}$ %, and my proceeds? *Ans.* Brokerage, $3.47; proceeds, $459.33

14. A sells a house and lot for me at $3850, and charges $\frac{5}{8}$ % brokerage: what is his fee? *Ans.* 24.06\frac{1}{4}$

15. I have a lot of tobacco on commission, and sell it through a broker for $4642.85: my commission is $2\frac{1}{2}$ %, the brokerage $1\frac{1}{8}$ %: what do I pay the broker, and what do I keep? *Ans.* I keep $63.84; and brok. $52.23

16. What does a tax gatherer get for collecting a tax of $37850, at 3 %, and how much does he pay over? *Ans.* $1135.50 rec'd; $36714.50 paid over.

17. A tax collector is paid $4\frac{1}{2}$ % for collecting a tax of $218096.75: what is his fee, and the net proceeds? *Ans.* $9814.35 fee, $208282.40 paid over.

CASE II.

ART. **266**. Given, the commission, and the amount of the sale, purchase, or collection, to find the rate of commission.

1. An auctioneer's commission for selling a lot was $50, and the sum paid the owner was $1200: what was the rate of commission? *Ans.* 4 %.

SUGGESTION.—Find the amount of the sale, $1250; then find what % $50 is of $1250, as in Case II of Percentage.

2. A commission merchant sells 800 bbl. of flour, at 6.43\frac{3}{4}$ a bbl., and remits the net proceeds, $5021.25: what is his rate of commission? *Ans.* $2\frac{1}{2}$ %.

3. The cost of a building was $19017.92, including the architect's commission, which was $553.92: what rate did the architect charge? *Ans.* 3 %.

4. Bought flour for A; my whole bill was $5802.57, including charges, $76.85, and commission, $148.72: find the rate of commission. *Ans.* $2\frac{2}{3}$ %.

5. Charged $52.50 for collecting a debt of $1050: what was my rate of commission? *Ans.* 5 %.

6. An agent gets $169.20 for selling property for $8460: what was his rate of brokerage? *Ans.* 2 %.

7. My commission for selling books was \$6.92, and the net proceeds, \$62.28: what rate did I charge?
Ans. 10 %.

8. Paid \$38.40 for selling goods worth \$6400: what was the rate of brokerage? *Ans.* $\frac{3}{5}$ %.

9. Paid a broker \$24.16, and retained as my part of the commission \$42.28, for selling a consignment at \$2416: what was the rate of brokerage, and my rate of commission? *Ans.* Brok. 1 %; Com. $1\frac{3}{4}$ %.

10. A tax gatherer is paid \$3711 for collecting a tax of \$74220: what rate is allowed him? *Ans.* 5 %.

11. A tax collector retains \$6826.45, and pays over \$539289.55: what % is his commission? *Ans.* $1\frac{1}{4}$ %.

CASE III.

ART. 267. Given, the commission, and rate of com., to find the sum on which commission is charged.

NOTE.—After finding the sum on which commission is charged, subtract the commission, to find the *net proceeds*, or add it, to find the *whole cost*, as the case may be.

1. My commissions in 1 year, at $2\frac{1}{2}$ %, are \$3500: what were the sales, and the whole net proceeds?
Ans. \$140000 and \$136500.

SUGGESTION.—$2\frac{1}{2}$ % = \$3500; find 1 %, then 100 %; as in Case III of Percentage.

2. An insurance agent's income is \$1733.45, being 10 % on the sums received for the company: what were the company's net receipts? *Ans.* \$15601.05

3. A pork merchant charged 15 % commission, and cleared \$2376.15, after paying out \$1206.75 for all expenses of packing: how many pounds of pork did he pack, if it cost $4\frac{1}{2}$ ct. a pound? *Ans.* 530800 lb.

4. An agent purchased, according to order, 10400 bu. of wheat; his commission, at $1\frac{1}{4}$ %, was \$156, and charges for storage, shipping, and freight, \$527.10: what did he pay a bushel? and what was the whole cost?
Ans. \$1.20 a bu., and \$13163.10, whole cost.

5. Paid \$64.05 for selling coffee, which was $\frac{7}{8}$ % brokerage: what are the net proceeds? *Ans.* \$7255.95

REVIEW.—266. What is Case 2? 267. What is Case 3? How are the net proceeds found? How, the whole cost?

6. Received produce on commission at $2\frac{1}{4}$ %; my surplus commission, after paying $\frac{1}{2}$ % brokerage, is $107.03: what was the amount of the sale, the brokerage, and net proceeds? *Ans.* Sale, $6116; brok., $30.58; pro., $5978.39

7. A is paid $861.27 for collecting taxes at $2\frac{1}{2}$ %: what were the taxes, and the sum received by the State?
Ans. Tax, $34450.80, paid to State, $33589.53

8. Paid A $1952.64 for collecting, at $1\frac{3}{4}$ %: what were my net proceeds? *Ans.* $109626.79

CASE IV.

ART. **268**. Given, the rate of commission, and the net proceeds, or the whole cost, to find the sum on which commission is charged.

NOTES.—1. After the sum on which commission is charged is known, find the commission by subtraction.

2. When the divisor in this case is little less than 100, use the contracted method of Division. (Art. 69.)

1. A lawyer collects a debt for a client, takes 4 % for his fee, and remits the balance, $207.60: what was the debt and the fee? *Ans.* $216.25 and $8.65

SOLUTION.—The debt being the quantity on which commission is charged, is 100 %; if 4 % is taken, there is left 96 % = $207.60: find 1 % and then 100 %; as in Case IV of Percentage.

2. Sent $1000 to buy a carriage, commission $2\frac{1}{2}$ %: what must the carriage cost? *Ans.* $975.61

SUGGESTION.—100 % + $2\frac{1}{2}$ % = $102\frac{1}{2}$ % = $1000; find 1 %, then 100 %.

3. A buys per order a lot of coffee; charges, $56.85; commission, $1\frac{1}{4}$ %; the whole cost is $539.61: what did the coffee cost? *Ans* $476.80

SUGGESTION.—Take out the charges; the rest as before.

4. Buy sugar at $2\frac{1}{4}$ % commission, and $2\frac{1}{2}$ % for guaranteeing payment: if the whole cost is $1500, what was the cost of the sugar? *Ans.* $1431.98

SUGGESTION.—100 + $2\frac{1}{4}$ + $2\frac{1}{2}$ = $104\frac{3}{4}$ % = $1500; rest as before

REVIEW.—268. What is Case 4? How can the commission be found If the divisor is little less than 100, what may be done?

5. Sold 2000 hams (20672 lb.); commission, $2\frac{1}{2}$ %, guarantee, $2\frac{3}{4}$ %, net proceeds due consignor, $2448.34: what did the hams sell for a lb.? *Ans.* $12\frac{1}{2}$ ct.

6. What tax must be assessed, to yield $26782.45 net proceeds, and pay the collector $3\frac{1}{8}$ %; and what is the collector paid? *Ans.* Tax, $27646.40; col., $863.95

7. What tax must be raised, to yield $1044073.50 nd pay for collection, at $\frac{5}{8}$ %? *Ans.* $1050640.

8. A sells 1000 bbl. ($30468\frac{3}{4}$ gal.) of whisky; brokerage, $\frac{4}{5}$ %; proceeds, $7254: how much a gallon was it sold for? *Ans.* 24 ct.

9. Sold cotton on commission at 5 %; invested the net proceeds in sugar, commission, 2 %; my whole commission was $210: what was the value of the cotton and sugar? *Ans.* Cotton, $3060; sugar, $2850.

Suggestion.—Cotton = 100 %. Com. on Cotton = 5 %. Proceeds = 95 %. Com. on sugar = $\frac{2}{102}$ of 95 % (Art. 256) = $1\frac{44}{51}$ %. Whole Com. = 5 % + $1\frac{44}{51}$ % = $6\frac{44}{51}$ % = $210; find 1 %, &c.

10. Sold flour at $3\frac{1}{2}$ % commission; invested $\frac{2}{3}$ of its value in coffee, at $1\frac{1}{2}$ % commission; remitted the balance, $432.50: what was the value of the flour, the coffee, and my commissions? *Ans.* Flour, $1500; Coffee, $1000; 1st Com., $52.50; 2d Com., $15.

11. Sold a consignment of pork, and invested the proceeds in brandy, after deducting my commissions, 4 % for selling, and $1\frac{1}{4}$ % for buying. The brandy cost $2304.00: what did the pork sell for, and what were my commissions?

Ans. Pork, $2430; 1st Com., $97.20; 2 Com., $28.80

12. Sold 1400 bbl. of flour, at $6.20 a bbl.; invested the proceeds in sugar, as per order, reserving my commissions, 4 % for selling, and $1\frac{1}{2}$ % for buying, and the expense of shipping, $34.16: how much did I invest in sugar? *Ans.* $8176.

XVIII. STOCKS AND DIVIDENDS.

Art. **269.** A joint-stock company is an association of individuals, empowered to transact a specified business under certain restrictions.

The business transacted by them is generally such as to exceed the means of one person; as, Banking, Mining, Insurance, &c. Railroads, canals, steamboats, turnpikes, bridges, telegraphs, &c., are owned and managed by joint-stock companies.

The capital of a company is called its *stock*, and, for convenience is usually divided into *shares* of $100, or $50, for each share a *certificate* is issued.

Persons who own shares are called *stockholders*, or *shareholders;* stock can be transferred from one person to another, the certificates being evidence of ownership.

The *dividend* is the gain to be divided among the stockholders, in proportion to their amounts of stock. Hence, a joint-stock company is in the nature of a partnership.

On account of the great number of shares in such a concern, it is convenient to declare the dividend, as a certain rate per cent. of the whole stock; this rate per cent. may be called the *rate of dividend.*

CASE I.—*Given, the stock, and rate of dividend, to find the dividend for that stock.* (See Case I, Percentage.)

1\. I own 18 shares of $50 each, in the City Insurance Co., which has declared a dividend of $7\frac{1}{2}$ %: what do I receive? *Ans.* $67.50

2\. I own 147 shares of railroad stock ($50 each), on which I am entitled to a dividend of 5 %, payable in stock: how many additional shares do I receive?
Ans. 7 shares, and $17.50 toward another share.

3\. A has 218 shares bank stock ($100 each), and gets a dividend of 12 %: how much is that? *Ans.* $2616.

4\. The Cincinnati Gas Co. declare a dividend of 18 %: what do I get on 50 shares ($100 each)? *Ans.* $900.

5\. The Western Stage Co. declare a dividend of $4\frac{1}{2}$ per cent: if their whole stock is $150000, how much is distributed to the stockholders? *Ans.* $6750.

6\. A railroad Co., whose stock account is $4256000, declared a dividend of $3\frac{1}{2}$ %: what sum was distributed mong the stockholders? *Ans.* $148960.

REVIEW.—269. What is a joint-stock company? What kind of business do they transact? What is the stock? How is it divided? What are the stockholders? What is the dividend? How is it divided among the stockholders? What is the rate of dividend? What is Case I?

7. A Telegraph Co., with a capital of $75000, declares a dividend of 7 %, and has $6500 surplus: what has it earned? *Ans.* $11750.

8. I own 24 shares of stock ($25 each) in a Fuel Co. which declares a dividend of 6 %; I take my dividend in coal, at 8 ct. a bu.: how much do I get? *Ans.* 450 bu.

ART. **270.** CASE II.—*Given, the stock, and dividend, to find the rate of dividend.* (See Case II, Percentage.)

1. My dividend on 72 shares bank stock, ($50 each), is $324: what was the rate of dividend? *Ans.* 9 %.

2. A Turnpike Co., whose stock is $225000, earns $16384.50: what rate of dividend can it declare? *Ans.* 7 %, and $634.50 surplus.

3. The receipts of a Canal Co., whose stock is $3650000, in one year are $256484; the outlay is $79383: what rate of dividend can it declare? *Ans.* $4\frac{1}{2}$ %, and $12851 sur.

ART. **271.** CASE III.—*The dividend, and rate of dividend, given, to find the stock corresponding.* (See Case III, Percentage.)

1. An Insurance Co. earns $18000, and declares a 15% dividend: what is its stock account? *Ans.* $120000.

2. A man gets $94.50 as a 7 % dividend: how many shares of stock ($50 each) has he? *Ans.* 27.

3. Received 5 shares ($50 each), and $26 of another share, as an 8 % dividend on stock: how many shares had I? *Ans.* 69.

ART. **272.** CASE IV.—(*See Case IV, Percentage.*)

1. Received, 10 % stock dividend, and then had 102 shares ($50 each), and $15 of another share: how many shares had I before the dividend? *Ans.* 93.

2. Having received two dividends in stock, one of 5 %, another of 8 %, my stock has increased to 567 shares: how many had I at first? *Ans.* 500.

XIX. PAR, DISCOUNT, AND PREMIUM.

ART. **273.** PAR, DISCOUNT, and PREMIUM, are mercantile terms, applied to *money*, *stocks*, *bonds*, and *drafts*.

Money is the circulating medium of trade, in the form of gold and silver coins, and bank notes.

Stocks, are money invested in Banks, Insurance companies, &c., in the form of *shares*, $50 or $100 each.

Drafts, *bills of exchange*, or *checks*, are written orders for money. *Bonds* are written obligations to pay money at a future time.

ART. **274**. All *money*, *stocks*, *bonds*, and *drafts*, have value on their *face*, called *nominal* or *par* value.

Their *real value*, is what they are intrinsically worth and sell for.

When their *real* is the same as their *nominal* value, they are said to be *par*, (the word *par* meaning *equal*, in Latin). When they sell for less than their nominal value, they are *below par*, or *at a discount*. When they sell for more than their nominal value, they are *above par*, or *at a premium*.

ART. **275**. *Discount* is how much *less*, money, stocks, drafts, &c., are worth, than their *face*. *Premium* is how much *more*, they are worth, than their *face*.

Rate of premium, or *rate of discount*, is the rate per cent. the *premium* or *discount* is of the *face*.

ART. **276**. The *face* or *par value* of money, stocks, drafts, &c., is the standard of comparison: hence, this

GENERAL RULE.

Represent the face by 100 *per cent; and then proceed by such rule of Percentage as the nature of the question requires.*

This subject has 4 cases, solved like the 4 corresponding cases of Percentage.

CASE I.—*Given, the par value, and the rate of premium or discount, to find the premium or discount.* (See Case I, Percentage.)

NOTE.—If the result is a *premium*, it must be *added* to the *par* to get the *real* value; if it is a *discount*, it must be subtracted.

REVIEW.—270. What is Case 2? 271. Case 3? 272. What does Case 4 correspond to? 273. What are Par, Discount, and Premium? What is money? What are stocks? What are drafts? What other names do they have? What are bonds? 274. What is the par value of money, stocks, &c.? What is their real value? When are they par? Why so called? When are they below par, or at a discount? When are they above par, or at a premium? 275. What is discount? Premium? Rate of discount? Rate of premium?

1. Buy 18 shares stock ($100) at 8 % discount: find the discount and cost. *Ans.* $144, and $1656.

2. Sell the same at $4\frac{1}{2}$ % prem.: find the prem., the price, and gain. *Ans.* $81, $1881, and $225.

3. Bought 62 shares Railroad stock ($50) at 28 % premium: what did they cost? *Ans.* $3968.

4. What is the cost of 47 shares Railroad stock ($50) at 30 % discount? *Ans.* $1645.

5. Bought $150 in gold at $\frac{3}{4}$ % premium: what is the premium, and cost? *Ans.* 1.12\frac{1}{2}$ and 151.12\frac{1}{2}$

6. Sold a draft on New York of $2568.45, at $\frac{1}{2}$ % premium: what do I get for it? *Ans.* $2581.29

7. Sold $425 uncurrent money at 3 % discount: what did I get, and lose? *Ans.* $412.25 and $12.75

8. What is a $5 bank note worth at 6 % discount? *Ans.* $4.70

9. Exchanged 32 shares Bank stock ($50), 5 % premium, for 40 shares Railroad stock ($50), 10 % discount, and paid the difference in cash: what was it? *Ans.* $120.

10. Bought 98 shares stock ($50) at 15 % discount; gave in payment a bill of exchange on New Orleans for $4000 at $\frac{5}{8}$ % premium, and the balance in cash: how much cash did I pay? *Ans.* $140.

11. Bought 56 shares Turnpike stock ($50) at 69 %; sold them at $76\frac{1}{2}$ %: what did I gain? *Ans.* $210.

REMARK.—The cost of stocks is generally given as so many % of the *face*, instead of so many % discount or premium.

12. Bought telegraph stock at 106 %; sold it at 91 %: what was my loss on 84 shares ($50)? *Ans.* $630.

13. What is the difference between a draft on Philadelphia of $8651.40, at $1\frac{1}{4}$ % premium, and one on N. Orleans for the same amount, at $\frac{1}{2}$ % discount? *Ans.* $151.40

14. Bought 18 shares Railroad stock ($50) at 12 % discount, paying $\frac{1}{2}$ % brokerage: what did it cost me? *Ans.* $795.96

15. Find the cost of 9 Ohio State bonds ($500) at 8 % premium, $\frac{3}{4}$ % brokerage. *Ans.* $4896.45

REVIEW.—276. What is the standard of comparison? What is the general rule for buying or selling money, stocks, drafts, &c.? How many cases in this subject? What do they correspond to? What is Case 1? How is the real value found after the premium or discount is known?

16. I change $380 bank notes for gold at $1\frac{1}{2}$ % premium; $25 of the notes are 1 % discount; $40 are 2 % discount; $65 are 5 % discount; and $10 are 3 % discount: if I receive $370 in gold, how much change should be given the broker? *Ans.* 15 ct.

17. Buy 364 shares stock ($50), at 16 % discount, brokerage $\frac{3}{8}$ %; sell them at 10 % premium, brokerage $\frac{3}{4}$ %: what is my gain? *Ans.* $4524.52

ART. **277**. CASE II.—*Given, the face, and the discount or premium, to find the rate of discount or premium.* (See Case II, Percentage.)

NOTES.—1. If the *face* and the *real value* are known, take their difference for the discount or premium; then, apply the rule.

2. If the rate of *gain* or *loss* is required, the *real value* or *cost* is the standard of comparison, not the *face*.

1. Paid $2401.30 for a draft of $2360 on New York: what was the rate of premium? *Ans.* $1\frac{3}{4}$ %.

2. Paid $2508.03 for 26 shares stock ($100), and brokerage, $25.03: what the rate of discount? *Ans.* $4\frac{1}{2}$ %.

3. Bought 112 shares Railroad stock ($50) for $3640: what was the rate of discount? *Ans.* 35 %.

4. If the stock in last example yields 8 % dividend, what is my rate of *gain?* *Ans.* $12\frac{4}{13}$ %.

5. I sell the same stock for $5936: what rate of premium is that? what rate of *gain?* *Ans.* 6 %; $63\frac{1}{13}$ %.

6. If I count my dividend as part of the gain, what is my rate of gain? *Ans.* $75\frac{5}{13}$ %.

7. Exchanged 12 Ohio bonds ($1000), 7 % premium, for 280 shares of Railroad stock ($50): what rate of discount were the latter? *Ans.* $8\frac{2}{7}$ %.

8. Gave 266.66\frac{2}{3}$ of notes, 4 % discount, for $250 of gold: what rate of premium was the gold? *Ans.* $2\frac{2}{5}$ %.

9. Bought 58 shares Mining stock ($50) at 40 % premium, and gave in payment a draft on Boston for $4000: what rate of premium was the draft? *Ans.* $1\frac{1}{2}$ %.

10. Received $4.60 for an uncurrent $5 note: what was the rate of discount? *Ans.* 8 %.

REVIEW.—277. What is Case 2? What case of Percentage does it correspond to? What, if the *face* and the real value are known? What, if the rate of gain or loss is required?

ART. 278. CASE III.—*Given, the discount or premium, and the rate of discount or premium, to find the face.* (See Case III, Percentage.)

NOTES.—1. After the *face* is obtained, add to it the premium, or subtract the discount, to get the real value or cost.

2. If the *gain* or *loss* is given, and the rate per cent. of the *face* corresponding to it, work by Case III, Percentage.

1. Paid 36 ct. premium for gold $\frac{3}{4}$ % above par: how much gold was there? *Ans.* $48.

2. Took stock at par; sold it for $2\frac{1}{4}$ % discount, and lost $117: how many shares ($50) had I? *Ans.* 104.

3. The discount, at $7\frac{1}{2}$ %, on stocks was $93.75: how many shares ($50) were sold? *Ans.* 25.

4. Buy stock at $4\frac{1}{2}$ % premium; sell at $8\frac{1}{4}$ % premium; gain, $345: how many shares ($100)? *Ans.* 92.

5. Buy stocks at 14 % discount; sell at $3\frac{1}{2}$ % prem.; gain, $192.50; how many shares ($50)? *Ans.* 22.

6. The premium on a draft, at $\frac{7}{8}$ %, was $10.36: what was the face? *Ans.* $1184.

7. Buy stocks at 6 % discount; sell at 42 % discount; loss, $666: how many shares ($50)? *Ans.* 37.

8. Stocks at 12 % discount, brokerage, $6.03, cost $239.97 less than the face: how many shares ($50)? *Ans.* 41.

9. Bonds, at 20 % premium, brokerage $\frac{5}{8}$ %, cost 300.87\frac{1}{2}$ more than the face: what is the face? *Ans.* $1450.

10. Buy uncurrent bank notes at 10 % discount, $2\frac{1}{2}$ % brokerage; sell them at par, and gain $348.75: what was the face of the notes? *Ans.* $4500.

11. Buy stocks at 40 % discount, brokerage $1\frac{1}{4}$ %; sell them at 20 % discount, brokerage $1\frac{1}{4}$ %, and gain 574.87\frac{1}{2}$: how many shares ($50)? *Ans.* 63.

ART. 279. CASE IV.—*Given, the real value, and the rate of premium or discount, to find the face of the drafts, stock, &c.* (See Case IV, Percentage.)

NOTES.—1. After the *face* is known, take the difference between it and the real value, to find the discount or premium.

REVIEW.—278. What is Case 3? What does it correspond to? How can the real value or cost be found? When the *gain* or *loss* is given, and the rate per cent. of the *face*, how proceed?

2. Bear in mind that the rate of premium or discount, and the rate of gain or loss, are entirely different things; the former is referred to the par value or *face*, as a standard of comparison, the latter to the real value or *cost.*

1. What is the face of a draft on Baltimore costing \$2861.45, at $1\frac{1}{2}$ % premium? *Ans.* \$2819.16

2. Invested \$1591 in stocks at 26 % discount: how many shares (\$50) did I buy? *Ans.* 43.

3. Bought a draft on New Orleans at $\frac{1}{2}$ % discount, for \$6398.30: what was its face? *Ans.* \$6430.45

4. Notes at 65 % discount, 2 % brokerage, cost \$881.79: what is their face? *Ans.* \$2470.

5. Exchanged 17 Railroad bonds (\$500), 25 % below par, for bank stock at $6\frac{1}{4}$ % premium: how many shares (\$100), did I get? *Ans.* 60.

6. How much gold, at $\frac{5}{8}$ % premium, will pay a check for \$7567? *Ans.* \$7520.

7. How much silver, at $1\frac{1}{4}$ % premium, can be bought for \$3252.96 of currency? *Ans.* \$3212.80

8. How large a draft, at $\frac{1}{4}$ % premium, is worth 54 city bonds (\$100), at 12 % discount? *Ans.* \$4740.15

9. Exchanged 72 Ohio State bonds (\$1000), at $6\frac{1}{4}$ % premium, for Indiana bonds (\$500), at 2 % premium: how many of the latter did I get? *Ans.* 150.

XX. INSURANCE.

ART. **280**. INSURANCE is of two kinds—Insurance on property and Life Insurance; the latter will be explained under the head of Annuities.

Insurance on Property is of two kinds—Fire and Marine; the former takes effect upon fixed property, as houses and their contents; the latter applies to property transported by water, as vessels and their cargoes.

REVIEW.—279. What is case 4? What does it correspond to? How is the discount or premium found? What is the distinction between the rate of premium or discount, and the rate of gain or loss? What is the standard of comparison for the former? What for the latter? 280. How many kinds of insurance? How is insurance on property divided? What is Fire insurance? Marine?

Property conveyed by railroads is insured after the manner of marine insurance.

ART. **281.** Insurance on property, then, is security against loss by fire or the dangers of transportation.

The companies or individuals that guarantee against such loss, are called *underwriters*, or *insurers*. If an individual insures, it is called *out-door* insurance.

The written contract between the insurers and the insured is called the *policy*.

The sum charged for insuring is called the *premium*, and is a certain rate per cent. of the amount insured. This rate per cent. is called the *rate of insurance*.

Insuring property is called *taking a risk*.

ART. **282.** Insurance has 4 cases solved like the 4 corresponding cases of Percentage. The amount insured is the standard of comparison; hence, this

GENERAL RULE FOR INSURANCE.

Represent the amount insured by 100 *per cent., and then proceed by such rule of Percentage as the nature of the question requires.*

CASE I.—*Given, the rate of insurance, and the amount insured, to find the premium.* (See Case I, Percentage.)

1. Insured a house for \$2500, and furniture for \$600, at $\frac{6}{10}$ %: what is the premium? *Ans.* \$18.60

2. Insured $\frac{5}{8}$ of a vessel worth \$24000, and $\frac{2}{3}$ of its cargo worth \$36000, the former at $2\frac{1}{4}$ %, the latter at $1\frac{1}{3}$ %: what is the premium? *Ans.* \$607.50

3. What is the premium on a cargo of railroad iron worth \$28000, at $1\frac{3}{4}$ %? *Ans.* \$490.

4. Insured goods invoiced at \$32760, for 3 mon., at $\frac{8}{10}$ %: what is the premium? *Ans.* \$262.08

5. My house is permanently insured for \$1800, by a deposit of 10 annual premiums, the rate per year being $\frac{3}{4}$ %:

REVIEW.—280. What is said of property conveyed by railroads, &c.? 281. What is insurance on property? Who are underwriters? What is out-door insurance? What is the policy? The premium? The rate of insurance? What is insuring property called? 282. How many cases in Insurance? What is the standard of comparison? What is the general rule? What is Case I? What does it correspond to?

how much did I deposit? and if, on terminating the insurance, I receive my deposit less 5 %, how much do I get?
Ans. $135 deposited; $128.25 received.

6. A shipment of pork costing $1275, is insured at $\frac{5}{9}$ %, the policy costing 75 ct.: what does the insurance cost? *Ans.* $7.83

7. An Insurance company having a risk of $25000 at $\frac{9}{10}$%, re-insured $10000 at $\frac{4}{5}$% with another office, and $5000 at 1% with another: how much premium did it clear above what it paid? *Ans.* $95.

ART. **283.** CASE II.—*Given, the amount insured, and premium, to find the rate of insurance.* (See Case II, Percentage.)

1. Paid $19.20 for insuring $\frac{2}{3}$ of a house worth $4800: what was the rate? *Ans.* $\frac{3}{5}$ %.

2. Paid $234, including cost of policy, $1.50, for insuring a cargo worth $18600: what was the rate?
Ans. $1\frac{1}{4}$ %.

3. Bought books in England for $2468; insured them for the voyage for $46.92, including the cost of the policy, $2.50: what was the rate? *Ans.* $1\frac{4}{5}$ %.

4. A vessel is insured for $42000; $18000 at $2\frac{1}{2}$ %, $15000 at $3\frac{4}{5}$ %, and the rest at $4\frac{2}{3}$ %: what is the rate on the whole $42000? *Ans.* $3\frac{3}{7}$ %.

5. I took a risk of $45000; re-insured at the same rate, $10000 each, in 3 offices, and $5000 in another; my share of the premium was $262.50: what was the rate? *Ans.* $2\frac{5}{8}$ %.

6. I took a risk at $1\frac{1}{2}$ %; re-insured $\frac{2}{5}$ of it at 2 %, and $\frac{1}{4}$ of it at $2\frac{1}{2}$ %: what rate of insurance do I get on what is left? *Ans.* $\frac{3}{14}$ %.

SUGGESTION.—Find what per cent. of the risk is paid for re-insurance; take this from my rate, and find what per cent. the remainder is of the partial risk taken by me.

ART. **284.** CASE III.—*Given, the premium, and rate of insurance, to find the amount insured.* (See Case III, Percentage.)

1. Paid $118 for insuring, at $\frac{4}{5}$ %: what was the amount insured? *Ans.* $14750.

REVIEW.—283. What is Case 2? What does Case 2 correspond to? 284. What is Case 3? What does it correspond to?

2. Paid $\$411.37\frac{1}{2}$ for insuring goods, at $1\frac{1}{2}$ %: what was their value? *Ans.* \$27425.

3. Paid \$42.30 for insuring $\frac{5}{8}$ of my house, at $\frac{9}{10}$ %: what is the house worth? *Ans.* \$7520.

4. Took a risk at $2\frac{1}{4}$ %; re-insured $\frac{3}{5}$ of it at $2\frac{1}{2}$ %; my share of the premium was \$197.13: how large was the risk? *Ans.* \$26284.

5. Took a risk at $1\frac{3}{5}$ %; re-insured half of it at the same rate, and $\frac{1}{3}$ of it at $1\frac{1}{2}$ %; my share of the premium was \$58.11: how large was the risk? *Ans* \$19370.

6. Took a risk at 2 %; re-insured \$10000 of it at $2\frac{1}{8}$ %, and \$8000 at $1\frac{3}{4}$ %; my share of the premium was \$207.50: what sum was insured? *Ans.* \$28000.

ART. **285.** CASE IV.—*Given, the rate of insurance, and the amount of property to be insured, to find the amount to be insured so as to cover both property and premium.*

In this case, the amount insured is *made up of the property and premium;* so that if a loss occurs, both the value of the property insured, and the premium, shall be recovered.

The value of the property is a certain rate per cent. *less* than the amount insured, since it is less than that amount by the premium, which is always a certain rate per cent. of the amount insured; hence, the amount insured is found as in Case IV of Percentage.

What sum must be insured, to cover property worth \$4800, and premium at the rate of $\frac{1}{2}$ %?

ANALYSIS.—The amount insured being the standard of comparison, is 100 %; the premium is $\frac{1}{2}$ %; the property, (\$4800), is the remaining $99\frac{1}{2}$ %: then 1 % = \$4800 ÷ $99\frac{1}{2}$ = \$48.2412+, and 100 % is 100 times this, = \$4824.12, the amount to be insured.

It generally happens in this case, that the divisor wants but little of being a unit, and the division can be performed with advantage by Case III of Contracted Division of Decimals (Art. 159). Hence, this mode of operation, when applied to this case of Insurance, is generally stated in the following Rule.

```
$4800.
   24.
     .12
--------
$4824.12
```

REVIEW.—285. What is Case 4? What does it correspond to? Why? Analyze the example. What is said of the divisor? How may the division be performed? Apply it to the example.

PRACTICAL RULE,

FOR INSURING BOTH PROPERTY AND PREMIUM.

Find the premium on the property to be insured; then, the premium on this premium: then, the premium on the 2d premium, if necessary, and so on, neglecting all figures below mills; the sum of the successive premiums will be the whole premium; to find the amount to be insured, add the property to the premium.

1. What sum must be insured to cover property worth $2600, and prem. at $\frac{7}{10}$ %? *Ans.* $2618.33

2. Insured to *cover* a shipment of pork, valued at $12368.50, at 1 %: find the amt. insured, and prem. *Ans.* $12493.43, am't ins'd; $124.93, prem.

3. Insured to *cover* my library, $1856.20, at $\frac{6}{10}$ %: what was the premium? *Ans.* $11.20

4. Insured to *cover* property to the value of $4840, at $\frac{3}{4}$ %: what was the premium? *Ans.* $36.57

XXI. TAXES.

ART. **286**. TAXES are money paid by the subjects of a government, to defray its expenses; and are either *direct* or *indirect*.

Direct taxes consist of a *property-tax*, and *poll-tax* or *capitation-tax*, and are generally collected once a year; a *property-tax* is levied on all property, but that exempt by law; it is estimated as a certain rate per cent. of the assessed value of the property: this rate per cent. is called the *rate of taxation*.

A *poll-tax* or *capitation-tax* is a fixed sum, charged on every citizen, without regard to his property.

This subject has four cases, solved like the four corresponding cases of Percentage: the taxable property is the standard of comparison. Hence, this

GENERAL RULE.

Represent the taxable property by 100 %; *then proceed by such rule of Percentage as the nature of the question requires.*

REVIEW.—What is the practical rule? 286. What are taxes? What are direct taxes? What is a property-tax? How is it estimated? What is a poll-tax? What is the general rule?

CASE I.—*Given, the taxable property, and the rate of taxation, to find the property-tax.* (See Case I, Percentage.)

NOTE.—If there is a poll-tax, the sum produced by it should be added to the property-tax, to give the whole tax.

1. The taxable property of a county is $486250, and the rate of taxation is 78 ct. on $100; that is, $\frac{78}{100}$ %: what is the tax to be raised? *Ans.* $3792.75

REM.—The rate of taxation being usually small, is expressed most conveniently as so many cents on $100, or as so many mills on $1.

2. A's property is assessed at $3800; the rate of taxation is 96 ct. on $100 ($\frac{96}{100}$ %): what is his whole tax, if he pays a poll-tax of $1? *Ans.* $37.48

3. The taxable property of Cincinnati, in 1855, was $85330880, on which was charged a tax of $\frac{32}{100}$ % for the State, $\frac{414}{1000}$ % for the county, and $\frac{746}{1000}$ % for the township and city: in all, $1\frac{48}{100}$ %; how much was raised for each and for all?

Ans. State tax, $273058.82; county, $353269.84; township and city, $636568.36; total, $1262897.02

In making out bills for taxes, a table is used, containing the units, tens, hundreds, thousands, &c., of property, with the corresponding tax opposite each.

To find the tax on any sum by the table, take out the tax on each figure of the sum, and add the results. In this table, the rate is $1\frac{1}{4}$ %, or 125 ct. on $100.

TAX TABLE.

Prop.	Tax.	Prop.	Tax.	Prop.	Tax.	Prop.	Tax.	Prop.	Tax.
$1	.0125	$10	.125	$100	$1.25	$1000	$12.50	$10000	$125
2	.025	20	.25	200	2.50	2000	25.	20000	250
3	.0375	30	.375	300	3.75	3000	37.50	30000	375
4	.05	40	.50	400	5.	4000	50.	40000	500
5	.0625	50	.625	500	6.25	5000	62.50	50000	625
6	.075	60	.75	600	7.50	6000	75.	60000	750
7	.0875	70	.875	700	8.75	7000	87.50	70000	875
8	.10	80	1.	800	10.	8000	100.	80000	1000
9	.1125	90	1.125	900	11.25	9000	112.50	90000	1125

4. What will be the tax by the table, on property assessed at $25349?

REVIEW.—What is Case 1? To what does it correspond?

SOLUTION.—The tax for 20000 is 250; for 5000 is 62.50; for 300 is 3.75; for 40 is .50; for 9 is .1125, which, added, give the tax for \$25349 = \$316.86.

5. Find the tax for \$6815.30. *Ans.* \$85.19

NOTE.—The tax for cents is not in the table, as being too insignificant; if desired, it may be found from the dollars, by moving the point to the left, 2 figures: the tax for 30 ct. = \$.00375.

6. What is the tax on property assessed at \$10424.50 and two polls, at \$1.50 each? *Ans.* \$133.31

ART. 287. CASE II.—*Given, the taxable property, and the tax, to find the rate of taxation.* (See Case II, Percentage.)

NOTE.—If there is a poll-tax, find what it produces, and subtract it from the whole tax; the remainder is the property-tax, and is used to find the rate of taxation.

1. Property, assessed at \$2604, pays \$19.53 tax: what is the rate of taxation? *Ans.* $\frac{3}{4}$ % = 75 ct. on \$100.

2. The taxable property in a town of 1742 polls, is \$6814320. A tax of \$66913.54 is proposed. If a poll-tax of \$1.25 is levied, what should be the rate of taxation? (See Note.) *Ans.* $\frac{95}{100}$ % = 95 ct. on \$100.

3. An estate of \$350000 pays a tax of \$5670: what is the rate of taxation? *Ans.* $1\frac{31}{50}$ % = \$1.62 on \$100.

4. A's tax is \$53.46; he pays for 3 polls, at \$1.50 each, and owns \$8704 taxable property: what is the rate of taxation? *Ans.* $\frac{9}{16}$ % = $56\frac{1}{4}$ ct. on \$100.

ART. 288. CASE III.—*Given, the tax, and the rate of taxation, to find the assessed value of the property.* (See Case III, Percentage.)

NOTE.—If any part of the tax arises from polls, it should be first deducted from the given tax.

1. What is the assessed value of property taxed \$66.96, at $1\frac{4}{5}$ %? *Ans.* \$3720.

2. A corporation pays \$564.42 tax, at the rate of $\frac{46}{100}$ %, or 46 ct. on \$100: find its capital. *Ans.* \$122700.

REVIEW.—If there is a poll-tax, what should be done? 287. What is Case 2? What does it correspond to? If any part of the tax is produced from polls, what should be done?

3. A is taxed $71.61 more than B; the rate is $1\frac{8}{25}$ %, or $1.32 on $100: how much is A assessed more than B? *Ans.* $5425.

4. A tax of $4000 is raised in a town containing 1024 polls, by a poll-tax of $1, and a property-tax of $\frac{24}{100}$ % (24 ct. on $100): what is the value of the taxable property in it? *Ans.* $1240000.

5. A's income is 16 % of his capital; he is taxed $2\frac{1}{2}$ % of his income, and pays $ 26.04: what is his capital? *Ans.* $6510.

ART. **289.** CASE IV.—(See Case IV of Percentage.)

1. A pays a tax of $1\frac{7}{20}$ % ($1.35 on $100) on his capital, and has left $125127.66: what was his capital, and his tax? *Ans.* Cap., $126840; tax, $1712.34

2. Sold a lot for $7599, which was cost and 2 % beside paid for tax: what was the cost? *Ans.* $7450.

XXII. DUTIES OR CUSTOMS.

ART. **290.** DUTIES or CUSTOMS, are taxes upon foreign merchandise, paid by the importer; they are of two kinds, *ad valorem* and *specific*.

Ad valorem duties are so much on the value of the goods as shown by the invoice.

Ad valorem is Latin for *on the value.* An *invoice* is a bill of the goods, showing the kinds, quantities, and cost.

Specific duties are a fixed sum for a fixed quantity, without regard to cost; as, $10 a cwt., $30 a hhd., &c.

ART. **291.** In specific duties, certain allowances are usual, called *draft, tare, leakage,* and *breakage.*

Draft is an allowance made, that the goods may hold out when retailed. It is calculated as follows:

REVIEW.—288. What is Case 3? What does it correspond to? If any part of the tax is produced from polls, what should be done? 289. What does Case 4 correspond to? 290. What are Duties or Customs? How many kinds? What are ad valorem duties? Why so called?

On each parcel weighing

112 lb., or less,	it is 1 lb.	from 336 lb. to 1120 lb.,	it is 4 lb.
from 112 lb. to 224 lb.	.. 2 lb.	from 1120 lb. to 2016 lb.,	.. 7 lb.
from 224 lb. to 336 lb.	.. 3 lb.	over 2016 lb.,	.. 9 lb.

Tare is an allowance for the weight of the box, bale, cask, or whatever contains the goods. It is generally estimated at a certain per cent. on the quantity remaining after the draft has been deducted.

Tare is sometimes estimated at a certain quantity for each cask package, etc., or a certain weight for each cwt., tun, &c.

The weight of the goods, *before* draft and tare have been allowed, is called *gross* weight; *after* they have been allowed, it is called *net* weight.

Leakage is an allowance of 2 %, on all liquors in casks paying duty by the gallon.

Breakage, usually 10 %, is allowed on ale, beer, and porter in bottles; on other liquors in bottles, it is only 5 %.

The common sized bottles are estimated at $2\frac{2}{3}$ gal. per dozen.

RULE FOR CALCULATING SPECIFIC DUTIES.

ART. **292**. *Find the net weight of the goods in that denomination for which the rate of duty is given, and multiply it by the given rate.*

NOTE.—Observe that in custom-house business, 28 lb. make 1 qr.. 112 lb. make 1 cwt., and 2240 lb. make 1 tun.

ART. **293**. In ad valorem duties, the calculation is one of Percentage, and since the invoiced value of the goods is the quantity on which the rate of duty takes effect, it is 100 per cent.

RULE FOR ALL QUESTIONS IN DUTIES AD VALOREM.

Represent the invoiced value of the goods by 100 *per cent., and proceed by such rule of Percentage as the case requires.*

REVIEW.—290. What is an invoice? What are specific duties? 291. What allowances are made in specific duties? What is draft? How is it estimated? What is tare? How is it generally estimated? What is it calculated on? How is tare sometimes estimated? What is gross weight? Net weight? What is leakage? Breakage? How is the quantity of liquor in bottles estimated? 292. What is the rule for calculating specific duties? In custom-house business, how many lb. in a qr.?

REMARK.—According to the last tariff-bill of the U. S., passed 1846, none but ad valorem duties are allowed.

CASE I.—*See Case I of Percentage.*

WHAT IS THE DUTY

1. On 45 casks of wine, of 36 gal. each, invoiced a $1.25 per gal., at 40 % ad valorem? *Ans.* $793.80

Allow for leakage.

2. On 12 boxes fancy soaps, each $98\frac{1}{2}$ lb., tare 10 %, at 5 ct. a lb.? *Ans.* $52.65

3. On 36 boxes sugar, each weighing 6 cwt. 2 qr. 18 lb., tare 16 lb. per cwt., at $2\frac{1}{2}$ ct. a lb.? *Ans.* $572.40

SUGGESTION.—Tare = 16 lb. per cwt. = $\frac{16}{112} = \frac{1}{7}$ of the lb.

4. On 460 E. E. of broadcloth, at $3.20 per ell, at 30 % ad valorem? and if I pay $41.40 freight, how much per yard should I charge to gain 20 %?
Ans. Duty, $441.60; price per yd., $4.08

5. On 50 drums figs, each weighing 57 lb., draft as usual, tare 20 lb. to the cwt., at $12 a cwt.? *Ans.* $246.43

6. On 8 hhd. of sugar, each 12 cwt. 3 qr. 14 lb., tare 12 lb. to the cwt., at $1\frac{1}{2}$ ct. a lb.? *Ans.* $153.75

7. On an invoice of carpets, worth $1859.60, at 30 % ad valorem? *Ans.* $557.88

8. On 680 boxes of raisins, invoiced at $1.25 a box, at 40 % ad valorem? *Ans.* $340.

9. On 26 chests of tea, each 118 lb., tare 10 lb. per chest, at 10 ct. a lb.? *Ans.* $275.60

10. On 42 hhd. of sugar, each 13 cwt. 3 qr., and the tare 7 lb. per cwt., at $2 a cwt.? *Ans.* $1077.89

11. On 30 bags of rice, each 7 cwt. 2 qr. 10 lb., tare 12 %, at $1.20 a cwt.? *Ans.* $239.30

12. On oil-cloth, 40 yd. long, and 3 yd. 2 ft. 8 in. wide, worth 75 ct. a sq. yd., at 30 % ad valorem? *Ans.* $35.

ART. **294.** CASE II.—*See Case II, Percentage.*

1. If goods invoiced at $3684.50 pay a duty of $1473.80, what is the rate of duty? *Ans.* 40 %.

REVIEW.—293. What is the general rule for questions in ad valorem duties? What kind of duties only are allowed in this country? What does Case 1 correspond to?

2. If laces, invoiced at $7618.75, cost, when landed, $9142.50, what is the rate of duty? *Ans.* 20 %.

3. If 40 hhd. (63 gal. each) of molasses, invoiced at $12\frac{1}{2}$ ct. a gal., pay $92.61 duty, after allowing 2 % for leakage, what is the rate of duty? *Ans.* 30 %.

ART. **295.** CASE III.—*See Case III, Percentage.*

1. Paid $806.12 duty on watches, at 35 %: what were they invoiced at, and what did they cost in store?

Ans. Invoiced, $2303.20; cost, $3109.32

2. The duty on 1800 yd. of silk was $337.50, at 25 % ad valorem: what was the invoice price per yd.? and what must I charge per yd. to clear 20 %?

Ans. Invoiced at 75 ct. per yd.; selling price $$1.12\frac{1}{2}$

3. The duty on 15 gross London porter, allowing 10 % breakage, was $40.50, at 20 % ad valorem: how much a dozen were they invoiced at? *Ans.* $1.25

ART. **296.** CASE IV.—*See Case IV, Percentage.*

1. French cloths, after paying 30 % duty, and other charges, $73.80, cost in store $7389.03: what were they invoiced at? *Ans.* $5627.10

2. Imported from Havre 80 baskets of champagne, 12 bottles each, 5 % breakage, duty 40 %, freight and other charges $67.20, and the whole cost $729.60: what did it cost a bottle at Havre, what in store, and how much a bottle should I charge to clear 35 %?

Ans. 50 ct. at Havre; 80 ct. in store; selling price $1.08

3. The cost in store of 20 puncheons Jamaica rum, 84 gal. each, is $631.43; duty 15 %, leakage 2 %, charges $53.34: what did it cost a gal. in Jamaica?

Ans. 30 ct.

XXIII. INTEREST.

ART. **297.** INTEREST is money charged for the use of money.

The profits accruing at regular periods on permanent investments, such as dividends or rents, are called interest, since they are the growth of capital, unaided by labor.

REVIEW.—297. What is interest? What are sometimes called interest?

The *principal* is the sum on which interest is charged.

The principal is either a sum loaned; money invested to secure an income; or a debt, which not being paid when due, is allowed by agreement or law to draw interest.

The *amount* is the principal, with its interest added.

ART. **298.** Interest is payable at regular intervals *yearly*, *half-yearly*, or *quarterly*, as may be agreed: if there is no agreement, it is understood to be *yearly*.

ART. **299.** The *rate of interest* is the rate per cent. of the yearly interest to the principal: it is called *rate per cent. per annum; per annum* meaning *by the year*, in Latin.

If interest is payable half-yearly, or quarterly, the rate is still the rate *per annum*, or rate per year. In short loans, the rate per month is generally given; but the rate per year, being 12 times the rate per month, is easily found: thus, 2 % a month = 24 % a year.

If the rate of interest is not specified, the rate established by the law of the country where the contract is made, prevails, called the *legal rate*, which is generally, but not always, the highest rate allowed by the law.

If a higher rate of interest is charged than the law allows, it is called *usury*, and the person offending is subject to a penalty.

ART. **300.** The following table shows the legal rate in each of the States of the Union.

LEGAL RATES.	STATES.
8 %	in Georgia, Alabama, Mississippi and Florida.
7 %	in New York, So. Carolina, Michigan, Wiscon. and Iowa.
5 %	in Louisiana.
10 %	in Texas.
6 %	in all other places in the U. S., and the U. S. courts.

ART. **301.** Interest is either *Simple* or *Compound*.

REVIEW.—297. What is the principal? What may it sometimes be? What is the amount? 298. How is interest payable? If there is no agreement, how is it payable? 299. What is the rate of interest? For what time is it given? When is it given per month? How is the rate per year then found? If no rate of interest is mentioned, what one prevails? What is usury? 301. What two kinds of interest?

Simple Interest is interest which, even if not paid when due, is not convertible into principal, and therefore can not draw interest itself and accumulate in the hands of the debtor, however long it may be retained.

Compound Interest is interest which, not being paid when due, is convertible into principal, and from that time draws interest itself, and accumulates in the hands of the debtor, according to the time it is retained.

Compound interest is more favorable to the creditor than simple interest, as shown in this example:

If I lend $100 at the rate of 10 % per annum, what should I receive at the end of three years, presuming no interest paid in the mean time?

SOLUTION BY SIMPLE INTEREST.—The interest for 1 yr. is 10 % of $100 = $10; for 3 yr. it is $30, and the whole sum due is $130.

SOLUTION BY COMPOUND INTEREST.—The interest for the 1st year is $10; which, not being paid, makes $110 then due and drawing interest: 10 % of $110 = $11, the interest for the 2d year; which, not being paid, makes $121 then due and drawing interest: 10 % of $121 = $12.10, the interest for the 3d year; which makes the whole sum then due, $133.10, being $3.10 more than by Simple Interest.

XXIV. SIMPLE INTEREST.

ART. **302**. SIMPLE INTEREST differs from the other applications of Percentage by taking the time into consideration, which they do not.

All questions in Percentage involve three different quantities, and may be solved by a simple proportion. All questions in interest contain *four* different quantities, and may be solved by a *compound* proportion.

ART. **303**. The four quantities embraced in every question of interest are, 1st, the *principal;* 2d, the *interest;*

REVIEW.—301. What is simple interest? Compound interest? Which is more favorable to the creditor? Show the difference by the example. 302. How does simple interest differ from the other applications of percentage? If the time were left out of view, how could questions in simple interest be solved? How may they be solved as it is? 303. What four quantities are involved in every question of simple interest?

3d, the *rate*; 4th, the *time*: of these four, *the principal being the standard of comparison, is* 100 *per cent.*

Any three of these quantities being given, find the 4th by this

GENERAL RULE.

Represent the principal by 100 *per cent., and proceed as in questions of compound proportion.*

NOTE.—If the *amount* is given or required, it may be necessary to perform an addition or subtraction before applying the rule, or after the result has been obtained.

Though the general rule is sufficient for all the ordinary problems in Simple Interest, it is *too* general for practical purposes, and a special rule will be given for each case.

CASE I.

ART. **304**. Given, the principal, rate of interest, and time, to find the interest.

RULE.—*Find the yearly interest by Case I of Percentage; then ascertain from it the interest for the given time by aliquot parts.*

NOTE.—In calculations of interest, every month is regarded as 30 days, and every year as 12 months, or 360 days.

What is the simple interest of $354.80 for 3 yr. 7 mon. 19 da. at 6 % ?

SOL. — The yearly interest, being 6 % of the principal, is found, by Case I of Percentage; the interest for 3 yr. 7 mon. 19 da. is then obtained by aliquot parts; each item of interest is carried no lower than mills, the next figure being neglected if less than 5; but if 5 or over it is counted 1 mill.

		$354.80
		6
1 yr.	=	21.2880
3 yr.	= 3	63.864
6 mon.	= $\frac{1}{2}$	10.644
1 mon.	= $\frac{1}{6}$	1.774
18 da.	= $\frac{1}{10}$	1.064
1 da.	= $\frac{1}{30}$	.059
		$77.41 *Ans.*

REVIEW.—303. How many must be known? What is the general rule for questions of simple interest? If the *amount* is given or required, what may be necessary? 304. What is Case 1? The rule? How many days are counted a month in interest? How many a year? Solve the example. What custom prevails in computing interest? Why?

If the rate of interest were 5 %, $7\frac{1}{2}$ %, 10 %, &c. the process would be similar; but, as the prevailing rate is either 6 %, or a simple aliquot part greater or less than 6 %, *it is customary, whatever be the rate, to compute the interest, first at 6 %; then increase or diminish the result by such a part of itself, as may be necessary to obtain the interest at the rate given.*

A SHORT METHOD.

ART. **305.** As 6 % a year is the same as 1 % for 2 months, take 1 % of the principal, by pointing off the two right hand figures of dollars; the result is the interest for 2 mon., or 60 da.; and the interest for the given time can be found by aliquot parts.

Applying this method to the last example, the operation will appear thus:

$ 3	54.80
70	96
1	774
1	064
	059
$77	41

SOLUTION.—Write the principal, $354.80; cut off the 2 right hand figures of dollars by a vertical line; $3.548 is the interest for 2 mon.: $70.96 is the int. for 40 mon., being 20 times the int. for 2 mon. just above; $1.774 is the int. for 1 mon., being $\frac{1}{2}$ the top number; $1.064 is the int. for 18 da., bein $\frac{3}{10}$ of the top number; (see Rem., Art. 234), and the $.059 is the int. for 1 day, being $\frac{1}{30}$ of $1.774, the int. for 1 mon. The sum of these is the int. for 43 mon. 19 da. = 3 yr. 7 mon. 19 da.

ART. **306.** After finding the Int. at 6%, observe, that the Interest at 5 % = interest at 6 %, — $\frac{1}{6}$ of itself. And the Interest

at $4\frac{1}{2}$ % = Int. at 6 % — $\frac{1}{4}$ of it.	at 1 % = $\frac{1}{6}$ Int. at 6 %.
at 4 % = Int. at 6 % — $\frac{1}{3}$ of it.	at 7 % = Int. at 6 % + $\frac{1}{6}$ of it.
at 3 % = $\frac{1}{2}$ Int. at 6 %.	at $7\frac{1}{2}$ % = " 6 % + $\frac{1}{4}$ of it.
at 2 % = $\frac{1}{3}$ Int. at 6 %.	at 8 % = " 6 % + $\frac{1}{3}$ of it.
at $1\frac{1}{2}$ % = $\frac{1}{4}$ Int. at 6 %.	at 9 % = " 6 % + $\frac{1}{2}$ of it.

at 12 %, 18 %, 24 % = 2, 3, 4 times Interest at 6 %.

at 5 %, 10 %, 15 %, 20 % = $\frac{1}{12}$, $\frac{1}{6}$, $\frac{1}{4}$, $\frac{1}{3}$ Interest at 6 %, after moving the point one figure to the right.

ART. **307.** ANOTHER METHOD. Take the example already solved.

REVIEW.—305. Explain the short method. 306. After the interest at 6 % has been found, how is the interest at 5 % obtained? At $4\frac{1}{2}$ %? 4 %, &c. 7 %? $7\frac{1}{2}$ %, &c. 307. Explain "another" method.

ANALYSIS.—The interest of any sum ($354.80) at 6 % equals the interest of half that sum ($177.40) at 12 %. But 12 % a year equals 1 % a mon., and for 3 yr. 7 mon. = 43 mon., the rate is 43 %, and for 19 da., which is $\frac{19}{30}$ of a mon., the rate is $\frac{19}{30}$ %; hence, the rate for the whole time is $43\frac{19}{30}$%, and $43\frac{19}{30}$ % of the principal will be the interest. To get $43\frac{19}{30}$%, multiply by $43\frac{19}{30}$ hundredths = $.43\frac{19}{30}$ = $.436\frac{1}{3}$ (Art. 147).

By using the contracted multiplication (explained in Art. 153), the work may be shortened, and the answer obtained correctly to cents.

```
 $354.80
 $177.40
    .436⅓
 70960
  638640
    5913
$77.40553
$77.41 Ans.

$17.74(0 | .436⅓
 7096
  639
    6
$77.41
```

In the multiplier $.436\frac{1}{3}$, the hundredths (43) are the number of months, and the thousandths ($6\frac{1}{3}$) are $\frac{1}{3}$ of the days (19 da.), in the given time, 3 yr. 7 mon. 19 da.

TWO PRACTICAL RULES.

RULE 1.—*Express the years, if any, in months, and write the whole number of months as decimal hundredths; after which, place $\frac{1}{3}$ of the days, if any, as thousandths: multiply half the principal by this number.*

RULE 2.—*Or, point off from the principal two more decimals than it already has. This gives the interest for 2 months, or 60 days, from which the interest for the given time can be found by aliquot parts.*

Either of these rules gives the interest at 6 %, from which the interest at any other rate can be found by aliquot parts.

NOTE.—In the 1st rule, when the number of days is 1 or 2, place a cipher to the right of the months, and write the $\frac{1}{3}$ or $\frac{2}{3}$: otherwise, they will not stand in the thousandths' place: thus, if the time is 1 yr. 4 mon. 1 da., the multiplier is $.160\frac{1}{3}$; for 9 mon. 2 da., $.090\frac{2}{3}$.

REVIEW.—307. How can the work be shortened? What are the hundredths of the multiplier equal to? The thousandths? What is the first rule for computing interest? The 2d rule? In using the 1st rule, when the days are 1 or 2, what is necessary? When the fraction in the multiplier is $\frac{2}{3}$, how do we proceed? Which rule is generally most convenient for short times? In taking aliquot parts for the days in the 2d rule, what should we make use of?

If the fraction in the multiplier is $\frac{2}{3}$, it is better to take $\frac{1}{3}$ of the *whole* principal, than $\frac{2}{3}$ of *half* the principal.

REMARK.—The 2d rule will generally be found most convenient for short times, and will not require more than *two* aliquot parts for the days, by using the *tenths*, as well as the *halves*, *fourths*, &c. (See Art. 234, Rem.)

ART. **308**. In New York and some other parts of the U. S., and in Great Britain, the interest for days is calculated at 365 days to the year. In those places, 1 day's interest = $\frac{1}{365}$ of a year's interest = $\frac{1}{365}$ of 360 da. interest obtained by the rule = $\frac{360}{365}$ or $\frac{72}{73}$ of a day's interest, as commonly obtained; but $\frac{72}{73}$ is $\frac{1}{73}$ less than the whole:

Hence, *the interest for any number of days, counting* 365 *da. to a year, is* $\frac{1}{73}$ *less than the interest for the same number of days, counting* 360 *da. to a year.*

FIND THE SIMPLE INTEREST OF ANS.

1. $178.63 for 2 yr. 5 mon. 26 da., at 7 %. = $31.12
2. $6084.25 for 1 yr. 3 mon., at 4½ %. = $342.24
3. $64.30 for 1 yr. 10 mon. 14 da., at 9 %. = $10.83
4. $1052.80 for 28 da., at 10 %. = $8.19
5. $419.10 for 8 mon. 16 da., at 6 %. = $17.88
6. $1461.85 for 6 yr. 7 mon. 4 da., at 10 %. = $964.01
7. $2601.50 for 72 da., at 7½ %. = $39.02

REMARK.—72 da. = 2 mon. 12 da.

8. $8722.43 for 5½ yr., at 6 %. = $2878.40
9. $326.50 for 1 mon. 8 da., at 8 %. = $2.76
10. $1106.70 for 4 yr. 1 mon. 1 da., at 6 %. = $271.33
11. $10000 for 1 da., at 6 %. = $1.67
12. $4642.68 for 5 mon. 17 da., at 15 %. = $323.05
13. $13024 for 9 mon. 13 da., at 10 %. = $1023.83
14. $615.38 for 4 yr. 11 mon. 6 da., at 20 %. = $607.17
15. $2066.19 for 3 yr. 6 mon. 2 da., at 30 %. = $2172.94
16 $92.55 for 3 mon. 22 da., at 5 %. = $1.44

REVIEW.—307. Into how many aliquot parts can the days always be divided? 308. What is said of interest for days in New York, Great Britain, &c.? How is interest for any number of days obtained in those places? Why?

Find the simple interest of

ANS.

17. \$1532.45 for 9 yr. 2 mon. 7 da., at 12 %. = \$1689.27
18. \$78084.50 for 2 yr. 4 mon. 29 da., at 18 %. = \$33927.72
19. \$512.60 for 8 mon. 18 da., at 7 %. = \$25.72
20. \$3278.12 for 1 yr. 6 mon. 3 da., at 4 %. = \$197.78
21. \$8408.46 for 11 mon. 5 da., at 3 %. = \$234.74
22. \$126.75 for 2 yr. 24 da., at 8 %. = \$20.96
23. \$1363.20 for 39 da., at 1¼ % a month. = \$22.15
24. \$402.50 for 100 da., at 2 % a month. = \$26.83
25. \$6919.32 for 7 yr. 6 mon., at 6 %. = \$3113.69
26. \$990.73 for 9 mon. 19 da., at 7 %. = \$55.67

Find the amount of

27. \$757.35 for 117 da., at 1½ % a mon. = \$801.65
28. \$1883 for 1 yr. 4 mon. 21 da., at 6 %. = \$2040.23
29. \$5000 for 10 yr. 10 mon. 20 da., at 9 %. = \$9900.
30. \$4212.45 for 5 yr. 5 mon. 25 da., at 5 %. = \$5367.95
31. \$262.70 for 53 da. at 1 % a month. = \$267.34
32. \$584.48 for 133 da., at 7½ %. = \$600.67
33. \$8291.56 for 294 da., at 6 %. = \$8697.85
34. \$16372.05 for 3 yr. 9 mon., at 8 %. = \$21283.67
35. \$2001.25 for 86 da., at 6 %. = \$2029.93
36. \$392.28 for 71 da., at 2½ % a mon. = \$415.49
37. \$3032.90 for 7 mon. 7 da., at 7 %. = \$3160.87

38\. Find the interest of \$7302.85 for 365 da., at 6 %, counting 360 da. to a year. *Ans.* \$444.26

39\. Of \$10000 for 360 da., at 6 %, counting 365 days to a year. (Art. 308.) *Ans.* \$591.78

40\. Same as last, 360 da. to a yr. *Ans.* \$600.

41\. If I borrow \$1000000 in New York, at 7 %, and lend it at 7 % in Ohio, what do I gain in 180 da.? *Ans.* \$479.45

42\. Find the interest of \$5064.30 for 7 mon. 12 da., at 7 %, in New York. (Art. 308). *Ans.* \$218.45

SUGGESTION.—Find the interest for years and months in New York as elsewhere; but for days, find the interest as usual, and diminish it by $\frac{1}{73}$ of itself, before adding it in.

Find the simple interest of

43. \$681.75 for 98 da., 7 % in N. Y. *Ans.* \$12.81

44. \$1353.10 for 2 yr. 8 mon. 29 da., at 7 % in New York. *Ans.* \$260.10

45. \$6786.24 for 1 yr. 10 mon. 16 da., at 10 % in New York. *Ans.* \$1273.89

46. If I borrow \$12500 at 6 %, and lend it at 10 %, what do I gain in 3 yr. 4 mon. 4 da.? *Ans.* \$1672.22

47. If I borrow \$23275 at 12 %, and invest it at 7 %, what do I lose in 2 yr. 1 mon. 23 da.? *Ans.* \$2498.83

48. What is the interest of \$3416.20, at 6 %, from Feb. 3, 1847, to Aug. 9, 1851? *Ans.* \$925.79

REMARK.—Find the time by Art. 325, *Rem.* 3.

49. If \$4603.15 is loaned July 17, 1853, at 7 % what is due March 8, 1855? *Ans.* \$5130.34

50. Find the interest at 8 % of \$13682.45, borrowed from a minor 13 yr. 2 mon. 10 da. old, and retained till he is of age (21 yr). *Ans.* \$8543.93

51. What is the amount of \$5772 from Oct. 26, 1850, to April 12, 1855, at 10%? *Ans.* \$8348.56

52. In one year, a broker loans \$876459.50 for 63 da., at $1\frac{1}{2}$ % a mon., and pays 6 % on \$106525.20 deposits: what is his gain? *Ans.* \$21216.96

53. What is a broker's gain in 1 yr., on \$100 deposited at 6 %, and loaned 11 times for 33 da. at 2 % a mon.? *Ans.* \$18.20

54. Find the simple interest of £493 16s. 8d. for 1 yr. 8 mon. at 6 %. *Ans.* £49 7s. 8d.

SUGGESTION.—Find the interest of sterling money by Rule in Art. 304, multiplying and dividing as in compound numbers; or, reduce the principal to one denomination, as pounds, and apply a Rule in Art. 307.

55. Find the simple interest of £24 18s. 9d. for 10 mon. at 6 %. *Ans.* £1 4s. $11\frac{1}{4}$d.

56. Of £25 for 1 yr. 9 mon. at 5 %. *Ans.* £2 3s. 9d.

57. Of £651 for 7 mon. at $4\frac{1}{2}$ %. *Ans.* £17 1s. $9\frac{1}{4}$d.

58. Of £648 15s. 6d. from June 2 to November 25, at 5 %. (Art. 308.) *Ans.* £15 12s. 10d.

59. Of £14 from March 23 to Nov. 2, at 6 %. *Ans.* 10s. $3\frac{3}{4}$d.

60. Find the simple interest of £66 8s. from May 6 to Aug. 21, at $5\frac{1}{2}$ %. *Ans.* £1 1s. 5d.

61. Of £98 for 3 yr. 122 da. at 6 %. *Ans.* £19 12s. $1\frac{1}{4}$d.

62. Of £374 5s. from April 1 to Dec. 29, at 4 %. *Ans.* £11 3s. $1\frac{1}{2}$d.

CASE II.

ART. **309.** Given, the principal, interest, and time, to find the rate.

RULE.—*Assume* 1 *per cent. for the rate; determine the interest on this supposition, and divide the given interest by it.*

PROOF.—Calculate the interest at the rate thus found; if it agrees with the given interest, the work is right.

NOTE.—If the amount is given instead of the interest, the principal may be subtracted from it, to obtain the interest.

If I loan $6875 at simple interest, and in 2 yr. 3 mon. 18 da. receive $7823.75, what is the rate?

SOLUTION.—Assume 1 % for the rate; the interest of $6875 for 2 yr. 3 mon. 18 da., at 1 %, is $158.125, (Case I); the given interest = $7823.75 — $6875 = $948.75, and since this contains $158.125 *six* times, it must have accumulated at a rate *six* times as great; that is, 6 %.

1. What is the rate of interest, when I pay $119.70 for the use of $3325 for 10 mon. 24 da.? *Ans.* 4 %.

2. If $65.47 be paid for a loan of $844.75 for 93 days, what is the rate per month? *Ans.* $2\frac{1}{2}$ %.

SUG.—Find the rate per year by the rule, and divide it by 12.

3. If I borrow $5000 for 7 yr. 6 mon. 28 da., and return $10000, what is the rate? *Ans.* $13\frac{67}{341}$ %.

4. At what rate per annum will any sum of money double itself by simple interest, in 5, 6, 7, 8, 9, 10, 12, 15, 20, 25 years, respectively?

Ans. 20, $16\frac{2}{3}$, $14\frac{2}{7}$, $12\frac{1}{2}$, $11\frac{1}{9}$, 10, $8\frac{1}{3}$, $6\frac{2}{3}$ 5, 4 %.

SUGGESTION.—Take $100 for the principal, and $100 for the interest. *If* 100 *be divided by the time in years, the quotient will be the rate at which any sum will double itself at simple interest in that time.*

REVIEW.—309. What is Case 2? The rule? The proof? If the amount is given instead of the interest, what is necessary? Solve the example.

5. At what rate per annum will any sum treble itself at simple interest, in 5, 10, 15, 20, 25, 30 years, respectively? *Ans.* 40, 20, $13\frac{1}{3}$, 10, 8, $6\frac{2}{3}$ %.

6. At what rate per annum will any sum quadruple itself at simple interest, in 6, 12, 18, 24, 30 years, respectively? *Ans.* 50, 25, $16\frac{2}{3}$, $12\frac{1}{2}$, 10 %.

7. What is the rate of interest, when $35000 yields an income of $175 a month? *Ans.* 6 %.

8. When $29200 produces $6.40 a day? *Ans.* 8 %.

9. When $12624.80 draws $315.62 interest quarterly? *Ans.* 10 %.

10. When stock bought at 40 % discount, yields a semi-annual dividend of 5 %? *Ans.* $16\frac{2}{3}$ % per annum.

11. A house that cost $8250, rents for $750 a year; the insurance is $\frac{6}{10}$ %, and the repairs $\frac{1}{2}$ %, every year: what rate of interest does it pay? *Ans.* 8 %—.

CASE III.

ART. **310.** Given, the interest, rate, and time, to find the principal.

RULE.—*Assume* $1 *for the principal; determine the interest on this supposition, and divide the given interest by it.*

PROOF.—Calculate the interest on the principal found; if it agrees with the given interest, the work is right.

NOTES.—1. After the principal has been found, the interest may be added to it, and the amount thus obtained.

2. The contracted method of division in Art. 158 may be generally used.

What sum will yield $228.80 interest in 4 mon. 23 da., at 15 % per annum?

SOLUTION.—Assume $1 for the principal; the interest of it for 4 mon. 23 da., at 15 % is $\$.059\frac{7}{12}$ (Case I): as the given interest, $228.80, contains this 3840 times, the principal producing it must be 3840 times as large, that is, $3840.

WHAT PRINCIPAL WILL PRODUCE

1. $1500 a year, at 6 %? *Ans.* $25000.

2. $1830 in 2 yr. 6 mon., at 5 %? *Ans.* $14640.

REVIEW.—310. What is Case 3? The rule? The proof? How can the amount be afterward found?

What principal will produce

3. $45 a mon., at 9 %? *Ans.* $6000.
4. $17 in 68 da., at 1 % a mon.? *Ans.* $750.
5. $86.15 in 9 mon. 11 da., at 10 %? *Ans.* $1103.70
6. $313.24 in 112 da. at 7 %? *Ans.* $14383.47
7. $146.05 in 7 mon. 14 da. at 6 %? *Ans.* $3912.05
8. $79.12 in 5 mon. 25 da. at 7 % in N. Y.? *Ans.* $2329.72

CASE IV.

ART. **311.** Given, the amount, rate, and time, to find the principal.

RULE.—*Assume* $1 *for the principal; determine the amount on that supposition, and divide the given amount by it.*

PROOF.—Calculate the amount on the principal found; if the same as the given amount, the work is right.

NOTES.—1. After finding the principal, subtract it from the amount, to get the interest.

2. The contracted method of division (Art. 158) may be used.

What sum, drawing simple interest at 5 %, will amount to $819.45 in 1 yr. 8 mon. 5 da.?

SOLUTION.—Assume $1 for the principal; the amount of $1 for 1 yr. 8 mon. 5 da., at 5 %, is $\$1.084\frac{1}{36}$ (Case I); as the given amount $819.45, contains this 755.93 times, it must have arisen from a principal 755.93 times as large; that is, $755.93.

1. What principal in 2 yr. 3 mon. 12 da., at 6 %, will amount to $1367.84? *Ans.* $1203.03

2. What principal in 10 mon. 26 da., will amount to $2718.96, at 10 % interest? *Ans.* $2493.19

3. What principal, at $4\frac{1}{2}$%, will amount to $4613.36 in 3 yr. 1 mon. 7 da.? *Ans.* $4048.14

4. What principal, at 7%, will amount to $562.07 in 79 da. (365 da. to a year)? *Ans.* $553.68

PRESENT WORTH

ART. **312.** Is an important application of Case IV.

REVIEW.—310. What method of division may be generally used? Solve the example. 311. What is Case 4? The rule? The proof? How is the interest afterward obtained? What method of division may be used? Solve the example.

Present Worth is a phrase used in speaking of a debt before it is due, and is the *sum which at the prevailing rate of interest, will amount to that debt when it is due.*

The difference between the *present worth* of a debt and the *debt itself*, is the interest of the *present worth* from the present time until the time the debt is due; it is called the *discount*, since it is the sum which must be deducted from the *debt* or *nominal value*, to give the *present worth* or *real value.*

REMARK.—The *discount* in this case, like that in Art. 275, is *an abatement or deduction from the apparent value;* in fact, *the difference between the real and the nominal value:* but the latter is a certain per cent. of the *nominal value*, while this is a certain per cent., not of the *nominal value* (the debt), but of the *real value* (the Present Worth).

1. Find the present worth and discount of \$5101.75 due in 1 yr. 9 mon. 19 da., rate of interest, 6 %.
Ans. Pres. wor., \$4603.775; Dis., \$497.975

2. Also, of \$1476.81, due in 4 mon. 11 da., rate, 6 %.
Ans. Pres. wor., \$1445.26; Dis., \$31.55

3. Find the present worth of \$2906.30, due in 103 days, rate, 8 %. *Ans.* \$2841.27

4. Find the discount of \$6344.25, due in 23 days, rate of interest, 5 %. *Ans.* \$20.20

5. Find the present worth of \$12720.40, due in 9 da., at 7 % in New York. *Ans.* \$12698.48

6. I can sell property for \$7500 cash, or for \$4250 payable in 6 mon. and \$4000 payable in 1 yr.: which should I prefer? and what do I gain by it if money is worth 12 % to me? *Ans.* The latter; \$80.86

CASE V.

ART. **313.** Given, the principal, rate, and interest, to find the time.

RULE.—*Assume* 1 *year for the time; determine the interest on this supposition, and divide the given interest by it.*

PROOF.—Calculate the interest for the time thus found; if it is the same as the given interest, the work is right.

REVIEW.—312. Why is this case important? What is present worth? What is the difference between a debt not due and its present worth? What is it called? Why? 313. What is Case 5? The rule? The proof?

NOTES.—1. The quotient will be the time in years; if it contain a fraction, reduce it to months and days by Art. 222.

2. If the amount is given, subtract the principal from it, to get the interest; then, apply the rule.

In what time will \$830 amount to \$1000, at 6 % simple interest?

SOLUTION.—Assume 1 yr. for the time; the interest of \$830 for 1 yr. at 6 % is \$49.80 (Case I): the given interest = \$1000 — \$830 = \$170, and as this contains \$49.80, $3\frac{103}{249}$ times, it must have taken $3\frac{103}{249}$ times as long to accumulate, that is, $3\frac{103}{249}$ yr. = 3 yr. 4 mon. 29 da., by reduction. (Art. 222.)

IN WHAT TIME WILL

1. \$1200 amount to \$1800, at 10 %? *Ans.* 5 yr.
2. \$415.50 to \$470.90, at 10 %? *Ans.* 1 yr. 4 mon.
3. \$3703.92 to \$4122.15, at 8 %? *Ans.* 1 yr. 4 mon. 28 da.

NOTE.—A part of a day is omitted in the answer, not being recognized in interest; it must be taken account of, however, in the proof.

4. In what time will any sum, as \$100, double itself by simple interest at $4\frac{1}{2}$, 5, 6, 7, $7\frac{1}{2}$, 8, 9, 10, 12, $12\frac{1}{2}$, 15, 18, 20, 25, 30 %?

Ans. $22\frac{2}{9}$, 20, $16\frac{2}{3}$, $14\frac{2}{7}$, $13\frac{1}{3}$, $12\frac{1}{2}$, $11\frac{1}{9}$, 10, $8\frac{1}{3}$, 8, $6\frac{2}{3}$, $5\frac{5}{9}$, 5, 4, $3\frac{1}{3}$ yr. 100 *divided by the rate of interest, will give the number of years in which any sum will double itself.*

5. In what time will any sum treble itself by simple interest at 4, $4\frac{1}{2}$, 5, 6, 7, $7\frac{1}{2}$, 8, 9, 10, 12 %?

Ans. 50, $44\frac{4}{9}$, 40, $33\frac{1}{3}$, $28\frac{4}{7}$, $26\frac{2}{3}$, 25, $22\frac{2}{9}$, 20, $16\frac{2}{3}$ yr.

6. In what time will any sum quadruple itself by simple Int., at 5, 6, 7, 8, 10, 12, 15, 20, 30, 40, 50, 100 %?

Ans. 60, 50, $42\frac{6}{7}$, $37\frac{1}{2}$, 30, 25, 20, 15, 10, $7\frac{1}{2}$, 6, 3 yr.

7. How long must I keep on deposit \$1374.50, at 10 %, to pay a debt of \$1480.78? *Ans.* 9 mon. 8 da.

8. How long will it take \$3642.08 to amount to \$4007.54, at 12 %? *Ans.* 10 mon. 1 da.

9. How long would it take \$175.12 to produce \$6.43 interest at 6 %? *Ans.* 7 mon. 10 da.

REVIEW.—313. What will the quotient be? If it contains a fraction, what is necessary? What, if the amount is given? Solve the example. What is said of a part of a day?

10. How long would it take $415.38 to produce $10.69 interest in New York at 7 %? *Ans.* 134 da.

REMARK.—In this example, multiply the fraction of the year by 365, instead of 12 and 30. (Art. 308.)

ART. **314**. All the Cases of Simple Interest, except the 4th, which treats of Present Worth, can be solved by Compound Proportion, after the following form:

$100.	Principal.			
:		: :	Rate.	: Interest.
1 yr.	Time in yr.			

ART. **315**. The last 4 cases show a striking similarity in their rules and modes of operation, and can be put into a

GENERAL RULE FOR THE LAST FOUR CASES.

Assume 1 *for the quantity required; determine the interest* (*or amount if it be necessary*), *on this supposition, and divide the given interest* (*or amount*) *by it.*

XXV. BANKING.

ART. **316**. The most important application of Simple Interest is *Banking*, including the Discounting of Notes, Exchange, and the settlement of depositors' accounts.

BANKS are corporations which deal in money.

A bank of *issue* is one which issues its own notes as money. A bank of *discount* is one which makes loans. A bank of *deposit* is one which takes charge of money belonging to others.

REMARK.—A bank is generally controlled by a board of directors, elected by the stockholders; the principal officers are the president and cashier.

REVIEW.—314. By what compound proportion may all the cases of interest, except the 4th, be solved? Why can not the 4th be solved thus? 315. What general rule serves for the last 4 cases? In which case only will it be necessary to use the amount? Why? 316. What is the most important application of simple interest? What does it include? What are banks? What kinds? Define each. How is a bank generally controlled? What are the principal officers?

PROMISSORY NOTES.

ART. 317. A PROMISSORY NOTE is a written promise by one party to pay a named sum to another.

The person by whom the note is signed is the *maker* of the note. The person to whom the money is promised is the *payee*. The owner of a note is the *holder*.

A promissory note is *negotiable*, when it is payable to *bearer*, or to the *order* of the payee; otherwise, it is not negotiable.

A negotiable note may pass from hand to hand; the payee or holder *endorsing* it by writing his name on its back, thus becoming liable for its payment, if the note is payable "to order."

If the note is payable "to bearer," no indorsement is required on transferring it, and only the maker is responsible.

It is essential to a valid promissory note, that it contain the words "value received," and that the sum of money to be paid should be written in words.

The *face* of a note is the sum promised to be paid.

If a note contain the words "with interest," it draws interest from date, and if no rate is mentioned, the legal rate prevails.

The face of such notes is the sum mentioned with its interest from date to the day of payment.

If a note does not contain the words "with interest," and is not paid when due, it draws interest from the day of maturity, at the legal rate, till paid.

REMARK.—The *day* of maturity is when the note is legally due.

If a note is not paid at maturity, the indorsers must be notified of the fact, in writing, when it falls due, or they will not be liable. This notification is generally made by a Notary, and is called a *Protest*.

REVIEW.—317. What is a promissory note? Who is the maker of a note? The payee? When is a promissory note negotiable? What may be done with a negotiable note? If it is payable to order, what is necessary? What is indorsing? What effect does it have? What kind of a note needs no indorsement when transferred? What are essential to a valid promissory note? What is the face of a note? What is the face, when the note contains the words "with interest?"

ART. 318. If a note is payable "on demand," it is legally due when presented.

Bank notes are of this sort, being payable "to bearer on demand."

If a day of payment is specified in the note, it is due the 3d day afterward; in some places, it is due on the day specified.

If a note is payable a certain time "after date," proceed thus,

TO FIND THE DAY A NOTE IS LEGALLY DUE.

RULE.—*Add to the date of the note, the number of years and months to elapse before payment; if this gives the day of a month higher than that month contains, take the last day in that month; then, count the number of days mentioned in the note and* 3 *more: this will give the day the note is legally due; but if it is a Sunday or a national holiday, it must be paid the day previous.*

NOTES.—1. When counting in the days, do not reckon the one *from* which the counting begins. The three additional days are called "days of grace"; in some countries they are 4, 5, or more. The day before "grace" begins, the note is *nominally* due; it is *legally* due on the last day of grace.

2. The months mentioned in a note are calendar months. Hence, a 3 mon. note would run longer at one time than at another; one dated Jan. 1st, will run 93 days, one dated Oct. 1st, will run 95 days: to avoid this irregularity, the time of short notes is generally given in days instead of months; as, 30, 60, 90 days, instead of 1, 2, 3 months.

DISCOUNTING NOTES.

ART. 319. DISCOUNTING NOTES is buying them *at a discount*, and is chiefly done by banks and brokers.

Notes to be discounted must be payable to order, and indorsed by the payee.

The *proceeds* or *cost* of a note is the sum paid for it.

The difference between the cost and the face of the note is the *discount;* it is more or less, according to the time the note has to run.

REVIEW.—317. If a note does not contain these words, and is not paid when due, what is the consequence? What is the day of maturity? 318. If a note is payable "on demand," when is it legally due?

The *time to run* is the number of days *from* the day the note is discounted, to the day the note is legally due, counting the latter, but not the former.

REMARK.—The time to run is taken in days by banks, because it is to their advantage; if it were taken in months and days, each month would have to be considered 30 days in calculating the discount, whereby a day would be lost in each month of 31 days.

ART. 320. In determining the *day of maturity* and *the time to run*, it is convenient to use this

TABLE

SHOWING THE NUMBER OF DAYS FROM ANY DAY OF

Jan.	Feb.	Mar.	Apr.	May.	June.	July.	Aug.	Sept.	Oct.	Nov.	Dec.	To the same day of next
365	334	306	275	245	214	184	153	122	92	61	31	Jan.
31	365	337	306	276	245	215	184	153	123	92	62	Feb.
59	28	365	334	304	273	243	212	181	151	120	90	Mar.
90	59	31	365	335	304	274	243	212	182	151	121	April.
120	89	61	30	365	334	304	273	242	212	181	151	May.
151	120	92	61	31	365	335	304	273	243	212	182	June.
181	150	122	91	61	30	365	334	303	273	242	212	July.
212	181	153	122	92	61	31	365	334	304	273	243	Aug.
243	212	184	153	123	92	62	31	365	335	304	274	Sept.
373	242	214	183	153	122	92	61	30	365	334	304	Oct.
304	273	245	214	184	153	123	92	61	31	365	335	Nov.
334	303	275	244	214	183	153	122	91	61	30	365	Dec.

REMARK.—In leap years, if the last day of February is included in the time, 1 day must be added to the number obtained from the table.

A note maturing Sept. 13, is discounted June 24 previous: what is the time to run?

SOLUTION.—From June 24 to Sept. 24, by the table, is 92 da.; the 13th being 11 days before the 24th, will give 92 — 11, = 81 da. *Ans.* for the time to run.

REVIEW.—318. If it is payable a certain time after date, what is the rule for finding when it is legally due? Which day is not counted? What are days of grace? Why so called? When is a note nominally due? When legally due? What are the months? What is said of notes drawn for months? How is this irregularity avoided in short notes?

ART. **321.** CASE I. To find the discount and proceeds of a Note of short date, custom has sanctioned this

RULE.

Take the face of the note as a principal, the rate of discount as the rate of interest, and calculate the interest for the time the note has to run; this will be the discount of the note, and if it is subtracted from the face of the note, the remainder will be the proceeds or cost of the note.

NOTES.—1. The discount on *short* notes, resembles the discount on money, stocks, bonds, &c., being calculated on the nominal value or face, but differs from the discounts on debts and *long* notes, (Art. 312): the former is called *Bank* discount; latter, *true* discount.

2. Since the face of every note is a debt due at a future time, its cost ought to be the present worth of that debt, and the bank discount to be the same as the true discount.

As it is, the former is greater than the latter; *for, bank discount is the interest on the face of the note, while true discount is the interest on the present worth, which is always less than the face.* Hence, their difference is the interest of the difference between the Present Worth and face; that is, *the interest of the true discount.*

3. In calculating the discount on notes, use generally the 2d rule (Art. 307). The operation may often be shortened, by recollecting that when the two right hand figures of dollars are pointed off, the result is the interest of the principal, for 60 da. at 6 %, for 72 da. at 5 %, for 45 da. at 8 %, for 36 da. at 10 %, or for any other number of days and rate, whose product is 360; so that the interest at any rate, can be frequently got in this way by aliquot parts, without first getting it at 6 %.

Find the day of maturity, the time to run, and the proceeds of the following notes.

1. $\$792\frac{50}{100}$ CINCINNATI, Jan. 3d, 1854.

Six months after date, I promise to pay to the order of Willis & Markham, seven hundred ninety-two $\frac{50}{100}$ dollars, at the Commercial Bank, value received. H. WHITTAKER.

Discounted, Feb. 18, at 6 %.

Ans. Due, July $^{3}|_{6}$; 138 da. to run; Pro. \$774.27

REVIEW.—319. What is discounting notes? By whom is it done? What kind of notes are discounted? What is the proceeds or cost of a note? The discount? What is the time to run? 320. Show the use of the table. What is said of leap years? 321. What is the rule for dis counting short notes? What does bank discount resemble?

REMARK.—The date before the line, in July $^{3}|_{6}$, is when the note is *nominally* due, the other when it is *legally* due.

2. 1066\frac{75}{100}$ LOUISVILLE, May 19, 1855.
Value received, ninety days after date, I promise to pay Thomas Beatty, or order, one thousand sixty-six $\frac{75}{100}$ dollars, at the City Bank. G. W. ALEXANDER.

Discounted June 8, at 6 %.

Ans. Due, Aug. $^{17}|_{20}$; 73 da. to run; Pro. $1053.77

3. 1962\frac{45}{100}$ NEW YORK, July 26, 1850.
Value received, four months after date, I promise to pay B. Thoms, or order, one thousand nine hundred sixty-two $\frac{45}{100}$ dollars, at the Chemical Bank. E. WILLIAMS.

Discounted Aug. 26, at 7 %.

Ans. Due, Nov. $^{26}|_{29}$; 95 da. to run; Pro. $1926.70

4. 543\frac{68}{100}$ CINCINNATI, Oct. 30, 1855.
Thirty days after date, I promise to pay to the order of Baker & Goodal, five hundred forty-three $\frac{68}{100}$ dollars, value received. T. H. SHORT.

Discounted Oct. 31, at 1 % a month.

Ans. Due, Nov. 29 | Dec. 2; 32 da. to run. Pro. $537.88

5. 2672\frac{18}{100}$ PHILADELPHIA, March 10, 1852.
Nine months after date, for value received, I promise to pay Edward H. King, or order, two thousand six hundred seventy-two $\frac{18}{100}$ dollars. JEREMIAH BARTON.

Discounted July 19, at 6 %.

Ans. Due, Dec. $^{10}|_{13}$; 147 da. to run. Pro. $2606.71

6. 804\frac{39}{100}$ COLUMBUS, Aug. 12, 1854.
Three months after date, I promise to pay at the City Bank of Columbus, eight hundred four $\frac{39}{100}$ dollars to the order of Irwin & Lee, value received. JOSIAH NEEDHAM.

Discounted September 3, at 6 %.

Ans. Due, Nov. $^{12}|_{15}$; 73 da. to run. Pro. $794.60

REVIEW.—321. What kind of notes are discounted by this rule? How are long notes running for one or more years discounted? What ought the cost of a note to be? What would the bank discount then be? Which is greater, bank discount or true discount? Why? How much? What rule of interest is generally used in notes? How may the operation often be shortened?

7. $3886. ST. LOUIS, Jan. 31, 1853.

One month after date, we jointly and severally promise to pay C. McKnight, or order, three thousand eight hundred eighty-six dollars, value received. T. MONROE, I. FOSTER.

Discounted Jan. 31, at $1\frac{1}{2}$ % a month.

Ans. Due, Feb. 28 | Mar. 3; 31da. to run. Pro. $3825.77

REMARK.—A note, drawn by two or more persons, "jointly and severally," may be collected of either of the makers, but if the words "jointly and severally" are not used, it can only be collected of the makers as a firm or company.

8. $1425. NASHVILLE, April 11, 1853.

For value received, eight months after date, we promise to pay Henry Hopper, or order, one thousand four hundred twenty-five dollars, with interest from date at 6 per cent. per annum. NIXON & MARSH.

Discounted June 15, at 6 %.

Ans. Due, Dec. $^{11}|_{14}$; 182da. to run. Pro. $1437.73

N. B.—Observe that the face of this note is the amount of $1425 for 8 mon. 3 da. at 6 %.

9. 3703\frac{84}{100}$ BALTIMORE, June 6, 1850.

For value received, four months after date, we promise to pay to the order of Jones & Newcome, three thousand seven hundred and three dollars and eighty-four cents, at the Savings Bank. THOMAS SHARPE & CO.

Discounted June 18, 1850, at 1 % a month.

Ans. Due, Oct. $^{6}|_{9}$; 113da. to run. Pro. $3564.33

10. 813\frac{60}{100}$ DAYTON, May 31, 1856.

For value received, sixty days after date, I promise to pay to the order of Hiram Wells & Co., eight hundred thirteen dollars sixty cents. JAMES T. FISHER.

Discounted May 31, 1856, at 2 % a month.

Ans. Due, July 30 | Aug. 2; 63da. to run. Pro. $779.43

11. 737\frac{40}{100}$ BOSTON, Feb. 14, 1856.

Value received, two months after date, I promise to pay to J. K. Eaton, or order, seven hundred thirty-seven $\frac{4}{10}$, dollars, at the Suffolk Bank. WILLIAM ALLEN.

Discounted Feb. 23, at 10 %.

Ans. Due, April $^{14}|_{17}$; 54da. to run. Pro. $726.34

12. $\$4085\frac{20}{100}$ NEW ORLEANS, Nov. 20, 1855.
Value received, six months after date, I promise to pay John A. Westcott, or order, four thousand eighty-five $\frac{20}{100}$ dollars, at the Planters' Bank. E. WATERMAN.

Discounted Dec. 31, 1855, at 5 %.

Ans. Due May $^{20}|_{23}$, 1856; 144 da. to run. Pro. \$4003.50

13. $\$2623\frac{50}{100}$ CINCINNATI, Aug. 7, 1854.
For value received, eighteen months after date, I promise to pay to the order of Jonathan Evans, two thousand six hundred twenty-three $\frac{50}{100}$ dollars, with interest at ten per cent. per annum. MORRIS TALBOT.

Discounted June 24, 1855, at $1\frac{1}{4}$ % a month.

Ans. Due, Feb. $^{7}|_{10}$, 1856; 231 da. to run. Pro. \$2728.61

ART. **322.** It may be inquired, what rate of interest is paid, when a note is discounted. The proceeds of the note is the sum received or borrowed, and is, therefore, the *principal*, while the discount is its *interest* for the time the note runs; so that having the principal, time, and interest, the rate of interest can be found as in Art. 309.

It is simpler, however, to leave the face of the note out of view, and proceed thus:

TO FIND THE RATE OF INTEREST WHEN A NOTE IS DISCOUNTED.

RULE.—*Assume* \$100 *for the face of the note, and on this supposition determine the discount and proceeds for the time it has to run; the former will be the interest; the latter, the principal; and the time to run, the time: from which the rate of interest can be found by Case II of Simple Interest.*

What is the rate of interest, when a sixty day note is discounted at 2 % a month?

SOLUTION.—For every \$100 in the face of the note, the discount for 63 da. at 2 % a month, is \$4.20, and the proceeds \$95.80; then \$95.80 being the principal, and \$4.20 its interest for 63 da., the rate of interest is found by Art. 309 to be $25\frac{25}{479}$% per annum.

REVIEW.—322. When a note is discounted, what is the principal or sum borrowed? What its interest? How may the rate of interest then be found? What may be left out of view in this calculation?

WHAT IS THE RATE OF INTEREST,

1. When a 30 da. note is discounted at 1, $1\frac{1}{4}$, $1\frac{1}{2}$, 2 % a mon.? *Ans.* $12\frac{132}{989}$, $15\frac{55}{263}$, $18\frac{594}{1967}$, $24\frac{88}{163}$ % per annum.

2 When a 60 da. note is discounted at 6, 8, 10 % per annum? *Ans.* $6\frac{126}{1979}$, $8\frac{56}{493}$, $10\frac{70}{393}$ % per annum.

3. When a 90 da. note is discounted at 2, $2\frac{1}{2}$, 3 % a mon.? *Ans.* $25\frac{275}{469}$, $32\frac{64}{123}$, $39\frac{627}{907}$ % per annum.

4. When a note running 1 yr. is discounted at 5, 6, 7, 8, 9, 10, 12 %? *Ans.* $5\frac{5}{19}$, $6\frac{18}{47}$, $7\frac{49}{93}$, $8\frac{16}{23}$, $9\frac{81}{91}$, $11\frac{1}{9}$, $13\frac{7}{11}$ %.

REMARK.—It may seem unnecessary to regard the time the note has to run, in determining the rate of interest; but, a comparison of examples 1 and 3, shows that a 90 da. note, discounted at 2 % a month, yields a higher rate of *interest* then a 30 da. note of the same face, discounted at 2 % a month. The discount, at the same rate, on all notes of the same face, *varies* as the time to run, and if in each case, it was referred to the *same principal*, the rate of interest would be the same; but when the discount becomes *larger*, the proceeds or principal to which it is referred, becomes *smaller*, and therefore *the rate of interest corresponding to any rate of discount increases with the time the note has to run.* Hence, the profit of the discounter is greater proportionally on *long* notes than *short* ones, at the same rate.

ART. **323**. Discounting Notes is an application of Simple Interest: the *face of the note* is the *principal;* the *discount* is the *interest;* the *rate of discount* is the *rate of interest;* and the *time to run* is the *time.*

Hence, it may be divided into cases corresponding to, and solved like those of simple interest. The only case of sufficient importance to require distinct notice, is

CASE II.

Given, the proceeds, time, and rate of discount, to find the face of the note.

RULE.—*Assume* $1 *for the face of the note; determine the proceeds on this supposition, and divide the given proceeds by it.*

REVIEW.—322. Give the rule. Analyze the example. Can the time to run be left out of view in determining the rate of interest corresponding to any rate of discount? Why not? What does the rate of interest increase with? What sort of notes are most profitable to the discounter, then, provided the rates of discount are the same? 323. What is Case 2? The Rule? The proof? Solve the example.

PROOF.—Discount the note thus found; if it yields the given proceeds, the work is right.

For what sum must a 60 da. note be drawn, to yield $1000, when discounted, at 6 % per annum?

SOLUTION.—For every $1 in the face of the note, the proceeds, by Case I, (Art. 321), is $.9895; hence, there must be as many dollars in the face as this sum, $.9895, is contained times in the given proceeds, $1000: this gives $1010.61 for the face of the note.

1. Find the face of a 30 da. note, which yields $1650 when discounted at $1\frac{1}{2}$ % a mon. *Ans.* $1677.68

2. The face of a 60 da. note, which, discounted at 6 % per annum, will yield $800. *Ans.* $808.49

3. The face of a 90 da. note, which, discounted at 7 % per annum, yields $1235.40. *Ans.* $1258.15

4. The face of a 4 mon. note, which, discounted at 1 % a month, yields $3375. *Ans.* $3519.29

5. The face of a 6 mon. note, which, discounted at 10 % per annum, yields $4850. *Ans.* $5109.75

6. The face of a 60 da. note, discounted at 2 % a month, to pay $768.25. *Ans.* $801.93

7. The face of a 40 da. note, which, discounted at 8 %, yields $2072.60. *Ans.* $2092.60

8. The face of a 30 da. and 90 da. note, to net $1000 when discounted at 6 %.

Ans. $1005.53 at 30 da.; $1015.74 at 90 da.

ART. **324.** To find the rate of discount corresponding to a given rate of interest.

RULE.—*Assume $100 for the proceeds; find its interest and amount at the given rate, for the time the note runs; the latter is the face of the note, and the former is the discount, or interest on that face for the time the note runs; the rate of interest will then be found by Case II of Simple Interest, and will be the rate of discount required.*

If I wish to get interest at the rate of 20 % per annum, at what rate should I discount 60 da. notes?

SOLUTION.—For every $100 in the proceeds, I wish to get interest at 20 % per annum, which for 63 days is $3.50, interest, and

REVIEW.—324. What is the rule for finding the rate of discount corresponding to a given rate of interest? Analyze the example.

\$103.50, amount · take \$103.50 as the face of the note, or principal, and \$3.50 as the discount, or interest of that principal for 63 days, and find the rate by Case II of Simple Interest, $19\frac{67}{207}$%.

WHAT RATES OF DISCOUNT

1. On 30 da. notes, yield 10, 15, 20, 30, 40, 50 % interest? *Ans.* $9\frac{1101}{1211}$, $14\frac{646}{811}$, $19\frac{391}{611}$, $29\frac{27}{137}$, $38\frac{182}{311}$, $47\frac{203}{251}$ %.

2. On 60 da. notes, yield 6, 8, 10, 12, 18, 24 % interest?
Ans. $5\frac{1895}{2021}$, $7\frac{451}{507}$, $9\frac{337}{407}$, $11\frac{769}{1021}$, $17\frac{929}{2063}$, $23\frac{17}{521}$ %.

3. On 90 da. notes, yield 1, $1\frac{1}{2}$, 2, $2\frac{1}{2}$, 3, 4 % a mon. int.?
Ans. $11\frac{659}{1031}$, $17\frac{419}{2093}$, $22\frac{106}{177}$, $27\frac{363}{431}$, $32\frac{1024}{1093}$, $42\frac{198}{281}$ %.

4. On notes running 1 yr. without grace, yield 5, 6, 7, 8, 9, 10, 12, 15, 18, 20, 25 % interest?
Ans. $4\frac{16}{21}$, $5\frac{35}{53}$, $6\frac{58}{107}$, $7\frac{11}{27}$, $8\frac{28}{109}$, $9\frac{1}{11}$, $10\frac{5}{7}$, $13\frac{1}{23}$, $15\frac{15}{59}$, $16\frac{2}{3}$, 20 %.

PARTIAL PAYMENTS.

ART. **325**. A note, not paid at maturity, draws interest at the legal rate, from the day it is due till it is paid; but if it contains the words "with interest," it draws interest from date. Any sums paid *on account* of the note, are indorsed with their respective dates, and are called *Partial Payments.* In such cases, the balance due on settlement is found by the

U. S. RULE FOR PARTIAL PAYMENTS.

"*Apply the payment, in the first place, to the discharge of the interest then due; if the payment* EXCEEDS *the interest, the surplus goes toward discharging the principal, and the subsequent interest is to be computed on the balance of principal remaining due.*

"*If the payment be* LESS *than the interest, the surplus of interest must not be taken to augment the principal; but interest continues on the former principal, until the period when the payments taken together exceed the interest due, and then the surplus*

REVIEW.—325. If a note is not paid at maturity, what is the consequence? If it contains the words "with interest," when does interest commence? What are partial payments? What is the United States rule for casting interest on notes and bonds when partial payments have been made? On what principle is this rule founded?

is to be applied toward discharging the principal, and interest is to be computed on the balance as aforesaid."

REMARK.—1. This rule, is adopted by the Courts of the U. S. and of most of the States; it is on the principle that *neither interest nor payment shall draw interest.*

2. It is worthy of remark, that the whole *aim* and *tenor* of legislative enactments and judicial decisions on questions of interest, have been to favor the debtor, by disallowing compound interest, and yet *this very rule* fails to secure the end in view, and really maintains and enforces the principle of compound interest in a most objectionable shape; *for it makes interest due,* (not every year as compound interest ordinarily does) *but as often as a payment is made;* by which it happens that the *closer the payments are together, the greater the loss of the debtor,* who thus suffers a penalty for his very promptness.

To illustrate, suppose the note to be for $2000, drawing interest at 6 %, and the debtor pays every month $10, which just meets the interest then due; at the end of the year he would still owe $2000. But if he had invested the $10 each month, at 6 %, he would have had, at the end of the year, $123.30 available for payment, while the debt would have increased only $120, being a difference of $3.30 in his favor, and leaving his debt $1996.70, instead of $2000.

3. To find the difference of time between two dates on the note, reckon by years and months as far as possible, and then count the days; as in the rule Art. 318.

$850. CINCINNATI, April 29, 1850.

Ninety days after date, I promise to pay Stephen Ludlow, or order, eight hundred fifty dollars, with interest; value received. CHARLES K. TAYLOR.

Indorsements.—Oct. 13, 1850, $40; June 9, 1851, $32; Aug. 21, 1851, $125; Dec. 1, 1851, $10; March 16, 1852, $80.

What was due Nov. 11, 1852?

SOLUTION.—Interest on face ($850) from April 29 to Oct. 13, 1850, being 5 mon. 14 da., at 6 % per annum,	$23.233
	850.
Whole sum due Oct. 13, 1850, ...	873.233
Payment to be deducted, ...	40.
Balance due Oct. 13, 1850, ...	833.233

REVIEW.—225. How does the rule allow compound interest? Illustrate by an example. How find the difference of time between two dates on the note?

Interest on balance ($833.233) from Oct. 13, 1850, to June 9, 1851, being 7 mon. 27 da., ...	32.913
Payment not enough to meet the interest, ...	32.
Surplus interest not paid June 9, 1851, ...	.913
Interest on former principal ($833.233) from June 9, 1851, to Aug. 21, 1851, being 2 mon. 12 da., . . .	9.999
Whole interest due Aug. 21, 1851, ...	10.912
	833.233
Whole sum due Aug. 21, 1851, ...	844.145
Payment to be deducted, ...	125.
Balance due Aug. 21, 1851, ...	719.145
Interest on the above balance ($719.145) from Aug. 21, 1851, to Dec. 1, 1851, being 3 mon. 10 da., ...	11.986
Payment not enough to meet the interest, ...	10.
Surplus interest not paid Dec. 1, 1851, ...	1.986
Interest on former principal ($719.145) from Dec. 1, 1851, to March 16, 1852, being 3 mon. 15 da., . . .	12.585
Whole interest due March 16, 1852, ...	14.571
	719.145
Whole sum due March 16, 1852, ...	733.716
Payment to be deducted,.	80.
Balance due March 16, 1852, ...	653.716
Interest on Balance from March 16, 1852, to Nov. 11, 1852, being 7 mon. 26 da., ...	25.713
Balance due on settlement, Nov. 11, 1852, ...	$679.43

1. $\$304\frac{75}{100}$ CHICAGO, March 10, 1852.

For value received, six months after date, I promise to pay G. Riley, or order, three hundred and four $\frac{75}{100}$ dollars.

H. McMAKIN.

No payments.

What was due Nov. 3, 1853? *Ans.* $325.63

2. $2250. LOUISVILLE, Dec. 6, 1850.

For value received, one year after date, I promise to pay Albert Rogers, or order, two thousand two hundred and fifty dollars, with interest. NATHAN PHILIPS.

Indorsed: April 13, 1852, $1000.

What was due Aug. 19, 1854? *Ans.* $1634.63

3. $\$429\frac{30}{100}$ INDIANAPOLIS, April 13, 1853.

On demand, I promise to pay W. Morgan, or order, four hundred and twenty-nine $\frac{30}{100}$ dollars, value received.

R. WILSON.

Indorsed: Oct. 2, 1853, $10; Dec. 8, 1853, $60; July 17, 1854, $200.

What was due Jan. 1, 1855? *Ans.* $195.06

4. $1750. NEW YORK, Nov. 22, 1852.

For value received, two years after date, I promise to pay to the order of Spencer & Ward, seventeen hundred and fifty dollars, with interest at 7 per cent. JACOB WINSTON.

Indorsed: Nov. 25, 1854, $500; July 18, 1855, $50; Sept. 1, 1855, $600; Dec. 28, 1855, $75.

What was due Feb. 10, 1856? *Ans.* $879.71

5. 4643\frac{50}{100}$ DAYTON, Ohio, March 7, 1851.

For value received, three years after date, I promise to pay R. Banks, or order, forty-six hundred and forty three $\frac{50}{100}$ dollars, with interest at 10 per cent. W. G. BROOKS.

Indorsed: June 25, 1854, $1000; Nov. 1, 1854, $500; Jan. 12, 1855, $2500; Sept. 4, 1855, $1350; May 10, 1856, $150.

What was due July 1, 1856? *Ans.* $1189.86

6. $540. BALTIMORE, Jan. 11, 1849.

Eighteen months after date, we promise to pay Silas Greene, or order, five hundred and forty dollars, with interest, value received. EVANS & HART.

Indorsed: Nov. 23, 1850, $125; March 5, 1851, $35; Feb. 27, 1852, $25; May 31, 1852, $80; Oct. 16, 1852, $100.

What was due March 1, 1853? *Ans.* $291.60

7. $2500. PHILADELPHIA, Aug. 8, 1850.

For value received, nine months after date, I promise to pay Abijah Warren, or order, two thousand five hundred dollars, with interest at 6 per cent. HENRY CROSS.

Indorsed: Nov. 1, 1851, $400; Dec. 14, 1852, $100; July 6, 1853, $50; Oct. 21, 1854, $750; April 18, 1855, $500.

What was due July 1, 1855? *Ans.* $1362.04

8. $6875. BOSTON, Oct. 22, 1852.

For value received, four months after date, we promise to pay Augustus King, or order, six thousand eight hundred and seventy-five dollars. DAVIS & UNDERWOOD.

Indorsed: Feb. 25, 1853, $2000; Jan. 1, 1854, $245; April 1, 1854 $75; July 10, 1854, $1500; Dec. 26, 1854, $95; May 12, 1855, $1200 Sept. 29, 1855, $50.

What was due Jan. 1, 1856? *Ans.* $2376.64

9. $550. ST. LOUIS, June 17, 1848.

For value received, one year after date, I promise to pay to the order of Ross & Wade, five hundred and fifty dollars, with interest. TIMOTHY GORDON.

Indorsed: Aug. 19, 1848, $100; Dec. 1, 1848, $40; Feb. 8, 1850, $30; Sept. 27, 1850, $200; Jan. 4, 1852, $10; Mar. 1, 1852, $60.

What was due Aug. 28, 1852? *Ans.* $195.87

CONNECTICUT RULE.

ART. **326.** "*Compute the interest to the time of the first payment, if that be one year or more from the time that interest commenced; add it to the principal, and deduct the payment from the sum total.*

"*If there be after payments made, compute the interest on the balance due to the next payment, and then deduct the payment as above; and in like manner from one payment to another, till all the payments are absorbed:* Provided, *the time between one payment and another be one year or more.*

"But *if any payment be made before one year's interest hath accrued, then compute the interest on the principal sum due on the obligation, for one year, add it to the principal, and compute the interest on the sum paid, from the time it was paid, up to the end of the year; add it to the sum paid, and deduct that sum from the principal and interest added as above.* (See Note.)

"*If any payment be made of a less sum than the interest arisen at the time of such payment, no interest is to be computed, but only on the principal sum, for any period.*"

NOTE.—"*If a year does not extend beyond the time of payment; but if it does, then find the amount of the principal remaining unpaid up to the time of settlement, likewise the amount of the payment or payments from the time they were paid to the time of settlement, and deduct the sum of these several amounts from the amount of the principal.*

1. What is due on the 2d, 3d, and 4th of the preceding notes, by the Connecticut rule?

Ans. 2d, $1634.63; 3d, $194.54; 4th, $877.95

VERMONT RULE.

ART. **327.** *Find the simple interest of the principal from the time it begins to draw interest to the time of settlement, and add it to the principal. Do the same for each payment. Add together the amounts of the several payments, and deduct the sum from the amount of the principal. The remainder will be the balance due on settlement.*

NOTE.—This rule is of frequent use, when settlement takes place *a year or less* from the time interest begins.

1. What is due on the 2d, 3d, and 4th of the preceding notes, by the Vermont rule?

Ans. 2d, $1608.88; 3d, $193.50; 4th, $855.79

REMARK.—The following has been recommended as a more equitable RULE for Partial Payments, than any in use.

"*Starting at the time interest begins, find the Present Worth by Simple Interest of each payment; deduct the sum of these Present Worths from the Principal: the amount of the balance, by Simple Interest, to the day of settlement, will be the sum then due.*"

The principle of this rule is, *each payment discharges a part of the principal with its simple interest to the day the payment is made;* making interest and principal due at the same time, instead of the interest all due first, and the principal afterward.

XXVI. EXCHANGE.

ART. **328**. EXCHANGE is the method of transmitting money from one place to another by means of *Bills of Exchange.*

If the places are in different countries it is called *Foreign Exchange;* if not, *Home, Domestic* or *Inland Exchange.*

Bills of Exchange, also called *drafts* or *checks*, are written orders for the payment of money.

A *Sight Bill*, is one payable "at sight."

A *Time Bill* is payable a specified time after sight, or after date.

The signer of the bill, is the *maker* or *drawer.*

The one to whom the draft is addressed, and who is requested to pay it, is the *drawee.*

The one to whom the money is ordered to be paid, is the *payee.*

The one who has possession of the draft, is called the *owner* or *holder*; when he sells it, and becomes an *indorser*, he is liable for payment.

REVIEW.—328. What is exchange? Foreign exchange? Home exchange? What are Bills of Exchange? What is a sight bill? A time bill? Who is the drawer? The drawee? The payee? Who is the holder? What, if he becomes an indorser?

A *special indorsement* is an order to pay the draft to a particular person named, who is called the *indorsee*, as "Pay to F. H. Lee.—W. Harris." and no one but the *indorsee* can collect the bill.

When the *indorsement* is *in blank*, the payee merely writes his name on the back, and any one who has lawful possession of the draft can collect it.

If the drawee promises to pay a draft at maturity, he writes across the face the word "Accepted," with the date, and signs his name, thus: "Accepted, July 11, 1851.—H. Morton." The *acceptor* is first responsible for payment, and the draft is called an *acceptance*.

ART. **329.** A bill of exchange, like a promissory note, may be payable "to order," or "bearer," and is subject to protest in case the payment or acceptance is refused: it is also entitled to the 3 days grace, whether a *sight* or *time bill.*

In some States, drafts are not entitled to days of grace.

ART. **330.** The following are forms of drafts or bills of exchange.

INLAND DRAFT.

$1500. CINCINNATI, Feb. 8, 1855.

Please pay, at sight, to Williams & Baker, or order, fifteen hundred dollars, value received, and charge

To KING & CLARK, Brokers, New York. THOMAS ATKINS.

FOREIGN DRAFT.

Exchange £2000. New York, May 21, 1853.

At sixty days sight of this first exchange (second and third of the same date and tenor unpaid), pay to the order of H. Watts, two thousand pounds, without further advice.

To GEORGE SPENCE, Merchant, Liverpool. ELMER & BATES.

NOTES.—1. The words "value received" are not essential to a bill of exchange.

2. In foreign exchange, three separate bills are generally drawn, so that if one or two are lost, the other may reach its destination. When one bill of a set has been paid, the rest are void.

REVIEW.—328. What is special indorsement? Give an example. What is the consequence of a special indorsement? What is an indorsement in blank? Its consequence? What is an acceptance? Give an example. What is the effect of it? 329. How does a bill of exchange resemble a promissory note? What is said of protest? Of grace? Of sight-bills? 330. Give an example of an inland draft. Of a foreign draft. What is said of the words "value received"? What is a set of foreign exchange? When one of a set is paid, what is the consequence?

ART. 331. The *rate of exchange* is a rate per cent. of the face of the draft.

The rate of exchange between two places depends on their trade with each other; thus, at New Orleans, drafts on New York will be *at a premium*, more or less, according as the demand at New Orleans, for drafts to make payments in New York, *exceeds*, more or less, the supply furnished by parties drawing against sums due them in New York.

If the demand is less than the supply, exchange on New York is *at a discount* in New Orleans, more or less, according to the excess.

If the demand just *equals* the supply, exchange is at *par*.

ART. 332. The calculations connected with *inland exchange* have been explained in Art. 276–279.

The computation of *foreign* exchange is similar, except that it is necessary to express the money of one country in that of the other.

The sum mentioned in a draft on a foreign country, is expressed in the money of that country.

In calculating the *premium, discount,* or *cost* of such a draft, in the home currency, it is necessary not only to have tables of foreign money, but also to know the comparative values of the home and foreign currencies.

The *par of exchange* is the comparative value of the money of two countries, depending upon the amounts of pure gold or silver they contain: it must, therefore, remain fixed, as long as the coins retain the same weight and fineness; as, \$4.86 (*intrinsic* value,) = £1, and \$1 = $5\frac{1}{3}$ francs.

The *course of exchange* is the *par of exchange* after allowing for the *rate of exchange;* since the rate of exchange is variable according to the balance of trade, the course of exchange will also fluctuate within certain limits, being sometimes above and sometimes below the *par*, as, \$1 = 5.25 francs, \$4.87 (*exchangeable* value,) = £1.

REVIEW.—331. What is the rate of exchange? What does it depend on? Example. 332. To what subject do computations in inland Exchange belong? How does the computation of foreign exchange differ from that of inland exchange? How is the face of a foreign draft expressed? What must be known to find its value in our currency? What is the par of exchange between two countries? What does it depend on? Can it ever change?

Since the *course of exchange* allows for the *rate of exchange*, when the former is known, the cost of the bill is found by reduction, without using Percentage.

If the *rate of exchange* is given, a reduction of currencies, and a calculation in Percentage, are both required.

NOTES.—1. For Great Britain and her possessions, the RATE of exchange is given. For other foreign countries, the COURSE of exchange is given.

2. By comparing the pure metal in an English sovereign with that in a U. S. dollar, it is found that 20s. or £1 = \$4.86; formerly the silver coin of the U. S. was purer than at present, and then 20s. or £1, was equal to $\$4\frac{4}{9}$ or $\$4.44\frac{4}{9}$. Instead of comparing U. S. money with sterling money at the rate of £1 = \$4.86, it is customary still to reckon £1 = $\$4.44\frac{4}{9}$, or £9 = \$40, which is a convenient relation; and, consider the difference between \$4.86 and $\$4.44\frac{4}{9}$ ($41\frac{5}{9}$ ct., = $9\frac{7}{20}$ % of $\$4.44\frac{4}{9}$), as part of the rate of exchange.

Hence, sterling money is actually *par*, (£1 = \$4.86), when it is quoted $9\frac{7}{20}$ % premium, (on £1 = $\$4.44\frac{4}{9}$); it is actually *below par*, when it is quoted less than $9\frac{7}{20}$ % premium, and *above par*, only when it is above $9\frac{7}{20}$ % premium.

ART. 333. TABLE OF FOREIGN COINS AND MONIES.

Amsterdam and Antwerp.—1 florin = 100 cents = \$.40. Sometimes Flemish money is used as follows: 1 pound = 6 florins = 20 schillings = 120 stivers = 240 groats = 1920 pfennings.

Bombay.—1 rupee = 4 quarters = 400 reas = 16 annas = 50 pice = 45 ct. (ct. *in this table means Cents, U. S. Money.*)

Brazil.—Same as Lisbon.

Bremen.—1 thaler = 72 grootes = 360 swares = $78\frac{3}{4}$ ct.

Cadiz.—1 peso duro = $10\frac{5}{8}$ reals of old plate; 1 real = 16 quintas = 34 maravedis = 10 ct.; 1 ducat of plate = 11 reals; 1 real vellon = 5 ct.

Calcutta.—1 gold mohar = 16 sicca rupees = 64 cahauns = 256 annas = 3072 pice = 20480 gundas; 1 sicca rupee = 50 ct.; 1 current rupee = $44\frac{1}{2}$ ct.; a lac = 100000 rupees; a crore = 100 lacs.

REVIEW.—332. What is the course of exchange? Does it fluctuate? Why? When the course of exchange is given, how is the computation performed? How, when the *rate* of exchange is given? On which country is the rate of exchange given? On which the course of exchange? What is the actual par of exchange between the United States and England? What was the old par? Which is used in computations of sterling exchange? Why? What rate of exchange in favor of England makes sterling money really par? When is it at a real premium? When at a real discount?

Canton.—1 tael = 10 mace = 100 candarines = 1000 cash = $1.48.

Civita Vecchia.—1 scudo = 10 paoli = 100 bajocchi = $1.

Constantinople.—1 piastre = 40 paras = 120 aspers = 4 ct.

Copenhagen.—1 rix dollar = 6 marcs = 96 skillings; 1 rigsbank dollar = 52 ct.; the old rix dollar = $1.05.

Dantzic.—1 thaler = 30 silver groschen = 360 pfennings = 69 ct.; sometimes the following are used: 1 rix dollar = 3 florins = 90 groschen = 270 schillings = 1620 pfennings = 69 ct.

Genoa.—1 lira = 100 centesimi = $1.86. Old divisions: 1 lira = 20 soldi; 1 soldo = 12 denari.

Hamburg.—1 mark banco = 16 sols or schillings lubs = 192 pfennings lubs = $.35. Also 1 pound = $2\frac{1}{2}$ crowns = $3\frac{3}{4}$ thalers = $7\frac{1}{2}$ marks = 20 schillings, Flem. = 240 grotes, Flem. Lubs is a contraction for money of Lubec. Money is either *banco* or *current*; the former being at a premium of 23 % over the latter.

Havana.—1 dollar = 8 reals plate = 20 reals vellon = $1. A doubloon = $17.

La Guayra.—1 dollar = 8 reals = 75 ct.

Leghorn.—1 pezza of 8 reals = $5\frac{3}{4}$ lire = 20 soldi = 240 denari = 87 ct. The Tuscan corona or scudo = $1.05. Also, 1 lire = 20 soldi = 240 denari.

Lisbon.—1 milree = 1000 rees = $1.12; 1 milree of Azores = $83\frac{1}{3}$ ct.; 1 milree of Madeira = $1; 1 old crusado = 400 rees; 1 new crusado = 480 rees; 1 testoon = 100 rees.

Madras.—1 pagoda = $3\frac{1}{2}$ rupees = 42 fanams = 3360 cash = $1.84.

Naples.—1 ducato di regno = 10 carlini = 100 grani = $.80; 1 scudo = 12 carlini = $.95.

Palermo.—1 ducato = 10 piccioli = 100 bajocchi = $.80. Accounts are generally kept in the following: 1 oncia = 30 tari = 600 grani = 3 ducati = $2.40.

St. Petersburgh.—1 bank rouble = 100 copecks = $.214. Also, 1 silver rouble = 360 copecks = $.75.

Stockholm.—1 rix dollar banco = 48 skillings = 576 rundstycks = $.40; 1 silver rix dollar = $1.06.

Trieste.—1 florin = 60 kreutzers = 240 pfennings = $.48$\frac{1}{2}$.

Venice.—1 lira = 100 centesimi = 1000 millesimi = $.16.

REM.—The values of the foreign coins in this table are expressed in the U. S. gold and silver coins as they were previous to 1853. If their values in the new silver coinage since 1853 are required, the values given above most be increased $7\frac{27}{64}$ per cent. (See Art. 204.)

ART. 334. HOME, OR INLAND EXCHANGE.

1. What is paid for $3805.40 sight exchange on Boston, at $\frac{1}{2}$ % premium? *Ans.* $3824.43

2. What for a 30 da. bill on New Orleans for $7216.85, at $\frac{3}{8}$ % discount, interest off at 6 %? *Ans.* $7150.09

Interest and discount are calculated on the face of the draft.

3. What cost a check on St. Louis for $1505.40, at $\frac{1}{4}$ % discount? *Ans.* $1501.64

4. What cost a 60 da. draft on New York for $12692.50, at $\frac{3}{4}$ % premium, int. off at 6 %? *Ans.* $12654.42

5. What must be the face of a draft to cost $2000, at $\frac{5}{8}$ % premium? *Ans.* $1987.58

Assume $1 for the face, determine the cost on this supposition, and divide the given cost by it.

6. What must be the face of a draft to cost $4681.25, at $1\frac{1}{4}$ % discount? *Ans.* $4740.51

7. Of an 18 da. draft, costing $5264.15, at $\frac{1}{2}$ % premium, int. off at 6 %? *Ans.* $5256.27

Assume $1 for the face, &c. (See Ex. 5.) Prove by selling the draft as proposed, and see if it yields the given amount.

8. Of a 21 da. draft, costing $6836.75, at $\frac{7}{8}$ % discount, and int. off at 6 %? *Ans.* $6925.04

Art. 335. Foreign Exchange.

1. What is the cost in New York, of a draft on London for £2748 11s. 6d., at 10 % prem.? *Ans.* $13437.48

Suggestion.—Express the face in pounds, reduce to $, at the rate of £1 = $\$\frac{40}{9}$, increasing the result by 10 %.

2. In Liverpool, of a draft on Boston for $26550, at $9\frac{1}{2}$ % premium for sterling? *Ans.* £5455 9s. 7d.

As £1 = $\$\frac{40}{9} \times \frac{109\frac{1}{2}}{100}$, divide $26550 by this value, or multiply it by the value $1 = $£\frac{9}{40} \times \frac{100}{109\frac{1}{2}}$; reduce the decimal to shillings and pence.

3. What is the face of a draft on London, at $9\frac{3}{4}$ % premium, costing $6244.50? *Ans.* £1280 3s. $10\frac{1}{2}$d.

Assume £1 for the face, reduce it to $ at $9\frac{3}{4}$ % premium on $\$\frac{40}{9}$, and divide $6244.50 by it.

4. Of a draft on Philadelphia costing £1500, at $9\frac{1}{4}$ % premium for sterling? *Ans.* $7283.33

5. What is the cost of a draft on Paris for 6072.25f., at $1 = 5.35f.? *Ans.* $1135.

6. Of a draft on New York, at Havre, for $2918.56 at $1 = $5.37\frac{1}{2}$f.? *Ans.* 15687.26f.

7. What is the face of a draft on Bordeaux, to cost $4534.20, at $1 = 5.40f.? *Ans.* 24484.68f.

8. What cost a draft on Hamburgh, for 1231 marks 10 schil. 8 pfen., at 1 mark banco = 33 ct.? *Ans.* $406.45

9. What cost at Hamburgh, a draft for $1826.70, at 1 mark banco = 35 ct.?
Ans. 5219 marks banco 2 schil. $3\frac{3}{7}$ pfen.

10. A draft on Amsterdam, for 3422 florins 15 cents, at 1 florin = 38 ct. U. S.? *Ans.* $1300.42

11. What cost at Amsterdam, a draft for $6312.80, at 1 florin = 42 ct. U. S.? *Ans.* 15030 florins 48 cents.

12. What cost a draft on Lisbon, for 724 milrees 960 rees, at 1 milree = $1.25? *Ans.* $906.20

13. What cost at Lisbon, a draft on U. S. for $3542.60, at 1 milree = $1.28? *Ans.* 2767 milrees 656 rees.

14. A draft on Cadiz for 1252 reals 27 maravedis plate, at 1 piastre of 8 reals = 66 ct.? *Ans.* $103.36

15. What cost at Barcelona, a draft for $7564.12, at 1 real = $8\frac{1}{2}$ ct.? *Ans.* 88989 reals 22 maravedis.

16. What cost at Stockholm, a draft on U. S. for $3529.50, at 1 rix dollar = $1.01?
Ans. 3494 rix dollars 26 skillings 7 rundstycks.

17. What cost a draft on Stockholm, for 5643 rix dol. 7 skil. 4 rund., at 1 rix dol. = 98 ct.? *Ans.* $5530.29

18. What cost a draft on Moscow, for 8751 roubles 63 copecks, at 1 rouble = 27 ct.? *Ans.* $2362.94

19. What cost at Archangel, a draft for $3387.06, at 1 rouble = 24 ct.? *Ans.* 14112 roubles 75 copecks.

20. What cost a draft on Berlin, for 634 rix dol. 13 silver groschen 5 pfen., at 1 rix dol. = 65 ct.? *Ans.* $412.39

21. What cost at Dantzic, a draft for $4687.20 at 1 rix dol. = 66 ct.? *Ans.* 7101 rix dol. 24 silver gro. 7 pfen.

22. What cost a draft on Copenhagen, for 2934 rigsbank dol. 5 marks 14 skill. at 1 rigsbank dol. = 48 ct.?
Ans. $1408.79

23. What cost at Copenhagen, a draft for $4427.35 at 1 rigsbank dol. = 51 ct?
Ans. 8681 rigsbank dol. $7\frac{1}{2}$ skil.

24. What cost at Canton, a draft for $13653.85, at 1 tael = $1.55?
Ans. 8808 taels 9 mace 3 candarines 5 cash.

ARBITRATION OF EXCHANGE.

ART. **336.** Exchange may be either *direct* or *circular*.

Direct exchange is confined to two places, the residence of the drawer and of the drawee.

Circular exchange is effected by a succession of bills drawn *at*, and *on*, one or more intermediate points.

The circular method is called *Arbitration* of Exchange.

A New York merchant wishes to remit \$2240 to Lisbon: is it more profitable to buy a bill directly on Lisbon at 1 milree = \$1.15: or, to remit through London at 8 % premium, and through Paris at £1 = 25.20f., to Lisbon at 1 milree = 6f.?

SOLUTION.—The direct exchange gives the value of \$2240 = 1947.826 milrees, by dividing \$2240 by \$1.15, the value of 1 milree in exchange. The value of the remittance by the circular exchange is, $2240 \times \frac{9}{40 \times 1.08} \times 25.20 \times \frac{1}{6} = 1960$ milrees. The \$2240 is reduced to £, by multiplying it by the value of \$1, which is £$\frac{9}{40 \times 1.08}$; the £ are reduced to francs, by multiplying by the value of £1, which is 25.20f.; the francs are reduced to milrees, by multiplying by the value of 1 franc, which is $\frac{1}{6}$ milree. After indicating the operations, cancel.

2d SOL.—The same result can be obtained by a *chain* of equations, each expressing the course of exchange between two points in the circuit, except the first, which has the sum to be remitted equal to an unknown number ( ) of the denomination required (milrees); these equations are so arranged that the same denominations appear on both sides of the sign (=); the product of both sides being equal, the vacant term is found by dividing the product of those on the other side, by the product of the others on its own side, canceling first, to shorten the work.

56	\$2240	=	( ) milrees.
28	1 milree	=	6f. 2
7	25.20f.	=	£1
	£9	=	\$40
	3		×
	.09)176.40		1.08
	Cir. ex. 1960 mil.		.36
	Dir. ex. 1947.826		.09
	Gain 12.174 milrees.		

REVIEW.—336. What two kinds of exchange? What is direct exchange? Circular? What is the circular called? Solve the example.

ART. 337. Hence, to compare two quantities of different denominations, through the medium of other denominations, this rule, called the

CHAIN RULE.

RULE. I.—*Reduce the given quantity to the denomination with which it is compared; reduce this result to the denomination with which it is compared; and so on until the required denomination is reached, indicating the operations by writing as multipliers, the proper unit values. The compound fraction, thus obtained, reduced to its simplest form, will be the amount of the required denomination.*

RULE II.—*Form a series of equations, expressing the conditions of the question; the first containing the quantity given equal to an unknown number of the quantity required, and all arranged in such a way that the right hand quantity of each equation and the left hand quantity in the equation next following, shall be of the same denomination, and also the right hand quantity of the last and the left hand quantity of the first. Cancel all equal factors on the right and left, and divide the product of the quantities in the column which is complete by the product of those in the other column. The quotient will be the quantity required.*

NOTES.—1. This is called the *chain* rule, because each equation and the one following, as well as the last and first, are connected like a chain, having the right hand quantity in one of the same denomination as the left hand quantity of the next.

2. In applying this rule to exchange, if any currency is at a premium or a discount, its unit value in the currency with which it is compared, must be multiplied by such a number of decimal hundredths, as will increase or diminish it accordingly. (See Operation.)

3. If commission or brokerage is charged at any point of the circuit, for effecting the exchange on the next point, the sum to be transmitted must be diminished accordingly, by multiplying it by the proper number of decimal hundredths.

EXAMPLES FOR PRACTICE.

1. A New York merchant remits $1460 to Hamburgh, through London and Paris: what is the value when received, if £1 = $4.85; 25.20f. = £1; 1 mark banco = 1.80f.?
Ans. 4214 mark bancos 6 sols 11 pfennings.

REVIEW.—337. What is the chain rule? Why is this called the chain rule? Why should this chain or connection be made? If any of the currencies is at a premium or discount, what should be done? If commission or brokerage is charged at any point of the circuit, what should be done?

2. A, of Philadelphia, wishes to remit $4532.80 to St. Petersburgh: the direct exchange is $1 = 3 roubles 30 copecks, while through London, Paris, and Dantzic, it is as follows: 8 % in favor of London; 25.20f. = £1; 1 thaler = 4f.; 2 roubles 60 copecks = 1 thaler. Which is better?

Ans. Circular, by 509 roubles 94 copecks.

3. A, of Boston, owes in Amsterdam 10000 florins. How much in U. S. money will pay it, by remitting through London and Paris at the following rates: 9 % in favor of London; £1 = 25f.; 1 florin = 2.32f.; allowing 1 % brokerage in London, and ½ % in Paris? *Ans.* $4563.87

SUGGESTION.—Multiply the £ by .99, and the francs by .99½, to allow for the brokerage. (Note 3.)

4. A, of Memphis, has $6000 to pay in New York. The direct exchange is 1 % premium; but exchange on New Orleans is ½ % premium, and from New Orleans to New York is 1 % discount. By circular exchange, how much will pay his debt, and what is his gain?

Ans. $5969.70; $90.30 gain.

SUGGESTION.—In home exchange, the currency being the same throughout the circuit, annex to the $, the name of the place, as $ N. Y. or $ N. O., for New York money and New Orleans money.

5. A merchant of St. Louis wishes to remit $7165.80 to Baltimore. Exchange on Baltimore is ¾ % premium; but, on New Orleans it is ½ % premium; from New Orleans to Havana, ½ % discount; from Havana to Baltimore, 1 % discount. What will be the value in Baltimore by each method, and how much better is the circular?

Ans. Direct, $7112.46; circ., $7238.36; gain, $125.90

6. A Louisville merchant has $10000 due him in Charleston. Exchange on Charleston is ¼ % premium. Instead of drawing directly, he advises his debtor to remit to his agent in New York at ⅜ % premium, on whom he immediately draws at 12 da., and sells the bill at 1½ % premium, interest off at 6 %. What does he realize in this way, and what gain over the direct exchange?

Ans. Realizes $10087.17; gain, $62.17

SUGGESTION.—Interest at 6 % per annum is ½ % a month, or ¼ % for 15 days, which diminishes the rate of premium to 1¼ %.

7. A Cincinnati manufacturer receives, April 18th, an account of sales from N. Orleans; net proceeds $5284.67,

due June $^4|_7$. He advises his agent to discount the debt at 6 %, and invest the proceeds in a 7 da. bill on New York, interest off at 6 %, at $1\frac{1}{2}$ % discount, and remit it to Cincinnati. The agent does this, April 26. The bill reaches Cincinnati May 3, and is sold at $\frac{7}{8}$ % prem. What is the proceeds, and how much greater than if a bill had been drawn May 3, on New Orleans, due June 7, and sold at $\frac{3}{8}$ % prem., and int. off at 6 %?

Ans. Proceeds, \$5383.32; gain, \$109.66

8. Find the net value in Cincinnati, of a sum due in Paris, of 12600f., which is transmitted to London, at 1 % commission; exchange, 25.20f. = £1; a draft for the net amount in London, is drawn at New York at 7 % premium, after deducting $\frac{1}{2}$ % commission; the proceeds of this bill being remitted to Cincinnati, in a check at $\frac{1}{2}$ % discount. *Ans.* \$2354.

9. If 5 oranges are worth 8 lemons, 3 lemons worth 10 apples, 4 apples worth 1 melon, and 6 melons worth 75 ct.; how much money, are 12 oranges worth? *Ans.* \$2.

10. If 3 men do as much as 7 women, and 10 women as much as 27 boys, and 42 boys as much as 75 girls, and 36 girls can bind 500 sheaves in an hour, how many can 12 men bind in an hour? *Ans.* 1875.

11. What must 9 oz. of tea be sold for, if 45 lb. cost \$28, and the retail profit is 20 %? *Ans.* 42 ct.

12. What should coffee be sold at per pound, if an invoice of 253 bags, averaging $166\frac{2}{3}$ lb. each, cost \$4620, the freight and other charges being 15 % on the invoice, and the profit on the whole cost, 25 %? *Ans.* $15\frac{3}{4}$ ct.

13. If the freight at New Orleans, per U. S. bu. of corn (56 lb.), is 10d., what is the freight at Liverpool, per imperial quarter (480 lb.); primage, 5 %? *Ans.* 7s. 6d.

REMARK.—Primage is an allowance to the master and mariners of a ship for loading, and is charged on the *freight*.

14. What is the freight on a cargo of 2560 bales of cotton, averaging 450 lb. each, at $\frac{5}{16}$d. per pound, allowing 5 % primage, 8 % premium on sterling money, and 5 % commission for advancing the freight? *Ans.* \$23814.

15. If the freight per U. S. bu. of corn (56 lb.), is 26 ct., charges 4 ct. per bu. (as usual in New Orleans), making the freight 30 ct. per bu. on board, what is the freight per

imperial quarter (480 lb.), allowing 4 % commission for advancing the freight, sterling exchange 4 % premium? *Ans.* 11s. $6\frac{6}{7}$d.

16. If the freight is 66 ct. per U. S. bu., charges and commission as in last example, exchange 8 % prem., what is the freight per imperial quarter? *Ans.* 26s.

17. What is the cost in Liverpool, of corn per quarter, ought in New Orleans at 50 ct. per U. S. bu., charges 4 ct. per bu., exchange 8 % premium, freight 13d. per bu., commission (on *Freight only,*) 5 %? *Ans.* 29s. $\frac{3}{7}$d.

Find the cost per qr.; then, the freight per qr., and add.

18. At how many cents per U. S. bushel can an order be filled in New Orleans, limited at 28s. per imperial quarter, the freight being 15d. per U. S. bu., commission 5 %*, and exchange 6 % premium? *Ans.* 42 ct.

Find the freight per imperial qr., subtract it from 28s., and find the cost per U. S. bu. corresponding to the remainder. This cost, less 4 ct. the usual charges per bu., is the price per U. S. bu.

19. How many cents can be paid for corn in New Orleans, to fill an order from Liverpool, limited to 30s. per imperial quarter, the freight being 10d. per U. S. bu., commission* 5 %, exchange 7 % premium? *Ans.* $58\frac{5}{12}$ ct.

*In Examples 17, 18 and 19, Commission is charged on *Freight only.*

XXVII. ACCOUNTS CURRENT.

ART. **338**. THE account which a banker keeps of money received and paid out, is an *Account Current.*

Accounts Current are also kept by merchants, showing the value of goods delivered to their agents and customers, and the payments made thereon.

Here is an Account Current of a bank with a depositor:

REVIEW.—338. What is an account current? By whom kept? What does the Dr. side show? The Cr. side? What is settling an account current? What is it also called? What is necessary in making the balance? What is the direct rule?

Dr. *Henry Armor, in acc't with City Bank.* Cr.

1855.			$	ct.	1854.				$	ct.	Prod.
Jan	5	To check,	300		Dec.	31	By bal. old acc't,		500		2500
,,	20	,, ,,	500		1855			200			400
,,	21	,, ,,	100		Jan.	7	,, cash,		50		
,,	27	,, ,,	850		,,	15		250			2000
,,	31	,, ,,	75				,, ,,		400		
								650			3250
					,,	24		150			150
					,,	,,		50			150
							,, ,,		1000		
								1050			3150
								200			800
							Bal. items,	125			12.400
							Bal. int.,	2.07			$2.07
							Total bal.	$127.07			

The left hand, or Dr. side, shows, with their dates, the sums paid by the bank on the checks of Henry Armor, for which he is their debtor.

The right hand, or Cr. side, shows, with their dates, the sums deposited in the bank by Henry Armor, for which he is their creditor.

Settling or *closing* an account, is finding how much is due at a particular time, between the parties. It is sometimes called *striking a balance.*

Generally, in an account current, each item draws interest from its date to the day of settlement; hence,

TO SETTLE AN ACCOUNT DRAWING INTEREST.

Rule.—*Find the sum of the items on the Dr. side, with the interest of each from its date to the date of settlement; do the same for the Cr. side. The difference of these sums will be the balance in favor of the greater side.*

The following is a more convenient and

PRACTICAL RULE.

Multiply the 1st item by the number of days from the 1st to the 2d dates. Find the balance after the 2d item, and multiply it by the number of days from the 2d to the 3d dates, and so on to the last balance, which is multiplied by the number of days from the last date to the date of settlement.

Each product is put into a Dr. or Cr. column, according as the balance, from which it comes, is Dr. or Cr. Balance the

Review.—338. What is the practical rule? Show the reason o[f the] rule as applied to the account given.

two columns of products, and find the interest on this balance for 1 day, at the agreed rate. This will be the balance of interest in favor of the larger column, which being added to, or subtracted from, the last balance of items, according as they are on the same or opposite sides, will give the final balance required.

DEMONSTRATION.—Take the account already given. The 1st item, $500, being a balance in favor of Henry Armor from an old account, is on interest from its date (Dec. 31, 1854) to the date of the next item (Jan. 5), which is 5 days. The interest of $500 for 5 da. is the same as the interest of $500 × 5 = $2500 for 1 da.

Place $2500 in the column of products opposite; find the balance of items ($200), which is the difference between the sum on deposit ($500), and the sum ($300) paid out on the check. This balance may, for convenience' sake, be set down in pencil or red ink on the side of the account to which it belongs, as shown in the form. This balance is on interest from the second date (Jan. 5) to the third date (Jan. 7), = 2 days ; the interest of $200 for 2 da. is the same as the interest of $200 × 2 = $400 for 1 da.; set this product $400 in the column opposite, and keep on, taking a balance and forming a corresponding product, for every item, to the day of settlement.

In this example, the balance being always on the Cr. side, the products will all be in one column; but, if the balance were sometimes on one side, and sometimes on the other, a Dr., as well as a Cr. column for products, would be needed; adding each column, the difference of the sums would be the balance in favor of the larger column; the interest of this difference for 1 da., would be the excess of interest in favor of the larger column.

The balance of products in this example, is $12400; its interest for 60 da. at 6% is $124.00 (Art. 305); dividing by 60 gives the interest for 1 da., $2.07, in favor of the Cr. side: as the balance of items, $125, is also in favor of the Cr. side, the total balance to Henry Armor's credit, is the sum of the two, ($127.07).

NOTE.—In settling depositors' accounts in New York, Great Britain, &c., the interest, as ordinarily obtained, must be diminished by $\frac{1}{73}$ of itself, (Art. 308); but elsewhere, *this deduction should not be made;* as, there is no more reason for it in case of depositors' accounts, than in the discounts of the bank: it is manifestly unjust for the bank to make a deduction on the interest it pays, and not

REVIEW.—338. When will two columns for products be necessary? How do we then proceed? On which side will the balance of interest be? What should be done with it when obtained?

to allow it on the interest it receives, the circumstances being precisely the same: if it insists on taking the time by days in its discounts, it ought not to object to any advantage to the depositor in making up his interest on the same plan. It is true, if the interest on a depositor's account were calculated on each item separately to the time of settlement, using calendar months when possible, the result would be slightly less than by the last rule, a day being occasionally lost, *but it would not be* $\frac{1}{73}$ *less:* the difference would vary with the size and dates of the items.

FIND THE INT. DUE, AND BALANCE THESE ACCOUNTS.

1. DR. *F. H. Willis in acc't with E. S. Kennedy.* CR.

			Dol.	ct.				Dol.	ct.	Products.
1849.					1848.					
Jan.	6	To check,	170		Dec.	31	By bal. forw.	325		
,,	13	,, ,,	480		1849.					
,,	16	,, ,,	96		Jan.	7	,, cash dep.	800		
,,	25	,, ,,	500		,,	20	,, ,, ,,	175		
,,	28	,, ,,	50		,,	30	,, ,, ,,	240		

Interest to Jan. 31, at 6 per cent.

Ans. Int. due Willis, $2.33: bal. due Willis, $246.33

2. DR. *A. L. Morris in acc't with T. J. Fisher & Co.* CR.

			Dol.	ct.				Dol.	ct.	Products.
1854.					1853.					
Jan.	13	To check,	350		Dec.	31	By bal. from old acc't.	813	64	
,,	22	,, ,,	275		1854.					
Feb.	25	,, ,,	100		Feb.	4	,, cash,	120		
May.	1	,, ,,	400		Mar.	17	,, ,,	500		
Jun.	23	,, ,,	108	25	May	31	,, ,,	84	50	

Interest to June 30, at 6 per cent.

Ans. Int. due Morris, $13.34: bal. due Morris, $298.23

3. DR. *Wm. White in acc't with Beach & Berry.* CR.

			Dol.	ct.				Dol.	ct.	Products.
1855.					1855.					
Jul.	2	To check,	212	50	Jun	30	By bal. from old acct.	1102	50	
,,	20	,, ,,	66		Jul.	6	,, cash dep.,	50		
Aug.	7	,, ,,	235		,,	15	,, ,, ,,	95		
,,	25	,, ,,	300		Aug	9	,, ,, ,,	168	75	
Sep.	5	,, ,,	110		Sep.	18	,, ,, ,,	32		
,,	11	,, ,,	46	40	Oct.	3		79	90	
,,	27	,, ,,	454	25						

Interest to Oct. 12, at 10 per cent.

Ans. Int. due White, $19.68; bal. due White, $123.68

REVIEW. — 338. In New York, &c., what should be done in finding the interest? Should this be done elsewhere? Why not?

ART. **339**. The rule (Art. 338) is not so applicable to a merchant's account current, where the items, being sometimes notes, drafts, acceptances, sales on credit, &c., are not generally due in the order they are written.

TO BALANCE A MERCHANT'S ACCOUNT CURRENT,

RULE.—*Multiply each item by the number of days from the ime it is due as cash until the day of settlement. Set the products of the Dr. items in a Dr. column, and those of the Cr. items in a Cr. column.*

Add each of the columns, and find the difference of their sums; the interest of this for 1 day, at the rate agreed, will be the balance of interest in favor of the larger column, which is placed on the side of the account where it belongs; the difference between the two sides will be the whole balance in favor of the larger side.

REMARK.—This rule serves for closing an account sales, when the amount due at a particular date is required.

4. DR. *F. Archer, in acc't with T. & W. Day.* CR.

1855.			Dol.	ct.	Products.	1855.			Dol.	ct.	Products.
Jun	29	To 100 barrels pork, @ $12,	1200			Jul.	20	By 20 hhd. sugar, @ $66	1320		
Jul.	15	To 50 barrels flour, @ $9,	450			Aug	20	By 15 bales cotton, 6000 lb., @ 8 c.,	480		
Sep.	9	To 40 barrels whisky, 1360 gal., @ 25 c.	340			Nov	12	By 8 hhd. tobacco, 12000 lb., @ 6 c.,	720		
Dec.	18	To 18 barrels pork, @ $12,	216								

Interest at 6 per cent. to Dec. 31.

Ans. Int. due T. & W. Day, $3.95: bal. due Archer, $310.05 The 1st rule may be used here to advantage.

5. DR. *Oliver Mason, in acc't with Sharpe & Swift.* CR.

1856.		da	Dol.	ct.	Prod.	1856.		da	Dol.	ct.	Prod.
Jan. 6	To invoice, cash		874	75		Mar. 6	By cash,		372		
Feb. 3	,, ,, ,,		293			Apr. 8	,, ,,		495		
May 16	,, acceptance, May 31,		591	25		May 1	,, remittance, June 30,		2000		
July 5	To invoice, 3 m.		1600			May 2	,, remittance, May 9,		1426	50	
Aug. 20	,, ,, 2 m.		1150	10		Aug. 15	,, cash,		250		
Sept. 1	,, check,		925			Sept. 28	,, remittance, Oct. 15.		342		
" 25	,, cash,		540								

Ans. Int. due Mason, $30.73; bal. due Sharpe & Swift, $1057.87 Int. at 6 % to Sept. 30.

NOTE.—In this account, two items on the Dr. side and one on the Cr., fall due *after* Sept. 30, and must therefore suffer discount, or, what is the same, *draw interest for the other side:* hence, the products obtained from them, are put on the side of the account, opposite to that which contains the items.

6. *Account sales of* 500 *hhd. sugar, per ship Wave, from New Orleans, sold by order and for account of Perrot, Clarke & Co.*

Discount and interest 6 % to date.

1855.		500 hhd. sugar.	Dol.	ct.	Due	Da	Prod'cts.
May	10	To Jonas & King, 2 mon., 100 hhd., (107210 lb.), @ 5c.	5360	50	Jul.10	14	75047
,,	18	To H. Johnson, 3 mon., 150 hhd. (158930 lb.), @ 5c.	7946	50	Aug18	53	421165
J'ne	16	To R. Force, 3 mon., 80 hhd. (85374 lb.), @ 5½c.	4695	57	Sep.16	82	385037
,,	25	To E. Woodruff, 3 mon., 170 hhd. (190184 lb.), @ 5½c.	10460	12	Sep.25	91	951871
		Total, 500 hhd.	28462	69			
		CHARGES.					
May	1	Freight, 500 hhd. @ $6 ... $3000					
		Cartage, labor, cooperage ... 287					
		$3287			May 1	56	184072
J'ne	26	Commission 2½ p. ct., on $28462.69 711.57				60)	20171.92
		Guarantee, 1 ,, ,, ,, 284.63					
		Insurance, storage, &c. 115.28					336.199
		Discount on credit sales, and interest on charges, reduced to date, 336.20	4734	68			Interest
		Net proceeds this day	23728	01			
		(Errors and omissions excepted.)					
		NEW YORK, June 26, 1855.					
		EDWARD PARKER.					

REMARK.—The charges paid May 1, are on interest till June 26 (56 da.); the sales being on credit, the payments are not due until various times after June 26, and must be discounted for 14, 53, 82, and 91 days, respectively, to reduce them to cash on June 26.

7. DR. *E. S. Lee, in acc't current with Burns & Co.* CR.

*Interest 9 % to June 30.

1851.		da	Dol.	ct.	Prod.	1851.		da	Dol.	ct.	Prod.
Feb. 10	To invoice, cash		800			Feb. 20	By remit. Mar. 5		600		
Mar. 25	,, ,, ,,		900			Mar. 18	,, ,, Apr. 10		400		
Apr. 15	,, ,, 2 mon.		700			Apr. 5	,, cash,		200		
May 18	,, ,, 2 mon.		500			May 20	,, ,,		300		
May 31	,, acceptance, 15da. sight,		400			Jun. 20	,, remit. July 15		1600		
June20	,, invoice, cash		1000								

Ans. Int. due Burns & Co., $27.18; bal. due Burns & Co., $1227.18 (* Days of grace are not counted.)

REVIEW.—339. Why is the rule for bank accounts not so applicable to a merchant's account current? What is the rule for it?

8. DR. *Joshua Parkins, in acc't with H. K. Foster.* CR.

*Closed Dec. 31; interest 6 %.

1853.		da	Dol.	ct.	Prod.	1853.		da	Dol.	ct.	Prod.
Jan. 15	To invoice 3 m.		738			Feb. 6	By remittance, Feb. 18		250		
Feb. 20	,, acceptance, 15 da. sight,		400			Mar. 15	,, remittance, sight,		380		
Apr. 15	,, invoice, 3 m.		550	15		May 10	,, acc't sales,		215	30	
June 1	,, ,, 2 m.		632			Aug. 10	,, cash,		326	15	
Aug. 31	,, acceptance, 30 da. sight.		920	83		Oct. 9	,, ,,		184		
Sept. 16	,, invoice, cash		240			Oct. 20	,, remittance, Nov. 14,		203	90	
Oct. 19	,, ,, 2 m.		184	24		Dec. 1	,, remittance, Dec. 15.		500		
Dec. 21	,, draft, sight,		100								

Ans. Int. due Foster, $49.04; bal. due Foster, $1754.91

* In these Acts. current, and in Equation of Payments, days of grace are not counted.

STORAGE ACCOUNTS.

ART. **340**. Storage accounts are similar to bank accounts, one side showing how many bbl., packages, &c. are received, and at what times; the other side showing how many have been delivered, and at what times.

Storage is generally charged at so much per month of 30 da. on each bbl., package, &c.

TO FIND THE BALANCE DUE ON A STORAGE ACC'T,

RULE.—*Multiply the number of bbl. first set down by the number of days from the* 1*st to the* 2*d date of the account, and set the result in the column of products opposite.*

Find the number of bbl. on hand after the second item has been set down, and multiply it by the number of days from the 2*d to the* 3*d date, setting the result in the column of products opposite.*

Continue thus, multiplying the number of bbl. finally on hand, by the number of days from the last date to the date of settlement.

Add these products, and divide their sum by 30: *the quotient will be the number of bbl. stored per month, and this number multiplied by the price of storage on each bbl. per month, will give the balance due on the account.*

REVIEW.—340. How is storage charged? What is the rule for settling a storage account?

1. Storage to Jan. 31, at 5 ct. a bbl. per mon.

RECEIVED. DELIVERED.

1856.		bbl.	Balance on hand.	Days.	Products.	1856.		bbl.
January.	2	200				January.	10	110
	5	150					13	90
	7	30					17	20
	10	120					20	115
	14	80					25	140
	17	150					27	72
	20	75					30	100
	24	60					31	
	28	200						

Ans. Storage, $19.25; bbl. on hand, 418.

2. Storage to Feb. 20, at 5 ct. a bbl. per mon.

RECEIVED. DELIVERED.

		bbl.	Balance on hand.	Days.	Products.			bbl.
January.	31	418				February.	5	100
February.	4	250					10	80
	9	120					12	220
	12	100					14	140
	16	30					18	90
							20	288

Ans. Storage, $15.85; bbl. on hand, 0.

XXVIII. EQUATION OF PAYMENTS.

ART. **341.** *Equation of Payments*, or *averaging accounts*, is a method used by merchants, by which debts due at different times can be paid in one sum, and at one time, without loss to either party.

A buys an invoice of $250 on 3 mon. credit; $800 on 2 mon.; and $1000 on 4 mon.; and gives his note for the whole amount: how long should the note run?

SOLUTION.—The whole debt is $2050; and since

the discount on $250 for 3 mon. = discount on $750 for 1 mon.
and " " $800 " 2 " = " " $1600 " "
and " " $1000 " 4 " = " " $4000 " "

The discount on $2050 . . . . = discount on $6350 for 1 mon.

But the discount on $6350 for 1 mon. is the same as the discount on $2050 for $\frac{6350}{2050}$ mon., $= 3\frac{4}{41}$ mon. = 3 mon. 3 da.; which is, there-

REVIEW.—341. What is equation of payments? Solve the example.

fore, the time the note should run, since it is the time for which the debt, ($2050), must be discounted to be considered cash. The time thus found is called the *mean* or *average time*.

I buy of a wholesale dealer, at 3 mon. credit, as follows: Jan. 7, an invoice of $600; Jan. 11, $240; Jan. 13, $400; Jan. 20, $320; Jan. 28, $1200; I give a note at 3 mon. for the whole amount: when is it dated?

SOLUTION.—Start at the date of the 1st purchase, Jan. 7; then

Purchases.		Disc. Prod.
600 × 0	=	0
240 × 4	=	960
400 × 6	=	2400
320 × 13	=	4160
1200 × 21	=	25200
2760		)32720(12

Hence, 12 days after Jan. 7 = Jan. 19, the mean time or date of the note.

Any date will do to start with. If you start *at* or *before* the 1st payment, count *forward* for the multipliers and quotient; if you start *at* or *after* the last payment, count *backward*. If you start between the first and last payments, there will be two columns of products, one of the purchases before the assumed date, and the other of those after it; in that case, add each column, take the difference of the sums, and divide it by the sum total of debts: this quotient must be counted forward or backward from the assumed date, according as the products *after* or *before* that day preponderate.

REMARK.—The interest *gained* on the purchases made after the mean time, should be just equal to the interest *lost* on the purchases made before it; but, if the quotient which fixes the mean time is not an exact number of days, the two interests will slightly differ, according to the size of the fraction neglected; in the example above, the quotient is not exactly 12 days, but $11\frac{59}{69}$ days, on which account, the interest before Jan. 19, will be a little more than that after it, but as a part of a day is not recognized in dating notes, the answer must be considered correct.

TO AVERAGE AN ACCOUNT HAVING DEBITS ONLY,

RULE.—*Start at the earliest day a debt is due, and multiply each debt by the number of days (or months) its date is after the day fixed on; take the sum of these products, and divide it by the sum total of the debts; the quotient will be the number*

REVIEW.—341. What date is the starting point? Would any other time do? What, if you start at or before the first payment? What, if you start at or after the last payment? What, if you start between the first and last dates? What is the rule for averaging an account?

of days (or months) which, added to the day first fixed on, will give the mean time required.

PROOF.—Assume the day the last debt is due (or any other day); determine the mean time again, and see if it agrees with the one already obtained.

ART. **342.** If the account has credits as well as debits, it may be averaged, on the same principle, by this

RULE.

Start at the earliest day when a payment is made, or is due in cash; counting from this day, form the product for each item, both on the Dr. and Cr. sides, setting each product in a Dr. or Cr. column, according to the item multiplied. Balance the columns of products, and also the columns of items; the former divided by the latter will be the number of days to be added to the earliest date, to give the day the balance of items is due.

What is the balance in this account, and when is it due?

DR. *A in acc't with B.* CR.

1855.			Da.	Dol.	ct.	Prod.	1855.			Da.	Dol.	ct.	Prod.
Jan.	14	To invoice, 2m	13	400		5200	Mar	1	By cash,	0	500		0
Feb.	1	,, ,, ,,	31	200		6200	Apr	15	" remittance May 1,	61	250		15250
Mar.	10	,, ,, ,,	70	500		35000							
,,	25	,, ,, ,,	85	800		68000	May	20	,, cash,	80	375		30000
May	16	,, ,, ,,	137	300		41100	Jun	4	,, remittance June 25,	116	450		52200
				2200		155500					1575		97450
				1575		97450							
				625)		58050							

93 days after March 1 = June 2.

Ans. A owes B $625, due June 2.

SOLUTION.—Start with Mar. 1, the earliest day a payment is made or due. Find the products, both on the Dr. and Cr. side, using the dates when the items are due as cash. The balance of products, 58050, divided by the balance of items, 625, determines the mean time to be 93 days after Mar. 1 = June 2.

DEMONSTRATION.—In order to be due Mar. 1, the Dr. items must suffer a discount of $155500 for 1 day, and the Cr. items a discount of $97450 for 1 day. The Dr. side then must suffer a discount of $58050 for 1 day, more than the Cr. side. Therefore, the balance due Mar. 1 on the Dr. side, is $625, *less* the discount of

REVIEW.—341. What is the proof? 342. What is the rule for averaging an account that has credits? Solve the example. Demonstrate the rule.

$58050 for 1 day; but the disc:unt of $58050 for 1 day is the same as the discount of $625 for 93 days (Art. 341); and $625, due March 1, after being discounted for 93 days, is the same as $625 due 93 days after March 1 = June 2.

REMARK.—If the balance of items is on a different side of the account from the balance of products, it will *draw interest*, instead of *suffering discount*, for the time determined by the quotient; or, *it will be due, that long before the assumed date.*

ART. **343**. The rule for equation of payments is founded on Bank Discount, and assumes that the interest lost, by paying $100 five days *before* due, is exactly the same as the interest gained, by paying $100 five days *after* it is due. This is not strictly true; for, if I pay $100, 5 days before due, my exact loss is not 5 days' interest on $100, but on the *present worth*, which is less than $100; while, if I pay $100, 5 days after due, my gain is exactly 5 days' interest on $100; a little interest is thus gained by the debtor, or rather the mean time of payment is postponed, but the rule being confined to debts of short date, the error is small.

EXAMPLES FOR PRACTICE.

1. What is the mean time of the following invoices:

A to B. DR.

1851.				When due	Days after	Products.
May	15	To invoice at 4 months.	$800			
June	1	" " " 4 "	700			
"	10	" " " 4 "	900			
July	20	" " " 4 "	600			
Aug.	1	" " " 4 "	500			
"	15	" " " 4 "	1000			
		Sum total,	4500			

Ans. Oct. 30, 1851.

REMARK.—Start with Sept. 15, the day the first debt is due.

2. DR. *E in acc't current with F.* CR.

Feb.	4	To invoice 3 mon.	550	—	Prod.	May	8	By cash,	150	—	Prod.
Mar.	20	" " 3 "	260	—		"	26	" remit. June 5.	420	—	
Apr.	1	" " 3 "	150	—		Jun.	3	" note, 1 mon.	340	—	
"	5	" " 3 "	325	—		Jul.	1	" " 2 "	170	—	

Ans. E owes F $205, due Feb. 26.

REVIEW.—342. If the balance of items is on a different side of the accounts from the balance of products, what does it indicate? On what sort of discount is the rule founded? To whose advantage is this? Why?

3. *H. Wright to Mason & Giles.* Dr.

1856.			Dol.	ct.	When due		Days after April 8.	Products.
Feb.	1	To invoice 3 months.	900					
,,	20	,, ,, 3 ,,	700					
Mar.	10	,, ,, 3 ,,	600					
Apr.	8	,, ,, cash,	500					
May	10	,, ,, 3 ,,	900					
Jun.	15	,, ,, 3 ,,	400					
		Sum total,	4000					

Ans. Due June 13.

Remark.—Start with April 8, the time the first payment is due.

4. Dr. *A in acc't current with B.* Cr.

Mar.	19	To invoice cash,	900	—	Prod.	Feb.	20	By cash,	400	—	Prod.
Apr.	20	,, ,, ,,	800	—		Mar.	5	,, remit. Mar. 15.	300	—	
Jun.	15	,, ,, ,,	700	—		Jun.	20	,, cash,	200	—	
May	10	,, ,, ,,	600	—		Jul.	10	,, ,,	500	—	

Ans. A owes B a balance of $1600, due April 23.

Remark.—Start with Feb. 20, and date the remittance, Mar. 15, the day it is received.

5. Dr. *C in acc't current with D.* Cr.

Jan.	4	To invoice 2 mon.	250	—	Prod.	Mar.	10	By cash.	350	—	Prod.
Feb.	3	,, ,, 1 ,,	140	—		,,	21	,, ,,	200	—	
,,	15	,, ,, 2 ,,	450	—		Apr.	4	,, note 2 mon.	240	—	
Apr.	2	,, ,, cash,	100	—		May	20	,, remit., May 25.	120	—	
						Jun.	16	,, accep. 16 da. sight	500	—	

Ans. C is Cr. $470, due Aug. 12.

6. I owe $912, due Oct. 16, and $500, due Dec. 20. If I pay the first, Oct. 1, 15 days before due, when should I pay the last? *Ans.* Jan. 16, next, 27 da. after due.

7. I owe $2150, due Sept. 16; I pay $500, Aug. 4: when is the balance due? *Ans.* Sept. 29.

8. I owe a man the following notes: one of $800, due May 16; one of $660, due July 1; one of $940, due Sept. 29: he wishes to exchange them for two notes of $1200 each, and wants one to fall due June 1st; when should the other fall due? *Ans.* Sept. [illegible].

9. A owes $840, due Oct. 3; he pays $400, July 1; $200, Aug. 1: when will the balance be due? *Ans.* April 30, next yr.

10. I owe $3200, Oct. 25; I pay $400, Sept. 15; $800, Sept. 30: when will the balance be paid? *Ans.* Nov. 12

11. An account of \$2500 is due Sept. 16; \$500 are paid Aug. 1; \$500, Aug. 11; \$500, Aug. 21: when will the balance be due? *Ans.* Nov. 9.

12. Exchange the five following notes for six others, each for the same amount, and payable at equal intervals. One of \$1200, due in 41 days; one of \$1500, due in 72 days; one of \$2050, due in 80 days; one of \$1320, due in 110 days; one of \$1730, due in 125 days; total, \$7800.

Ans. The notes are \$1300 each, and run 25, 50, 75, 100, 125, 150 days respectively.

REMARK.—Since the notes of \$1300 are due at 1, 2, 3, 4, 5, 6 intervals respectively, use these numbers as multipliers: the quotient, 25 days, is the length of the interval.

13. I buy property for \$12000; $\frac{1}{3}$ in cash, and the balance in 2 equal payments, at 3 and 6 mon.; I pay $\frac{1}{4}$ down, and the balance in 3 equal payments, at equal intervals: what is the interval? *Ans.* 2 mon.

14. Exchange 3 notes, \$300 due in 10 da., \$500 due in 25 da., \$1000 due in 40 da., for \$600 cash, and 2 notes for \$550 and \$650 due at equal intervals; find the interval.

Ans. 30 da.

15. Account sales of lard oil, per steamboat Madison, for acc't of X and Y, Cincinnati.

1855.			Dol.	ct.	Days after Mar. 9.	Products.
Mar.	9	1 bbl. lard oil, 39 gal. at 80 c.—30 days.	31	20		
,,	10	1 ,, ,, ,, 39 ,, ,, 80 c.— ,, ,,	31	20		
,,	12	8 ,, ,, ,, 318 ,, ,, 80 c.—60 ,,	254	40		
	15	10 ,, ,, ,, 399 ,, ,, 78 c.— ,, ,,	311	22		
	19	1 ,, ,, ,, $40\frac{1}{2}$,, ,, 80 c.—30 ,,	32	40		
	21	16 ,, ,, ,, 638 ,, ,, 75 c.—60 ,,	478	50		
	22	3 ,, ,, ,, $119\frac{1}{2}$,, ,, 78 c.— ,, ,,	93	21		
	23	10 ,, ,, ,, 402 ,, ,, 77 c.— ,, ,,	309	54		
		50 bbl. = 1995 Due as cash, May 15 | 18.	1541	67		
		CHARGES.				
		To freight, \$45, drayage, \$3.25.................... 48 | 25				
		,, storage, labor, coop'age 8 |				
		,, fire insur. and adv....... 5 | 35				
		,, commission & guarantee, 4 per cent.......... — | —	123	27		
		Net proceeds, due May 15 | 18.	1418	40		
		Errors and omissions excepted.				
		EVANS & PRICE, New Orleans.				

REMARK.—The object in averaging an account sales, is that the consignor may draw a bill of exchange for the net proceeds, to fall due on the day of equation, without loss to either party.

16. Account sales of 1000 boxes star candles, for account of Messrs. A and B, of Cincinnati.

1854.			Dol.	ct.	Days after Apr. 21.	Products.
Apr.	21	Sold 5 bxs. 150 lb. $22\frac{1}{2}$ c. cash	33	75		
May	11	„ 1 box, 30 lb. $22\frac{1}{2}$ c. „	6	75		
Jun	28	„ 1 „ 30 lb. 25 c. „	7	50		
Jul.	13	„ 1 „ 30 lb. 23 c. „	6	90		
„	18	„ 10 bxs. 300 lb. 23 c. 2 per cent. off,	67	62		
Sept.	1	„ 500 „ 15000 lb. 23 c. at 6 mon.	3450	00		
„	2	„ 482 „ 14460 lb. 23 c. at 6 mon.	3325	80		
		1000 Due as cash, Feb. 25 | 28, 1855,	6898	32		
		CHARGES.				
Apr.	6	To freight on 1000 boxes . 216 | 23				
		„ drayage, storage, ins. 104 | 64				
		„ interest on freight to Feb. 28, 1855, ...				
		„ com. and guarantee 5 per ct. on $6898.32	677	39		
		Net proceeds due Feb. 25 | 28, 1855,	6220	93		
		Errors and omissions excepted.				
		HENRY WHITE, Philadelphia.				

ART. **344**. Accounts are averaged by tables of interest, as follows:

Take out the interest at any rate, on each item of the account, counting from the last day of the previous year, or any other convenient day: find the balance of interest, and refer to the table, under the head of the same rate, for the time corresponding to this balance: this will be the average time required.

REMARK.—Start at the last day of the previous year, because the date of each item will show how many months' and days' interest must be taken.

XXIX. COMPOUND INTEREST.

ART. **345**. Compound Interest differs from Simple Interest in allowing the Interest, as it accrues, to be converted into principal; thus producing interest on interest, and increasing the amount due, much more rapidly than simple interest, for the same time and rate.

REVIEW.—345. Does compound interest differ from simple interest?

ART. **346**. Compound, like Simple Interest, may be payable annually, semi-annually, quarterly, &c.; but the rate per cent. is given for a year, it being understood, as in simple interest, that the rate semi-annually is *one half*, and the rate quarterly is *one quarter*, of the annual rate.

Strictly speaking, the quarterly and semi-annual rates in compound interest are not one-fourth and one-half of the annual rate; for, the compound interest of

$1 for 1 year, at 8 % annually, is	.08 of a $.
" " " " 4 % semi-annually, is	.0816	"
" " " " 2 % quarterly, is	.08243216	"

ART. **347**. Compound, like Simple Interest, has 5 cases.

CASE I.

Given, the principal, time and rate, to find the compound amount and interest.

RULE.—*Find the amount of the principal for* 1 *interval, at the rate for that interval; then, the amount of this amount for another interval, and so on through the whole number of intervals; if there be, besides, a part of an interval, find the amount of the last amount for this time by the rule for simple interest. This will be the compound amount required, and the compound interest can be found by deducting the principal from it.*

NOTE.—By *interval* is meant a year, half-year, or quarter, as the interest may be payable.

Find the compound amount and interest of $4600 for 2 yr. 3 mon. 15 da., at 6 %, payable semi-annually.

SOLUTION.—The interval is 6 months, and the corresponding rate is 3 %. The number of intervals being 4, find the amounts of 4 successive principals at 3 %, and then the amount of the last amount, ($5177.34), for 3 mon. 15 da., at 6 % per annum: the result ($5267.94) is the compound amount; deducting the principal, $4600, the remainder, $667.94, is the compound interest.

REVIEW.—346. How is compound interest payable? What is said of the rate in compound interest? Show that rate quarterly and semi-annually are not $\frac{1}{4}$ and $\frac{1}{2}$ of the yearly rate in compound interest? 347 What is the rule for finding the comp. amount and interest of any sum for a given time and rate? What is meant by *interval?*

1. Find the compound amount, and interest, of $3850 for 4 yr. 7 mon. 16 da., at 5 %, payable annually.
Ans. $4826.59 and $976.59

2. The compound interest of $13062.50, for 1 yr. 10 mon. 12 da., at 8 %, payable quarterly. *Ans.* $2082.25

3. The compound amount of $1000 for 3 yr. at 10 %, payable semi-annually. *Ans.* $1340.10

4. Find the difference between the simple and comp. int. of $6320 for 5 yr. 8 mon. 6 da., at 6 %. *Ans.* $329.23

5. What sum, in 2 yr. 5 mon. 27 da., at 6 % simple int., will amount to the same as $2000 at comp. int., for the same time and rate, payable semi-annually? *Ans.* $2016.03

6. Find the compound int. of $1000000 for 10 yr. at 10 %, payable annually. *Ans.* $1593742.46

7. I have 80 shares of stock ($50), which pay a dividend every 6 mon. of 5 % in stock: how many shares will I have in 3 yr. 6 mon.?
Ans. 112 shares and $28.40 of another.

8. If I start with $5000, and increase my capital 15 % every year, what will it be in 6 years?
Ans. $11565.30

9. Find the comp. int. of $1200 invested at 8 % for 2 yr., and then at 10 % for 3 yr. *Ans.* $662.97

ART. **348**. As the operations in comp. interest are very laborious when the time is long, tables are used to facilitate the work.

To find the compound amount and interest of any principal, for any time and rate in the limits of the table,

RULE.

Observe at what intervals the interest is payable, also the number of such intervals, and the rate corresponding to each. Find from the table the compound amount of $1 for this rate and number of intervals, and multiply it by the given principal.

If there is any remaining time, calculate the amount on this product for that time, at the given rate: the result will be the compound amount for the whole time, and the compound interest can be found by deducting the principal from it.

REVIEW.—348. What are used to facilitate computations in compound interest? How do you find the compound amount and interest of any principal for any time and rate within the limits of the table?

Amount of $1 at Compound Interest in any number of years.

Yrs.	2 per cent.	2½ per cent.	3 per cent.	3½ per cent.	4 per cent.	4½ per cent.
1	1.0200 0000	1.0250 0000	1.0300 0000	1.0350 0000	1.0400 0000	1.0450 0000
2	1.0404 0000	1.0506 2500	1.0609 0000	1.0712 2500	1.0816 0000	1.0920 2500
3	1.0612 0800	1.0768 9062	1.0927 2700	1.1087 1787	1.1248 6400	1.1411 6612
4	1.0824 3216	1.1038 1289	1.1255 0881	1.1475 2300	1.1698 5856	1.1925 1860
5	1.1040 8080	1.1314 0821	1.1592 7407	1.1876 8631	1.2166 5290	1.2461 8194
6	1.1261 6242	1.1596 9342	1.1940 5230	1.2292 5533	1.2653 1902	1.3022 6012
7	1.1486 8567	1.1886 8575	1.2298 7387	1.2722 7926	1.3159 3178	1.3608 6183
8	1.1716 5938	1.2184 0290	1.2667 7008	1.3168 0904	1.3685 6905	1.4221 0061
9	1.1950 9257	1.2488 6297	1.3047 7318	1.3628 9735	1.4233 1181	1.4860 9514
10	1.2189 9442	1.2800 8454	1.3439 1638	1.4105 9876	1.4802 4428	1.5529 6942
11	1.2433 7431	1.3120 8666	1.3842 3387	1.4599 6972	1.5394 5406	1.6228 5305
12	1.2682 4179	1.3448 8882	1 4257 6089	1.5110 6866	1.6010 3222	1.6958 8143
13	1.2936 0663	1.3785 1104	1.4685 3371	1.5639 5606	1.6650 7351	1.7721 9610
14	1.3194 7876	1.4129 7382	1.5125 8972	1.6186 9452	1.7316 7645	1.8519 4492
15	1.3458 6834	1.4482 9817	1.5579 6742	1.6753 4883	1.8009 4351	1.9352 8244
16	1.3727 8570	1.4845 0562	1.6047 0644	1.7339 8604	1.8729 8125	2.0223 7015
17	1.4002 4142	1.5216 1826	1.6528 4763	1.7946 7555	1.9479 0050	2.1133 7681
18	1.4282 4625	1.5596 5872	1.7024 3306	1.8574 8920	2.0258 1652	2.2084 7877
19	1.4568 1117	1.5986 5019	1.7535 0605	1.9225 0132	2.1068 4918	2.3078 6031
20	1.4859 4740	1.6386 1644	1.8061 1123	1.9897 8886	2.1911 2314	2.4117 1402
21	1.5156 6634	1.6795 8185	1.8602 9457	2.0594 3147	2.2787 6807	2.5202 4116
22	1.5459 7967	1.7215 7140	1.9161 0341	2.1315 1158	2.3699 1879	2.6336 5201
23	1.5768 9926	1.7646 1068	1.9735 8651	2.2061 1448	2.4647 1555	2.7521 6635
24	1.6084 3725	1.8087 2595	2.0327 9411	2.2833 2849	2.5633 0417	2.8760 1383
25	1.6406 0599	1 8539 4410	2.0937 7793	2.3632 4498	2.6658 3633	3.0054 3446
26	1.6734 1811	1.9002 9270	2.1565 9127	2.4459 5856	2.7724 6979	3.1406 7901
27	1.7068 8648	1.9478 0002	2.2212 8901	2.5315 6711	2.8833 6858	3.2820 0956
28	1.7410 2421	1.9964 9502	2.2879 2768	2.6201 7196	2.9987 0332	3.4296 9999
29	1.7758 4469	2.0464 0739	2.3565 6551	2.7118 7798	3.1186 5145	3.5840 3649
30	1.8113 6158	2.0975 6758	2.4272 6247	3.8067 9370	3.2433 9751	3.7453 1813
31	1.8475 8882	2.1500 0677	2.5000 8035	2.9050 3148	3.3731 3341	3.9138 5745
32	1.8845 4059	2.2037 5694	2.5750 8276	3.0067 0759	3.5080 5875	4.0899 8104
33	1.9222 3140	2.2588 5086	2.6523 3524	3.1119 4235	3.6483 8110	4.2740 3018
34	1.9606 7603	2.3153 2213	2.7319 0530	3.2208 6033	3.7943 1634	4.4663 6154
35	1.9998 8955	2.3732 0519	2.8138 6245	3.3335 9045	3.9460 8899	4.6673 4781
36	2.0398 8734	2.4325 3532	2.8982 7833	3.4502 6611	4.1039 3255	4.8773 7846
37	2.0806 8509	2.4933 4870	2.9852 2668	3.5710 2543	4.2680 8986	5.0968 6049
38	2.1222 9879	2.5556 8242	3.0747 8348	3.6960 1132	4.4388 1345	5.3262 1921
39	2.1647 4477	2.6195 7448	3.1670 2698	3.8253 7171	4.6163 6599	5.5658 9908
40	2.2080 3966	2.6850 6384	3.2620 3779	3.9592 5972	4.8010 2063	5.8163 6454
41	2.2522 0046	2.7521 9043	3.3598 9893	4.0978 3381	4.9930 6145	6.0781 0094
42	2.2972 4447	2.8209 9520	3.4606 9589	4.2412 5799	5.1927 8391	6.3516 1548
43	2.3431 8936	2.8915 2008	3.5645 1677	4.3897 0202	5.4004 9527	6.6374 3818
44	2.3900 5314	2.9638 0808	3.6714 5227	4.5433 4160	5.6165 1508	6.9361 2290
45	2.4378 5421	3.0379 0328	3.7815 9584	4.7023 5855	5.8411 7568	7.2482 4843
46	2.4866 1129	3.1138 5086	3.8950 4372	4 8669 4110	6.0748 2271	7.5744 1961
47	2.5363 4351	3.1916 9713	4.0118 9503	5.0372 8404	6.3178 1562	7.9152 6849
48	2.5870 7039	3.2714 8956	4.1322 5188	5.2135 8898	6.5705 2824	8.2714 5557
49	2.6388 1179	3.3532 7680	4.2562 1944	5.3960 6459	6.8333 4937	8.6436 7107
50	2.6915 8803	3.4371 0872	4.3839 0602	5.5849 2686	7.1066 8335	9.0326 3627
51	2.7454 1979	3.5230 3644	4.5154 2320	5.7803 9930	7.3909 5068	9.4391 0490
52	2.8003 2819	3.6111 1235	4.6508 8590	5.9827 1327	7.6865 8871	9.8638 6463
53	2.8563 3475	3.7013 9016	4.7904 1247	6.1921 0824	7.9940 5226	10.3077 3853
54	2.9134 6144	3.7939 2491	4.9341 2485	6.4088 3202	8.3138 1435	10.7715 8677
55	2.9717 3067	3.8887 7303	5.0821 4859	6.6331 4114	8.6463 6692	11.2563 0817

Amount of $1 at Compound Interest in any number of years.

Yr.	5 per cent.	6 per cent.	7 per cent.	8 per cent.	9 per cent.	10 per cent.
1	1.0500 000	1.0600 000	1.0700 000	1.0800 000	1.0900 000	1.1000 000
2	1.1025 000	1.1236 000	1.1449 000	1.1664 000	1.1881 000	1.2100 000
3	1.1576 250	1.1910 160	1.2250 430	1.2597 120	1.2950 290	1.3310 000
4	1.2155 063	1.2624 770	1.3107 960	1.3604 890	1.4115 816	1.4641 000
5	1.2762 816	1.3382 256	1.4025 517	1.4693 281	1.5386 240	1.6105 100
6	1.3400 956	1.4185 191	1.5007 304	1.5868 743	1.6771 001	1.7715 610
7	1.4071 004	1.5036 303	1.6057 815	1.7138 243	1.8280 391	1.9487 171
8	1.4774 554	1.5938 481	1.7181 862	1.8509 302	1.9925 626	2.1435 888
9	1.5513 282	1.6894 790	1.8384 592	1.9990 046	2.1718 933	2.3579 477
10	1.6288 946	1.7908 477	1.9671 514	2.1589 250	2.3673 637	2.5937 425
11	1.7103 394	1.8982 986	2.1048 520	2.3316 390	2.5804 264	2.8531 167
12	1.7958 563	2.0121 965	2.2521 916	2.5181 701	2.8126 648	3.1384 284
13	1.8856 491	2.1329 283	2.4098 450	2.7196 237	3.0658 046	3.4522 712
14	1.9799 316	2.2609 040	2.5785 342	2.9371 936	3.3417 270	3.7974 983
15	2.0789 282	2.3965 582	2.7590 315	3.1721 691	3.6424 825	4.1772 482
16	2.1828 746	2.5403 517	2.9521 638	3.4259 426	3.9703 059	4.5949 730
17	2.2920 183	2.6927 728	3.1588 152	3.7000 181	4.3276 334	5.0544 703
18	2.4066 192	2.8543 392	3.3799 323	3.9960 195	4.7171 204	5.5599 173
19	2.5269 502	3.0255 995	3.6165 275	4.3157 011	5.1416 613	6.1159 090
20	2.6532 977	3.2071 355	3.8696 845	4.6609 571	5.6044 108	6.7275 000
21	2.7859 626	3.3995 636	4.1405 624	5.0338 337	6.1088 077	7.4002 499
22	2.9252 607	3.6035 374	4.4304 017	5.4365 404	6.6586 004	8.1402 749
23	3.0715 238	3.8197 497	4.7405 299	5.8714 637	7.2578 745	8.9543 024
24	3.2250 999	4.0489 346	5.0723 670	6.3411 807	7.9110 832	9.8497 327
25	3.3863 549	4.2918 707	5.4274 326	6.8484 752	8.6230 807	10.8347 059
26	3.5556 727	4.5493 830	5.8073 529	7.3963 532	9.3991 579	11.9181 765
27	3.7334 563	4.8223 459	6.2138 676	7.9880 615	10.2450 821	13.1099 942
28	3.9201 291	5.1116 867	6.6488 384	8.6271 064	11.1671 395	14.4209 936
29	4.1161 356	5.4183 879	7.1142 571	9.3172 749	12.1721 821	15.8630 930
30	4.3219 424	5.7434 912	7.6122 550	10.0626 569	13.2676 785	17.4494 023
31	4.5380 395	6.0881 006	8.1451 129	10.8676 694	14.4617 695	19.1943 425
32	4.7649 415	6.4533 867	8.7152 708	11.7370 830	15.7633 288	21.1137 768
33	5.0031 885	6.8405 899	9.3253 398	12.6760 496	17.1820 284	23.2251 544
34	5.2533 480	7.2510 253	9.9781 135	13.6901 336	18.7284 109	25.5476 699
35	5.5160 154	7.6860 868	10.6765 815	14.7853 443	20.4139 679	28.1024 369
36	5.7918 161	8.1472 520	11.4239 422	15.9681 718	22.2512 250	30.9126 805
37	6.0814 069	8.6360 871	12.2236 181	17.2456 256	24.2538 353	34.0039 486
38	6.3854 773	9.1542 524	13.0792 714	18.6252 756	26.4366 805	37.4043 434
39	6.7047 512	9.7035 075	13.9948 204	20.1152 977	28.8159 817	41.1447 778
40	7.0399 887	10.2857 179	14.9744 578	21.7245 215	31.4094 200	45.2592 556
41	7.3919 882	10.9028 610	16.0226 699	23.4624 832	34.2362 679	49.7851 811
42	7.7615 876	11.5570 327	17.1442 568	25.3394 819	37.3175 320	54.7636 992
43	8.1496 669	12.2504 546	18.3443 548	27.3666 404	40.6761 098	60.2400 692
44	8.5571 503	12.9854 819	19.6284 596	29.5559 717	44.3369 597	66.2640 761
45	8.9850 078	13.7646 108	21.0024 518	31.9204 494	48.3272 861	72.8904 837
46	9.4342 582	14.5904 875	22.4726 234	34.4740 853	52.6767 419	80.1795 321
47	9.9059 711	15.4659 167	24.0457 070	37.2320 122	57.4176 486	88.1974 853
48	10.4012 697	16.3938 717	25.7289 065	40.2105 731	62.5852 370	97.0172 338
49	10.9213 331	17.3775 040	27.5299 300	43.4274 190	68.2179 083	106.7189 572
50	11.4673 998	18.4201 543	29.4570 251	46.9016 125	74.3575 201	117.3908 529
51	12.0407 698	19.5253 635	31.5190 168	50.6537 415	81.0496 969	129.1299 382
52	12.6428 083	20.6968 853	33.7253 480	54.7060 408	88.3441 696	142.0429 320
53	13.2749 487	21.9386 985	36.0861 224	59.0825 241	96.2951 449	156.2472 252
54	13.9386 961	23.2550 204	38.6121 509	63.8091 260	104.9617 079	171.8719 477
55	14.6356 309	24.6503 216	41.3150 015	68.9138 561	114.4082 616	189.0591 425

10. Find the compound amount of $750, for 17 yr. at 6 %, payable annually. *Ans.* $2019.58

11. Of $5428, for 33 yr., at 5 % annually. *Ans.* $27157.31

12. The compound interest of $1800 for 14 yr., at 8 %, payable semi-annually. *Ans.* $3597.67

13. If $1000 is deposited for a child, at birth, and draws 7 % comp. int., payable semi-annually, till it is of age (21 yr.), what will it amount to? *Ans.* $4241.26

14. Find the compound amount and int. of $9401.50 for 19 yr. 4 mon., at 9 %, payable semi-annually. *Ans.* Am't. $51576.68, Int. $42175.18

15. The comp. int. of $1176.80 for 10 yr. 10 mon. 10 da., at 10 %, payable quarterly. *Ans.* $2263.75

16. The comp. amt. of $5000 for 12 yr. 5 mon. 22 da., at 12 %, payable semi-annually. *Ans.* $21405.37

17. The comp. int. of $8025 for 18 yr. 7 mon. 9 da., at 8 %, payable semi-annually. *Ans.* $26523.27

If the time is beyond the limits of the table, use this

RULE.

Separate the given time into two or more periods, each within the limits of the table. Find the compound amount of the principal for the first of these periods, then the compound amount of this amount for the second period, and so on. The last amount will be the compound amount required, from which the compound interest can be found as before.

18. Find the compound amount of $1000 for 100 yr., at 10 %, payable annually. *Ans.* $13780612.34

19. The comp. amt. and int. of $3600 for 15 yr., at 8 %, payable quarterly. *Ans.* Amt. $11811.71, Int. $8211.71

20. The comp. int. of $4000 for 40 yr. at 5 %, payable semi-annually. *Ans.* $24838.27

21. The comp. amt. of $2500 for 32 yr. 8 mon. 6 da., at 7 %, payable semi-annually. *Ans.* $23691.95

22. Find the comp. int. of $1200 for 27 yr. 11 mon. 4 da., at 12 %, payable quarterly. *Ans.* $31404.74

ART. **349**. CASE II.—Given, the principal, time, and interest or amount, to find the rate.

REVIEW.—348. What, if the time is beyond the limits of the table?

RULE.—*If the interest be given, add it to the principal to get the amount; then divide the amount by the principal; the quotient will be the amount of $1 for the given time and rate. Look in the table opposite the given number of intervals until this amount is found; the rate per cent. at the head of the column will be the one required.*

NOTE.—If the time contains a part of an interval, find the amounts, for that time, of the numbers in the table, before comparing them with the quotient.

AT WHAT RATE, BY COMPOUND INTEREST,

1. Will $1000 amount to $1593.85 in 8yr.? *Ans.* 6 %.
2. $3600 yield $6332.51 int. in 15 yr.? *Ans.* 7 %.
3. $13200 amt. to $48049.58, in 26 yr. 5 mon. 21 da.? *Ans.* 5 %.
4. $2813.50 amt. to $13276.03, in 17 yr. 7 mon. 14 da., int. payable semi-annually? *Ans.* 9 %.

SUGGESTION.—The rate found by the rule in this example is the semi-annual rate, and must be doubled to get the rate per annum.

5. $7652.18 yield $17198.67 int., payable quarterly, in 11 yr. 11 mon. 3 da.? *Ans.* 10 %.
6. At what rate will any sum double itself by compound int. in 8, 10, 12, 15, 20 yr.?

Ans. 1st, little over 9 %; 2d, between 7 and 8 %: 3d, not quite 6 %; 4th, not quite 5 %; 5th, little over $3\frac{1}{2}$ %.

ART. **350.** CASE III.—Given, the compound interest, the time, and rate, to find the principal.

RULE.—*Assume $1 for the principal; determine the compound interest on this supposition, and divide the given compound interest by it.*

PROOF.—With the principal thus found, calculate the compound interest for the given time and rate; if it agrees with the given compound interest, the work is right.

NOTE.—To get the amount, add the principal to the interest.

1. What principal will yield $52669.93 compound int. in 25 yr., at 6 % per annum? *Ans.* $16000.

REVIEW.—349. What is Case 2? The rule? If the time contain a part of an interval what is necessary? 350. What is Case 3? The rule? The proof?

2. What sum, in 6 yr. 2 mon., will yield $1625.75 compound int. at 7%, payable semi-annually? *Ans.* $3075.

3. What sum, at 10%, payable quarterly, will produce $3598.61 comp. int. in 3 yr. 6 mon. 9 da.? *Ans.* $8640.

4. What principal, in 37 yr., at 5% per annum, will produce $6891.61 compound interest? *Ans.* $1356.24

5. What sum, in $9\frac{1}{2}$ yr., at 8%, payable semi-annually, will yield $31005.76 compound int.? *Ans.* $28012.63

ART. **351**. CASE IV.—Given, the compound amount, the time, and rate, to find the principal.

RULE.—*Assume* $1 *for the principal; determine the compound amount on this supposition, and divide the given compound amount by it.*

PROOF.—With the principal thus found, calculate the compound amount at the given time and rate; if it agrees with the given compound amount, the work is right.

NOTES.—1. To get the interest, subtract the principal from the amount.

2. If the compound amount is a debt not yet due, the principal is its *present worth* by compound interest, and the difference between the debt and present worth is the *compound discount* of the *debt*, or the comp. int. of its present worth for the given time and rate.

1. What principal in 7 yr., at 4% compound interest, will amount to $27062.85? *Ans.* $20565.54

2. Find the present worth of $14625.70, due in 5 yr. 9 mon., at 6% comp. int., payable semi-annually. *Ans.* $10409.77

3. The discount on $8767.78, due in 12 yr. 8 mon. 25 da., at 5% comp. int. *Ans.* $4058.87

4. The present worth and comp. discount of $48941.28, due in 10 yr. 4 mon. 6 da., at 10%, payable quarterly. *Ans.* P. W. $17606.60, C. D. $31334.68

5. The present worth of $100000, due in 50 yr., at 9% comp. int. *Ans.* $1344.85

6. The difference between the present worth, at simple and at comp. int., of $34058.75, due in 20 yr., at 6%. *Ans.* $4861.57

REVIEW.—351. What is Case 4? The rule? The proof? How is the comp. int. then found?

ART. 352. CASE V.—Given, the principal, rate, and compound interest or amount, to find the time.

RULE.—*If the interest is given, add it to the principal to get the amount; then divide the amount by the principal; the quotient will be the compound amount of $1 for the time and rate.*

Look in the table under the head of the given rate until this number is found; the corresponding number in the left hand column will be the number of years (or intervals) required.

But if the number can not be found exactly in the table, take out the number of years (or intervals) corresponding to the smaller one of the two numbers between which it lies, and the remainder of the time will be such part of a year (or interval) as the difference between the number taken and the quotient, is of the difference between that smaller one and the next larger.

The part of a year (or interval) thus determined can be converted into months and days, and annexed to the whole number of years (or intervals) already taken out.

PROOF.—With the time thus found, calculate the compound amount of the given principal at the given rate; if it agrees with the given compound amount, the work is right.

1. In what time will $8000 amount to $12000, at 6 % compound interest? *Ans.* 6 yr. 11 mon. 15 da.

2. In what time will any sum of money double itself, by compound interest at 5, 6, 7, 8, 10 %?
Ans. 14 yr. 2 mon. 13 da.; 11 yr. 10 mon. 21 da.; 10 yr. 2 mon. 26 da.; 9 yr. 2 da.; 7 yr. 3 mon. 5 da.

3. In what time will $5200 draw $1308 comp. int. at 6 % payable semi-annually? *Ans.* 3 yr. 9 mon. 16 da.

4. In what time will $12500 amount to $18000, by comp. int. at 10 %, payable quarterly?
Ans. 3 yr. 8 mon. 9 da.

5. In what time will $9862.50 amount to $22576.15 by compound interest at 12 %, payable semi-annually?
Ans. 7 yr. 1 mon. 7 da.

XXX. ANNUITIES.

ART. 353. The principal application of Compound Interest is to Annuities.

REVIEW.—352. What is Case 5? The rule? The proof?

An Annuity is an estate which entitles its owner to the payment of a fixed sum, at regular intervals of time.

Annuities comprise Life-insurance, Rents, Dowers, Leases, Life-estates, Survivorships, Pensions, &c.

It is called Annuity, from the Latin *annus*, meaning a year, because the interval between the payments is generally a year, though sometimes it is half or quarter of a year.

A *Perpetuity* is an annuity which continues forever.

An *Annuity certain* begins and ends at fixed times.

A *contingent Annuity* begins or ends with the happening of an uncertain event; as, the birth or death of one or more persons.

ART. **354**. An *immediate* annuity begins at once.

A *deferred* annuity, or annuity *in reversion*, begins at a future time.

The *forborne* or *final value* of an annuity, is the sum of the compound amounts of all its payments at an assumed rate, from the time each is due, to the end of the annuity.

The *present value* of an annuity, is the present worth of the final or forborne value; that is, such a sum as put out at compound interest as proposed, will, at the expiration of the annuity, amount to its final value.

The value of a deferred annuity at the time it commences, may be called its *initial value;* its *present value*, is the present worth of its initial value, at an assumed rate of compound interest.

An annuity is said to be worth *as many years' purchase*, as its present value contains one of its payments.

An annuity begins, not at the time the first payment is made, but one interval before; if an annuity begins now, its first payment will be a year, half-year, or quarter of a year hence, according to the interval between the payments.

CASE I.

ART. **355**. Given, the rate, the payment, and the interval, to find the initial value of a perpetuity.

REVIEW.—353. What are annuities? What do they comprise? Why so called? What is a perpetuity? An annuity certain? A contingent annuity? 354. What is an immediate annuity? A deferred annuity? What is the forborne or final value of an annuity? Its present value? The initial value of a deferred annuity? 355. What is Case 1?

RULE.—*Find, by Case III of Simple Interest, the principal whose interest at the given rate, for the given interval, equals the given payment: this will be the initial value required.*

NOTE.—If the perpetuity is immediate, its initial value is its present value.

What is the initial value of a perpetual lease of $250 a year, allowing 6 % interest?

SOLUTION.—The initial value must be the principal, which, at 6 %, yields $250 interest every year; it is found by Art. 310. The operation amounts to *dividing the given payment by the interest of $1 for the interval and at the rate proposed.*

$$\begin{array}{l} \$1 \\ .06 \\ \hline .06)\underline{250.0000} \\ \textit{Ans. } \$4166.66\tfrac{2}{3} \end{array}$$

1. What is the initial value of a perpetual leasehold of $300 a year, allowing 6 % interest? *Ans.* $5000.

2. What must I pay for a perpetual lease of $756.40 a year, to secure 8 % interest? *Ans.* $9455.

3. Ground rents on perpetual lease, yield an income of $15642.90 a year: what is the present value of the estate, allowing 7 % interest? *Ans.* $223470.

4. What is the initial value of a perpetual leasehold of $1600 a year, payable semi-annually, allowing 5 % interest, payable annually? *Ans.* $32400.

SUGGESTION.—Here the yearly payment is $1620, by allowing 5 per cent. interest on the half-yearly payment first made.

5. What is the initial value of a perpetual leasehold of $2500 a year, payable quarterly, interest 6 % payable semi-annually; 6 % payable annually; 6 % payable quarterly?
Ans. $41979.16⅔; $42604.16⅔; $41666.66⅔

CASE II.

ART. **356**. To find the present value of a deferred perpetuity, when the payment, the interval, the rate, and the time the perpetuity is deferred, are known.

RULE.—*Find the initial value of the perpetuity by the last rule; then find the present worth of this sum for the time the perpetuity is deferred, by Case IV of Compound Interest; this will be the present value required.*

REVIEW.—355. What is the rule? If the perpetuity is immediate, what is the initial value? Solve the example.

Find the present value of a perpetuity of $250 a year, deferred 8 yr., allowing 6 % int.

SOLUTION.—Initial value of perpetuity of $250 a year, by last rule = $4166.66⅔. The present value of $4166.66⅔, due 8 yr. hence, at 6 % comp. int., = $4166.66⅔ ÷ 1.5938481 (Art. 351). Use the contracted method, reserving 3 decimal places: the quotient, $2614.22, is the present value of the perpetuity.

1. Find the present value of a perpetuity of $780 a yr., to commence in 12 yr., int. 5 %. *Ans.* $8686.66

2. Of a perpetual lease of $160 a year, to commence in 3 yr. 4 mon., int. 7 %. *Ans.* $1823.28

3. Of the reversion of a perpetuity of $540 a year, deferred 10 yr., int. 6 %. *Ans.* $5025.55

4. Of an estate which, in 5 yr. is to pay $325 a yr. for ever: int. 8 %, payable semi-annually. *Ans.* $2690.67

5. Of a perpetuity of $1000 a year, payable quarterly, to commence in 9 yr. 10 mon. 18 da., int. 10 %, payable semi-annually. *Ans.* $3858.88

CASE III.

ART. **357**. Given, the rate, the payment, the interval, and the time to run, to find the present value of an annuity certain.

RULE.—*Find the present value of two perpetuities having the given rate, payment, and interval, one of them commencing when the annuity does, and the other when the annuity ends. The difference between these values will be the present value of the annuity.*

NOTE.—This rule applies whether the annuity is immediate or deferred; in the latter, the time the annuity is deferred must be known, and used in getting the values of the perpetuities.

2. By using the *initial* instead of the *present* values of the perpetuities, the rule gives the *initial* value of the deferred annuity, which may be used in finding its final or forborne value. (Rem. I, Case IV.)

1. Find the present value of an immediate annuity of $250 continuing 8 years, 6 % interest.

Pres. val. of immediate perpetuity of $250, . . .	= $4166.67
Pres. val. of perpetuity of $250, deferred 8 yr. . . .	= 2614.22
Pres. val. of immediate annuity of $250, running 8 yr.	= $1552.45

REVIEW.—356. What is Case 2? The rule? Solve the example. 357. What is Case 3? The rule?

2. The present value of an annuity of $680, to commence in 7 yr. and continue 10 yr., 5 % int.

Pres. val. of perpetuity of $680, deferred 7 yr., at 5 %, = $9665.27
Pres. val. of perpetuity of $680, deferred 17 yr., at 5 %, = 5933.20
Pres. val. of annuity of $680, defer'd 7 yr., to run 10 yr. = $3732.07

3. The pres. val. of an annuity of $125, to commence in 12 yr. and run 12 yr., int. 7 %. *Ans.* $440.83

4. The present value of an immediate annuity of $400 running 15 yr. 6 mon., int. 8 %. *Ans.* $3484.41

5. The pres. val. of an annuity of $826.50, to commence in 3 yr. and run 13 yr. 9 mon., int. 6 %, payable semi-annually. *Ans.* $6324.69

6. The present value of an annuity of $60, deferred 12 yr. and to run 9 yr., int. $4\frac{1}{2}$ %. *Ans.* $257.17

7. Sold a lease of $480 a yr., payable quarterly, having 8 yr. 9 mon. to run, for $2500: do I gain or lose, int. 8 %, payable semi-annually? *Ans.* Lose $509.96

CASE IV.

ART. **358**. Given, the payment, the interval, the rate, and time to run, to find the final or forborne value of an annuuity.

RULE.—*Consider the annuity a perpetuity, and find its initial value, by Case I. The compound interest of this sum, at the given rate for the time the annuity runs, will be the final or forborne value.*

Find the final or forborne value of an annuity of $250, continuing 8 yr., int. 6 %.

SOLUTION.—The initial value of a perpetuity of $250, at 6 %, = 4166.66\frac{2}{3}$; its compound interest, at 6 % for 8 yr., = 4166.66\frac{2}{3}$ × .5938481 = $2474.37, the final or forborne value of the annuity.

REMARKS.—1. The final or forborne value of an annuity may be got by first getting its initial value by Case III, and finding its compound amount for the time the annuity runs, at the given rate.

2. The present value of an annuity can be obtained, by first getting its final or forborne value by the rule in this case, and finding its present worth for the time the annuity runs, at the given rate.

REVIEW.—357. What kind of annuities does the rule apply to? How can the initial value of a deferred annuity be found? Of what use is it? 358. What is Case 4? What is the rule of Case 4? How else may the final value of an annuity be found? How may the present value of an annuity be found from this rule?

ART. **359**. *The present value of* $1 *per annum in any number of years.*

Yrs.	4 per cent.	5 per cent.	6 per cent.	7 per cent.	8 per cent.	10 per cent.
1	.961538	.952381	.943396	.934579	.925926	.909091
2	1.886095	1.859410	1.833393	1.808018	1.783265	1.735537
3	2.775091	2.723248	2.673012	2.624316	2.577097	2.486852
4	3.629895	3.545951	3.465106	3.387211	3.312127	3.169865
5	4.451822	4.329477	4.212364	4.100197	3.992710	3.790787
6	5.242137	5.075692	4.917324	4.766540	4.622880	4.355261
7	6.002055	5.786373	5.582381	5.389289	5.206370	4.868419
8	6.732745	6.463213	6.209794	5.971299	5.746639	5.334926
9	7.435332	7.107822	6.801692	6.515232	6.246888	5.759024
10	8.110896	7.721735	7.360087	7.023582	6.710081	6.144567
11	8.760477	8.306414	7.886875	7.498674	7.138964	6.495061
12	9.385074	8.863252	8.383844	7.942686	7.536078	6.813692
13	9.985648	9.393573	8.852683	8.357651	7.903776	7.103356
14	10.563123	9.898641	9.294984	8.745468	8.244237	7.366687
15	11.118387	10.379658	9.712249	9.107914	8.559479	7.606080
16	11.652296	10.837770	10.105895	9.446649	8.851369	7.823709
17	12.165669	11.274066	10.477260	9.763223	9.121638	8.021553
18	12.659297	11.689587	10.827603	10.059087	9.371887	8.201412
19	13.133939	12.085321	11.158116	10.335595	9.603599	8.364920
20	13.590326	12.462210	11.469921	10.594014	9.818147	8.513564
21	14.029160	12.821153	11.764077	10.835527	10.016803	8.648694
22	14.451115	13.163003	12.041582	11.061241	10.200744	8.771540
23	14.856842	13.488574	12.303379	11.272187	10.371059	8.883218
24	15.246963	13.798642	12.550358	11.469334	10.528758	8.984744
25	15.622080	14.093945	12.783356	11.653583	10.674776	9.077040
26	15.982769	14.375185	13.003166	11.825779	10.809978	9.160945
27	16.329586	14.643034	13.210534	11.986709	10.935165	9.237223
28	16.663063	14.898127	13.406164	12.137111	11.051078	9.306567
29	16.983715	15.141074	13.590721	12.277674	11.158406	9.369606
30	17.292033	15.372451	13.764831	12.409041	11.257783	9.426914
31	17.588494	15.592811	13.929086	12.531814	11.349799	9.479013
32	17.873552	15.802677	14.084043	12.646555	11.434999	9.526376
33	18.147646	16.002549	14.230230	12.753790	11.513888	9.569432
34	18.411198	16.192904	14.368141	12.854009	11.586934	9.608575
35	18.664613	16.374194	14.498246	12.947672	11.654568	9.644159
36	18.908282	16.546852	14.620987	13.035208	11.717193	9.676508
37	19.142579	16.711287	14.736780	13.117017	11.775179	9.705917
38	19.367864	16.867893	14.846019	13.193473	11.828869	9.732651
39	19.584485	17.017041	14.949075	13.264928	11.878582	9.756956
40	19.792774	17.159086	15.046297	13.331709	11.924613	9.779051
41	19.993052	17.294368	15.138016	13.394120	11.967235	9.799137
42	20.185627	17.423208	15.224543	13.452449	12.006699	9.817397
43	20.370795	17.545912	15.306173	13.506962	12.043240	9.833998
44	20.548841	17.662773	15.383182	13.557908	12.077074	9.849089
45	20.720040	17.774070	15.455832	13.605522	12.108402	9.862808
46	20.884654	17.880067	15.524370	13.650020	12.137409	9.875280
47	21.042936	17.981016	15.589028	13.691608	12.164267	9.886618
48	21.195131	18.077158	15.650027	13.730474	12.189136	9.896926
49	21.341472	18.168722	15.707572	13.766799	12.212163	9.906296
50	21.482185	18.255925	15.761861	13.800746	12.233485	9.914814
51	21.617485	18.338977	15.813076	13.832473	12.253227	9.922559
52	21.747582	18.418073	15.861393	13.862124	12.271506	9.929599
53	21.872675	18.493403	15.906974	13.889836	12.288432	9.935999
54	21.992957	18.565146	15.949976	13.915735	12.304103	9.941817
55	22.108612	18.633472	15.990543	13.939939	12.318614	9.947107

1. Find the forborne value of an immediate annuity of \$300, running 18 yr., int. 5 %. *Ans.* \$8439.72

2. A pays \$25 a year for tobacco; how much better off would he have been in 40 yr., if he had invested it at 10 % per annum? *Ans.* \$11064.81

3. Find the forborne value of an annuity of \$75, to commence in 14 yr., and run 9 yr., int. 6 %. *Ans.* \$861.85

SUGGESTION.—The 14 yr. is not used.

4. A pays \$5 a year for a newspaper; if invested at 9 %, what will his subscription have produced in 50 yr.? *Ans.* \$4075.42

5. B pays \$150 a year, to have his life insured for \$5000: if he dies in 20 yr., does the Insurance Co. gain or lose? (int. 6 %.) *Ans.* Gains \$517.84

ART. **360**. By the table just given, some interesting and important cases in annuities can be solved, among which are the following three:

CASE V.

Given, the rate, time to run, and the present or final value of an annuity, to find the payment.

RULE.—*Assume* \$1 *for the payment; determine the present or final value on this supposition, and divide the given present or final value by it.*

NOTE.—Get the *final* value, if necessary, by the rule in Case IV.

1. An immediate annuity running 11 yr., can be purchased for \$6000: what is the payment, int. 6 % ?

SOLUTION.—The present value of an immediate annuity of \$1 for 11 yr., at 6 %, is \$7.886875; \$6000 divided by this gives \$760.76, the payment required.

2. How much a year should I pay, to secure \$15000 at the end of 17 yr., int. 7 % ? *Ans.* \$486.38

3. What is the payment of an annuity, deferred 4 yr., running 16 yr., and worth \$4800, int. 6%? *Ans.* \$599.64

CASE VI.

ART. **361**. Given, the payment, the rate, and present value of an annuity, to find the time it runs.

RULE.—*Divide the present value by the yearly payment; the quotient will be the present value of an annuity of* \$1, *similar to the one proposed; by looking in the table for this number, under the head of the given rate per cent., the corresponding number of years can be found.*

NOTE.—If the number is not in the table, take the nearest one below it; the balance remaining can be ascertained by finding the difference between the compound amount of the present value, and the final amount of the annuity for the number of years taken out.

REMARK.—This case is of use in finding the time required to liquidate a debt drawing interest, by means of a *sinking fund*, that is, by installments at regular intervals of a year, half-year, or quarter of a year; for example,

In what time will a debt of $10000, drawing interest at 6 %, be paid by installments of $1000 a year?

SOLUTION.—The $10000 may be considered the present value of an annuity of $1000 a year at 6 %; but $10000 ÷ $1000, = $10, the present value of an annuity of $1 for the same time and rate; by reference to the table, the time corresponding to this present value, under the head of 6 %, is 15 yr., a balance being then due of $689.61.

Comp. amt. of $10000 for 15 yr., at 6 %, (Art. 348.) .	=$23965.58
Final val. of annuity $1000 for 15 yr., at 6 %, (Art. 358.) =	23275.97
Bal. due at end of 15 yr.	$689.61

1. In how many years can a debt of $1000000, drawing interest at 6 %, be discharged by a sinking fund of $80000 a year? *Ans.* 23 yr., and $60083.43 then unpaid.

2. In how many years can a debt of $30000000, drawing interest at 5 %, be paid by a sinking fund of $2000000? *Ans.* 28 yr., and $798709.00 unpaid.

3. In how many years can a debt of $22000, drawing 7 % interest, be discharged by a sinking fund of $2500 a year? *Ans.* 14 yr., and $351.53 then unpaid.

CASE VII.

ART. **362**. Given, the payment, time to run, and present value of an annuity, to find the rate of interest.

RULE.—*Divide the present value by the yearly payment; the quotient will be the present value of an annuity of $1 similar to the one proposed; by looking in the table opposite the given*

REVIEW.—360. What is Case 5? The rule? How may the final value of the annuity of $1 be found if necessary? 361. What is Case 6? What is the rule of Case 6? If the number of intervals be not exact, how find the balance remaining for the partial interval?

number of years until this number is found, the rate per cent. at the head of the column which contains it, will be the rate required.

1. If \$9000 is paid for an immediate annuity of \$750 to run 20 yr., what is the rate? *Ans.* About $5\frac{1}{2}$ %.

2. If an immediate annuity of \$80, running 14 yr., sells for \$650, what is the rate? *Ans.* 8 %+

CONTINGENT ANNUITIES.

ART. **363**. Contingent Annuities comprise Life Insurance, Dowers, Pensions, &c.

A person who has his life insured pays a fixed sum, called the *annual premium*, every year he lives, and at his death, the sum insured is paid to the person designated in the *policy*.

The premium is a certain rate per cent. of the sum insured, varying with the age of the applicant.

Bills of Mortality are tables showing how many of a given number, say 10000, born in the same year, die at each age.

Expectation of life is the average number of years a person of any age lives, as deduced from bills of mortality.

A table of life-annuities shows the sum to be paid by a person of any age, to secure an annuity of \$1 for life.

CASE I.

ART. **364**. To find the *value* of a given annuity on the life of a person whose age is known.

RULE.—*Find from the table the value of a life-annuity of* \$1, *for the given age and rate of interest, and multiply it by the payment of the given annuity.*

NOTES.—1. To find the value of a life-estate or widow's dower (which is a life-estate in *one-third* of her husband's real estate): *Estimate the value of the property in which the life-estate is held; the yearly interest of this sum at an agreed rate, will be a life-annuity, whose value for the given age and rate, will be the value of the life-estate.*

2. The reversion of a life-annuity, life-estate, or dower, is found *by deducting its value from the value of the property.*

REVIEW.—362. What is Case 7? The rule? 363. What do contingent annuities comprise? What is life insurance? The annual premium? The policy? How is the premium estimated? What are bills of mortality? What is expectation of life? What is a table of life-annuities?

TABLE

Showing the values of Annuities on Single Lives, according to the Carlisle Table of Mortality.

Age.	4 per cent.	5 per ct.	6 per ct.	7 per ct.	Age.	4 per cent.	5 per ct.	6 per ct.	7 per ct.
0	14.28164	12.083	10.439	9.177	52	12.25793	11.154	10.208	9.392
1	16.55455	13.995	12.078	10.605	53	11.94503	10.892	9.988	9.205
2	17.72616	14.983	12.925	11.342	54	11.62673	10.624	9.761	9.011
3	18.71508	15.824	13.652	11.978	55	11.29961	10.347	9.524	8.807
4	19.23133	16.271	14.042	12.322	56	10.96607	10.063	9.280	8.595
5	19.59203	16.590	14.325	12.574	57	10.62559	9.771	9.027	8.375
6	19.74502	16.735	14.460	12.698	58	10.28647	9.478	8.772	8.153
7	19.79019	16.790	14.518	12.756	59	9.96331	9.199	8.529	7.940
8	19.76443	16.786	14.526	12.770	60	9.66333	8.940	8.304	7.743
9	19.69114	16.742	14.500	12.754	61	9.39809	8.712	8.108	7.572
10	19.58339	16.669	14.448	12.717	62	9.13676	8.487	7.913	7.403
11	19.46357	16.581	14.384	12.669	63	8.87150	8.258	7.714	7.229
12	19.33493	16.494	14.321	12.621	64	8.59330	8.016	7.502	7.042
13	19.20937	16.406	14.257	12.572	65	8.30719	7.765	7.281	6.847
14	19.08182	16.316	14.191	12.522	66	8.00966	7.503	7.049	6.641
15	18.95534	16.227	14.126	12.473	67	7.69980	7.227	6.803	6.421
16	18.83636	16.144	14.067	12.429	68	7.37976	6.941	6.546	6.189
17	18.72111	16.066	14.012	12.389	69	7.04881	6.643	6.277	5 945
18	18.60656	15.987	13.956	12.348	70	6.70936	6.336	5.998	5 690
19	18.48649	15.904	13.897	12.305	71	6.35773	6.015	5.704	5.420
20	18.36170	15.817	13.835	12.259	72	6.02548	5.711	5.424	5.162
21	18.23196	15.726	13.769	12.210	73	5.72465	5.435	5.170	4.927
22	18.09586	15.628	13.697	12.156	74	5.45812	5.190	4.944	4.719
23	17.95016	15.525	13.621	12.098	75	5.23901	4.989	4.760	4.549
24	17.80058	15.417	13.541	12.037	76	5.02399	4.792	4.579	4.382
25	17.64486	15.303	13.456	11.972	77	4.82473	4.609	4 410	4 227
26	17.48586	15.187	13.368	11.904	78	4.62166	4.422	4.238	4 067
27	17.32023	15.065	13.275	11.832	79	4.39345	4.210	4.040	3.883
28	17.15412	14.942	13.182	11.759	80	4.18289	4.015	3.858	3.713
29	16.99683	14.827	13.096	11.693	81	3.95309	3.799	3.656	3.523
30	16.85215	14.723	13.020	11.636	82	3.74634	3.606	3.474	3.352
31	16.70511	14.617	12.942	11.578	83	3.53409	3.406	3.286	3.174
32	16.55246	14.506	12.860	11.516	84	3.32856	3.211	3.102	2.999
33	16.39072	14.387	12.771	11.448	85	3.11515	3.009	2.909	2.815
34	16.21943	14.260	12.675	11.374	86	2.92831	2.830	2.739	2.652
35	16.04123	14.127	12.573	11.295	87	2.77593	2.685	2.599	2.519
36	15.85577	13.987	12.465	11.211	88	2.68337	2.597	2.515	2.439
37	15.66586	13.843	12.354	11.124	89	2.57704	2.495	2.417	2.344
38	15.47120	13.695	12.239	11.033	90	2.41621	2.339	2.238	2.198
39	15 27184	13.542	12.120	10.939	91	2.39835	2.321	2.248	2.180
40	15.07363	13.390	12.002	10.845	92	2.49199	2.412	2.337	2.266
41	14.88314	13.245	11.890	10.757	93	2.59955	2.518	2.440	2.367
42	14.69466	13.101	11.779	10.671	94	2.64976	2.569	2.492	2.419
43	14.50529	12.957	11.668	10.585	95	2.67433	2.596	2.522	2.451
44	14.30874	12.806	11.551	10.494	96	2.62779	2.555	2.486	2.420
45	14.10460	12.648	11.428	10.397	97	2.49204	2.428	2.368	2.309
46	13.88928	12.480	11.296	10.292	98	2.33222	2.278	2.227	2 177
47	13.66208	12.301	11.154	10.178	99	2.08700	2.045	2 004	1.964
48	13.41914	12.107	10.998	10.052	100	1.65282	1.624	1.598	1.569
49	13.15312	11.892	10.823	9.908	101	1.21005	1.192	1.175	1.159
50	12.86902	11.660	10.631	9.749	102	0.76183	0.753	0.744	0.735
51	12.56581	11.410	10.422	9.573	103	0.32051	0.317	0.314	0.312

1. What must be paid for a life-annuity of $650 a year, by a person aged 72 yr., int. 7 %? *Ans.* $3355.30

2. What is the life-estate and reversion in $25000, age 55 yr., int. 6 %? *Ans.* Life-estate, $14286; Rev. $10714.

3. The dower and reversion in $46250, age 21 yr., int 6 %? *Ans.* Dower, $12736.33; Rev. $2680.34

CASE II.

ART. **365**. To find how large a life-annuity can be purchased for a given sum, by a person whose age is known.

RULE.—*Assume $1 a year for the annuity; find from the table its value for the given age and rate of interest, and divide the given cost by it: the quotient will be the payment required.*

How large an annuity can be purchased—

1. For $500, age 26 yr., int. 6 %? *Ans.* $37.40
2. For $1200, age 43, int. 5 %? *Ans.* $92.61
3. For $840, age 58, int. 7 %? *Ans.* $103.03

CASE III.

ART. **366**. To find the present value of the *reversion* of a given annuity; that is, what remains of it, after the death of its possessor, whose age is known.

RULE.—*Find the present value of the annuity during its whole continuance; find its value during the given life: their difference will be the value of the reversion.*

NOTE.—It will save work, to *consider the annuity as* $1 *a year, then apply the rule, using the tables in* Art. 359 *and* Art. 364, *and multiply the result by the given payment.*

1. Find the present value of the reversion of a perpetuity of $500 a year, after the death of a person aged 47, int. 5 %. *Ans.* $3849.50

2. Of the reversion of an annuity of $165 a year for 30 yr., after the death of a person 38 yr. old, int. 6 %. *Ans.* $251.76

3. Of the reversion of a lease of $1600 a year, for 40 yr., after the death of A, aged 62, int. 7 %. *Ans.* $9485.93

CASE IV.

ART. **367**. To find the *single*, and *annual premium*, paid by one of a given age, to secure a given sum at death.

REVIEW.—364. What is Case 1? The rule? 365. What is Case 2? The rule? 366. What is Case 3? The rule?

RULE.—*Find the present value of an immediate perpetuity of* $1, *at the given rate; subtract from it the present value of an annuity of* $1 *for the given age and rate; divide the remainder by the value of the perpetuity increased by* 1: *the quotient will be the present value of* $1 *due at the expiration of the given life, and if it be multiplied by the given sum, the product will be the single premium.*

To get the annual premium, divide the single premium by 1 *more than the present value of an annuity of* $1 *for the given age and rate.*

NOTE.—The annual premiums being paid in advance, the *single premium* must exceed the present value of an ordinary life-annuity of the same payments, by one payment, viz: the one made when the insurance is obtained.

1. Find the single premium and annual premium necessary to secure $1000 at the death of a person aged 32, allowing 5 % interest. *Ans.* $261.62, and $16.87

2. To secure $1500 at the death of a person aged 66, allowing 6 % interest. *Ans.* $816.59, and $101.45

CASE V.

ART. **368.** If the holder wishes to sell his policy, or *surrender* it to the Insurance Co., proceed thus,

TO VALUE A POLICY OF LIFE INSURANCE,

RULE.—*Take the difference between the premium in the policy, and the premium for insuring the holder's life for the same sum, now; multiply it by* 1 *more than the present value of a life-annuity of* $1 *on the present age; this product will be the value of the policy.*

1. The premium in the policy is $2.04 per $100; the premium now, $2.50 per $100; age 27 yr., int. 6 %: value the policy for $5000. *Ans.* $328.33

2. Value the policy, when the premium in it is $74, the premium now, $82, age 42 yr., int. 5 %. *Ans.* $112.81

3. When the premium in the policy is $51.76, the premium now, $68.44, age 36 yr., int. 7 %. *Ans.* $203.68

REVIEW.—367. What is Case 4? The rule? How does life-insurance differ from ordinary life-annuities? 368. Give the rule for valuing a policy of life-insurance.

XXXI. PROPORTIONAL PARTS.

ART. 369. Divide the number 90 into 3 other numbers proportional to 2, 3 and 4.

SOLUTION.—Since the required numbers are to have the same ratios as 2, 3 and 4, they must contain a common factor 2, 3 and 4 times respectively. Call this common factor a *part;* then the 1st number is 2 parts, the 2d is 3 parts, the 3d is 4 parts; and the whole number $90 = 2 + 3 + 4$, or 9 parts. Hence, 1 part is $\frac{1}{9}$ of $90 = 10$, and the 1st number is $\frac{2}{9}$ of $90 = 20$, the 2d is $\frac{3}{9}$ of $90 = 30$, and the 3d is $\frac{4}{9}$ of $90 = 40$.

PROOF.—$20 + 30 + 40 = 90$; and 20, 30 and 40 have the same ratios as 2, 3 and 4, omitting the common factor 10.

TO DIVIDE A NUMBER INTO PROPORTIONAL PARTS,

RULE.—*Take separately such parts of the number to be divided, as each of the proportional numbers is of their sum; the results will be the required numbers.*

PROOF.—The sum of the parts must be equal to the number divided, and the ratio of each to its proper proportional must be the same.

NOTES.—1. If the proportional numbers are fractional, reduce them to a least common denominator, and use their numerators for the proportional numbers.

2. Any common factor may be canceled out of the proportional numbers, before using them.

1. Divide the number 84 into 2 parts proportional to 30 and 40. *Ans.* 36, 48.

SUG.—For 30 and 40, use 3 and 4, dropping the common factor 10.

2. Divide 60 apples among 3 boys in proportion to their ages, which are 7, 10 and 13. *Ans.* 14, 20, 26.

3. One part of a pole 16 ft. long, is $\frac{3}{5}$ of the other part: find each part. *Ans.* 6 ft., and 10 ft.

SUG.—$\frac{3}{5}$ and $\frac{5}{5}$, or 3 and 5 are the proportional numbers.

4. $\frac{2}{9}$ of A's money is equal to $\frac{3}{5}$ of B's, and both have $222; how much has each? *Ans.* A $162, B $60.

SUGGESTION.—A's $= \frac{27}{10}$ of B's; if B's is 10 parts, A's is 27 parts; hence, 27 and 10 are the proportional numbers.

REVIEW.—369. What is the rule for dividing a number into proportional parts? The proof? If the proportional numbers are fractional, what should be done? If they are large? Solve the example.

5. A tree, 51 ft. high, was broken by the wind; $\frac{2}{3}$ of the part that fell was equal to $\frac{3}{4}$ of the stump: how long was each? *Ans.* Stump 24 ft., other part 27 ft.

6. Divide 38 in proportion to $4\frac{1}{6}$ and $3\frac{3}{4}$ *Ans.* 20, 18.

SUG.—$4\frac{1}{6}$ and $3\frac{3}{4}$ are equal to $\frac{25}{6}$ and $\frac{15}{4}$, or $\frac{50}{12}$ and $\frac{45}{12}$; hence, 50 and 45—or, dividing by 5—10 and 9 are the proportional numbers.

7. Divide the cost of a supper, $4.50, among 3 persons in proportion to their money: A has $100, B $75, C $125.
Ans. A $1.50, B $1.12½, C $1.87½

8. A father left $7140 to his children, A, B and C; A's share was to B's as 2 to 3, and B's to C's as 4 to 5: what did each get? *Ans.* A $1632, B $2448, C $3060.

SUGGESTION.—If C's share is 5 parts, B's is 4 parts, and A's is $\frac{2}{3}$ of B's, that is, $\frac{2}{3}$ of 4 parts $= 2\frac{2}{3}$ parts; hence, $2\frac{2}{3}$, 4 and 5, or $\frac{8}{3}$, $\frac{12}{3}$, $\frac{15}{3}$, or 8, 12, and 15, are the proportional numbers.

9. A paid $3.60, and B, $2 for a bbl. of flour (196 lb.): how many lb. should each have? *Ans.* A 126 lb., B 70 lb.

10. Distribute $180 among 5 poor families in proportion to the number of children in each, which are 3, 4, 5, 2, 6
Ans. 1st $27, 2d $36, 3d $45, 4th $18, 5th $54.

11. Divide 1065 in proportion to 3, 5 and 7; also in proportion to $\frac{1}{3}$, $\frac{1}{5}$, $\frac{1}{7}$. *Ans.* 213, 355, 497; and 525, 315, 225.

12. How much copper and tin, 100 parts of the former to 11 of the latter, will make a cannon weighing 18 cwt. 3 qr. 12 lb.? *Ans.* 17 cwt. copper, 1 cwt. 3 qr. 12 lb. tin.

13. U. S. standard silver and gold are 9 parts pure metal to 1 part alloy: how much pure silver in the half-dollar, which weighs 8 pwt.? how much pure gold in the eagle, which weighs 10 pwt. 18 gr.?
Ans. 7 pwt. 4.8 gr., and 9 pwt. 16.2 gr.

14. Pewter is 112 parts tin, 15 lead, and 6 brass: how much of each ingredient in 2 lb. 1 oz. 4 dr. of pewter?
Ans. 1 lb. 12 oz. tin; 3 oz. 12 dr. lead; 1 oz. 8 dr. brass.

15. A man bequeaths $875 to A; $2450 to B; $6035 to C: if the estate yields only $8190, what will each get?
Ans. A $765.62½, B $2143.75, C $5280.62½

16. I owe $840, due Oct. 1; I pay part Aug. 15 (47 da. before due), the rest Jan. 1 (92 da. after due): what are the payments? *Ans.* $555.97, Aug. 15; $284.03, Jan. 1.

REMARK.—Divide $840 in proportion to 47 and 92.

17. C owes $1200, due Nov. 6; he pays part Aug. 1, and the rest Jan. 15; what are the payments?

Ans. $502.99, Aug. 1; $697.01, Jan. 15.

18. A owes $1680, due July 18; he pays $480 before, the rest after due; when were the payments made, if they were 49 days apart? *Ans.* June 13, and Aug. 1.

PARTNERSHIP.

ART. **370.** An important application of Proportional Parts, is in dividing the gains or losses of partners.

Partners are persons who join to carry on a business, and constitute a *firm*, *house*, or *company*.

The money used in the business, is called the *capital* or *stock*, and is contributed by the partners.

The gain or loss in the business, is called the *dividend*, because it is *to be divided*.

REMARK.—When each partner's stock is employed the same time, it is sometimes called *Simple Fellowship*, and when they are employed for different times, it is called *Compound Fellowship*.

CASE I.

To divide the gain or loss, when each partner's stock is employed for the same time.

RULE.—*Divide the gain or loss among the partners in proportion to their shares of the stock.*

A, B and C are partners, with $3000, $4000, and $5000 stock, respectively; if they gain $5400, what is each one's share?

SOL.—Each should have the same part of the gain as he has of the stock;			
3	$\frac{3}{12}$ of $5400 =	$1350	A's share.
4	$\frac{4}{12}$ of $5400 =	$1800	B's "
5	$\frac{5}{12}$ of $5400 =	$2250	C's "
12		$5400	whole gain.

hence, the gain should be divided in proportion to their stocks, using 3, 4, 5, for 3000, 4000, 5000.

REVIEW.—370. What are partners? What is the firm? What is the stock or capital? The dividend? Why so called? What is simple fellowship? Compound fellowship? What is Case 1 of partnership? The rule? Solve the example.

1. A and B gain in 1 year $3600; their store expenses are $1500. If A's stock was $2500 and B's $1875, how much does each gain? *Ans.* A $1200, B $900.

Take out the expenses, then divide.

2. A, B and C are partners; A puts in $5000, B $6400, C $1600. C is allowed $1000 a year for personal attention to the business; their store expenses for 1 year are $800, and their gain, $7000. Find A's and B's gain, and C's income. *Ans.* A $2000, B $2560, C $1640.

3. A pasture rents for $160; A puts in 24 cattle; B, 20; C, 60; D, 96; and they pay in proportion: what does each pay? *Ans.* A $19.20, B $16, C $48, D $76.80

4. A and B speculate together in flour; A contributes 800 bbl. at $5.40 per bbl., and B, 600 bbl. at $4.80 per bbl.; they lose 75 ct. a bbl., and pay storage $68.45: what is the loss of each? *Ans.* A $671.07, B $447.38

5. A, B and C form a partnership; A puts in $24000, B $28000, C $32000; they lose $\frac{1}{6}$ of their stock by a fire, but sell the remainder at $\frac{3}{5}$ more than cost: if all expenses are $8000, what is the gain of each?

Ans. A 5714.28\frac{4}{7}$, B 6666.66\frac{2}{3}$, C 7619.04\frac{16}{21}$

6. A, B and C are partners: A's stock is $5760, B's, $7200; their gain is $3920, of which C has $1120: what is C's stock, and A's and B's gain?

Ans. C's stock, $5184; A's gain, 1244.44\frac{4}{9}$; B's gain, 1555.55\frac{5}{9}$

7. A, B and C are partners: A's stock, $8000; B's, $12800; C's, $15200; A and B together gain $1638 more than C: what is the gain of each?

Ans. A $2340; B $3744; C $4446.

8. A and B buy a house and lease; A contributes $3480; B, $2900: the ground-rent, taxes, &c., are $108.30, and the property rents for $915.80: what does each get?

Ans. A, 440.45\frac{5}{11}$; B, 367.04\frac{6}{11}$

9. A, B and C, in partnership, have capitals respectively $19200, $24000, and $32400; they sell out for $100000; how much of this does each get?

Ans. A 25396.82\frac{34}{63}$; B 31746.03\frac{11}{63}$; C 42857.14\frac{2}{7}$

10. Four men rent 57 A. 2 R. 16 P. of land at $3.75 an acre: A puts in 72 sheep; B, 80; C, 96; D, 112: what should each pay? *Ans.* A $43.20; B $48; C $57.60; D $67.20

11. A's gain is $1750; B's, $1225; C's, $2275: what part of the capital, $18690, has each?

Ans. A $6230; B $4361; C $8099.

12. A, B and C have a joint capital of $27000: neither draws from the firm, and when they quit, A had $20000; B, $16000; C, $12000: what did each contribute?

Ans. A $11250; B $9000; C $6750.

CASE II.—PARTNERSHIP WITH TIME.

ART. **371** The rule for distributing gain or loss among partners, (Art. 370,) applies only when each partner's capital is employed for the same time; it needs some modification if such be not the case.

A, B and C are partners: A puts in $2500 for 8mon., B, $4000 for 6 mon.; C, $3200 for 10 mon.; their net gain is $4750: divide the gain.

SOLUTION.—A's capital ($2500), used 8 months, is equivalent to 8 × $2500, or $20000, used 1 month; B's capital ($4000), used 6 months, is equivalent to 6 × $4000, or $24000, used 1 month; C's capital ($3200), used 10 months, is equivalent to 10 × $3200, or $32000 used 1 month. Dividing the gain ($4750) in proportion to the stock equivalents, $20000, $24000, $32000, used for the same time (1 month), the results will be the gain of each.

Ans. A's $1250, B's $1500, C's $2000.

The *stock-equivalents* are obtained by multiplying each partner's stock by the time it is used; hence,

WHEN THE STOCKS ARE USED FOR DIFFERENT TIMES,

RULE.—*Multiply each partner's stock by the time it is used; and divide the gain or loss in proportion to the products so obtained.*

PROOF.—The sum of the separate gains or losses must equal the whole gain or loss, and the ratio of each partner's gain or loss to his stock-equivalent must be the same.

NOTE.—Express the times in the same denomination, before multiplying.

1. A begins business with $6000: at the end of 6 mon. he takes in B, with $10000: 6 mon. after, their gain is $3300: what is each share? *Ans.* A's $1800; B's $1500.

REVIEW.—371. What is Case 2? What are stock equivalents? The rule? The proof? In multiplying by the times, what is necessary?

2. A and B are partners: A puts in \$2500; B, \$1500. after 9 mon., they take in C with \$5000; 9 mon. after, their gain is \$3250: what is each one's gain?

Ans. A's \$1250; B's \$750; C's \$1250.

3. A and B rent a pasture for \$275: A puts in 80 sheep, and B, 100; after 6 mon. they each sell half their stock, and allow C to feed 50 sheep the rest of the year: how much should each pay?

Ans. A \$103.12$\frac{1}{2}$; B \$128.90$\frac{5}{8}$; C \$42.96$\frac{7}{8}$

4. A and B are partners, each contributing \$1000: after 3 months, A withdraws \$400, which B advances; the same is done after 3 months more; their year's gain is \$800: what should each get? *Ans.* A \$200, B \$600.

5. A works 9 hours a day; B remains idle the first two days of the week, and works 5$\frac{1}{4}$, 8$\frac{1}{2}$, 10$\frac{3}{4}$ and 11$\frac{1}{2}$ hr. respectively, on the other four: if their week's wages are \$38.75, what should each have? *Ans.* A \$23.25, B \$15.50

6. A, B and C are employed to empty a cistern by two pumps of the same bore: A and B go to work first, making 37 and 40 strokes respectively a minute; after 5 minutes, each makes 5 strokes less a minute; after 10 minutes, A gives way to C, who makes 30 strokes a minute until the cistern is emptied, which was in 22 min. from the start: divide their pay, \$2. *Ans.* A 46 ct.; B \$1.06; C 48 ct.

7. A and B are partners: A's stock is to B's, as 4 to 5: after 3 mon., A withdraws $\frac{2}{3}$ of his, and B $\frac{3}{4}$ of his: divide their year's gain, \$1675. *Ans.* A \$800, B \$875.

8. A, B and C join capitals, which are as $\frac{1}{2}$, $\frac{1}{3}$, $\frac{1}{4}$: after 4 mon., A takes out $\frac{1}{2}$ of his; after 9 mon. more, their gain is \$1988: divide it. *Ans.* A \$714, B \$728, C \$546.

9. A and B are partners: A's capital is \$4200; B's, \$5600: after 4 months, how much must A put in, to entitle him to $\frac{1}{2}$ the year's gain? *Ans.* \$2100.

10. A's capital is \$9750, B's \$10500, C's \$12250: after 3 months, A takes out \$1500, B \$1250, C \$1750: after 4 months more, what must A and B put in, to entitle each to $\frac{1}{3}$ of the year's gain? *Ans.* A \$5550, B \$3300.

11. A, B and C are in partnership, with capitals of \$10800, \$14400, and \$18000: after 2 mon., A draws out \$600, in 3 mon. more he draws out \$1200; and in 4 mon. more puts in \$1000. B draws out \$2400 in the first 6 mon.; \$800 more in 4 mon.; and in 1 mon. returns

$1600. C puts in $2000 at the end of 5 mon.; and 4 mon after, draws out $4000. Divide their year's gain, $14838.
Ans. A $3546; B $4752; C $6540.

12. A, B and C form a partnership, with capitals of $10000, $20000, $30000, respectively. A draws out $1000 a year, B $1600 a year, and C $1800 a year. In 5 years, their capital is $57200: how much of it does each own? *Ans.* A $8000; B $18300; C $30900.

13. A and B enter into partnership, with $2500 and $4000 capital, respectively. A draws out $600 after 3 months, and $600 after 6 months more; B draws $800 at each of these times. At the end of the year, they have $2680: divide it. *Ans.* A $920; B $1760.

14. A and B go into partnership, each with $4500. A draws out $1500, and B $500, at the end of 3 mon., and each the same sum at the end of 6 and 9 mon.: at the end of 1 yr. they quit with $2200: how must they settle?
Ans. B takes $2200, and has a claim on A for $300?

15. A's gain is $1800, B's $2250, C's $3200; A's capital was in 6 mon.; B's, 9 mon.; and C's, 1 year 4 mon.: how much of the capital, $27450, did each own?
Ans. A, $10980; B, $9150; C, $7320.

SUGGESTION.—Divide each one's gain by the number of months his capital was used, to get each one's gain for 1 month, divide the whole capital in proportion to these quotients.

16. A, B, C and D go in partnership: A owns 12 shares of the stock; B, 8 shares; C, 7 shares; D, 3 shares. After 3 mon., A sells 2 shares to B, 1 to C, and 4 to D; 2 mon. afterward, B sells 1 share to C, and 2 to D; 4 mon. afterward, A buys 2 from C and 2 from D. Divide the year's gain ($18000.)
Ans. A $4650, B $4650, C $4700, D $4000.

BANKRUPTCY.

ART. **372**. A *bankrupt*, or *insolvent*, is one who, from want of means, is unable to carry on his business.

The property of a bankrupt is usually entrusted to an *assignee*, who converts it as fast as possible into money, and pays the debts *pro rata;* that is, so that each creditor receives the *same part* of his claim.

TO DIVIDE THE PROPERTY OF A BANKRUPT,

RULE.—*Divide the whole property among the creditors in proportion to their claims.*

NOTES.—1. All necessary expenses, including assignee's fee (which is generally a certain rate per cent. on the whole amount of property), must be deducted, before dividing.

2. *The amount paid on a dollar can be found by taking such a part of $1 as the whole amount of the debts is of the whole property;* each creditor's proportion may be then found by multiplying his claim by the amount paid on the dollar.

1. A has a lot worth $8000, good notes $2500, and cash $1500; his debts are $20000: what can he pay on $1, and what will A receive, whose claim is $4500?

SOL.—$\$8000 + \$2500 + \$1500 = \$12000$, the amount of property which is $\frac{12000}{20000}$, or $\frac{3}{5}$ of the whole debts. Hence, $\frac{3}{5}$ of $\$1 = 60$ ct., the amount paid on $1, and $\$4500 \times .60 = \$2700$, the sum paid to A.

2. My assets are $2520; I owe A $1200; B, $720; C, $600; D, $1080: what does each get, and what is paid on each dollar?

Ans. A $840; B $504; C $420; D $756; 70 ct. on $1.

3. A bankrupt's estate is worth $16000; his debts, $47500. The assignee charges 5 %. What is paid on $1? and what does A get, whose claim is $3650?

Ans. 32 ct. on $1, and $1168.

GENERAL AVERAGE.

ART. **373.** *General Average* is the method of apportioning among the owners of a ship and its cargo, any loss or expense incurred during a voyage, for their general benefit, such as sacrificing a part of the cargo to save the rest, or making necessary repairs.

Such loss or expense is borne by the *ship*, the *freight*, and the *cargo*, in proportion to their values; the loss borne by the *cargo* is divided among the several shippers in proportion to the value of their goods.

REVIEW.—372. What is a bankrupt? What is done with a bankrupt's property? What is the rule for dividing a bankrupt's property among his creditors? What must be paid before distribution? How can the amount paid on a dollar be found? How can each creditor's share be then found?

In estimating these *contributory interests*, as they are called, it is customary to value the goods on board at the price they would bring at the port to which they are bound; while the *freight* is the amount of money received for *freight and passage, less* $\frac{1}{3}$ for seamen's wages, (in New York, less *one half*).

If the loss is for repairs, throw off $\frac{1}{3}$ from the cost of the new masts, rigging, &c., as they are considered that much better than the old.

RULE.—*After ascertaining the contributory interests, divide the loss among them, in proportion to their values.*

The goods thrown overboard (*jettison*) are reckoned a part of the cargo, and bear their proportion of loss.

1. The ship Dolphin, overtaken by a storm, throws overboard goods worth $4000, and puts into Fayal for repairs. The necessary expenses of detention were $250; repairs $1200. Divide the loss, estimating the vessel at $36000, the freight at $3750, the cargo at $52400. A's interest is $7500, B's, $16000, C's, $10500, D's, $12000, E's, $6400. The property lost was $2600 of A's, $900 of B's, and $500 of C's.

SOLUTION.

CONTRIBUTORY INTERESTS.		DAMAGES.	
Vessel, ...	$36000	Goods lost, ...	$4000
Freight (less $\frac{1}{3}$), . .	2500	Expense of detention,	250
Cargo, ...	52400	Repairs (less $\frac{1}{3}$), . .	800
Total, ...	$90900	Total loss, . . .	$5050

Then $\frac{5050}{90900} = \frac{1}{18}$, the part that each interest loses.

The owners of the vessel and freight pay $38500 $\times \frac{1}{18}$ = 2138\frac{8}{9}$

The owners of the cargo pay $52400 $\times \frac{1}{18}$ . . . . . = 2911$\frac{1}{9}$

Total loss, ... $5050

	RECEIVED.	PAID.
Owners of the vessel pay 2138\frac{8}{9}$, and receive $1050, or pay		1088$\frac{8}{9}$
A pays 416\frac{2}{3}$ and receives $2600, or receives	2183\frac{1}{3}$,	
B pays 888\frac{8}{9}$ and receives $900, or receives	11$\frac{1}{9}$,	
C pays 583\frac{1}{3}$ and receives $500, or pays		83$\frac{1}{3}$
D pays		666$\frac{2}{3}$
E pays		355$\frac{5}{9}$
	2194$\frac{4}{9}$,	2194$\frac{4}{9}$

REVIEW.—373. What is general average? How is loss at sea borne? Which are the contributory interests? How are the goods valued? How is the freight valued? What deduction is made from the cost of repairs? After the contributory interests have been estimated, how is the loss divided? Why should the goods lost bear their proportion of the loss?

2. The brig Adams, of New York, bound to New Orleans, suffered damage, $480, and loss of cargo, $5600. The vessel was valued at $18800; the freight, $3200; the cargo, $29600. Divide the loss; and settle A's account, who shipped $14400, and lost $1800.

Ans. Rate of loss $\frac{74}{625}$; ship pays $2415.36, and receives $320, making a balance paid, $2095.36; cargo pays $3504.64; A pays $1704.96, and receives $1800, making a balance received of $95.04

3. The schooner Washington, crippled by a storm, was relieved by the ship Leopard. The repairs cost $1350, salvage $4500. The vessel is worth $22000; the freight, $3450. The cargo was owned by A, B, C, and D: A, $10800; B, $16200; C, $19652; D, $7348. What is the rate of loss, and what does each pay?

Ans. $\frac{2}{29}$ rate of loss; A pays $744.83; B, $1117.24; C, $1355.31; D, $506.76; ship pays $1675.86, and receives $900, making a balance to be paid of $775.86

RATE BILLS FOR SCHOOLS.

ART. **374.** Another application of proportional parts is in making out rate-bills for public schools, in districts where they are not supported by general tax.

RULE.—*From the whole expenses of the school deduct the public money, if any, and divide the remainder among the families of the district, in proportion to the number of days of attendance of each.*

NOTE.—It is generally more convenient to find first the *rate per day*, by dividing the expense to be distributed by the whole number of days of attendance, and then make out each pupil's bill, by multiplying his number of days of attendance by this rate.

1. A school pays $500 for teacher's salary, $26.50 for repairs, and $12.30 for fuel, and draws $50 public money: the whole number of days of attendance is 3640. Find the rate per day, and the bill of A, who sends one pupil 175 days; of B, who sends 2 pupils, 220 days each: of C, who sends one pupil 108 days, one 76 days, and one 192 days.

Ans. Rate $13\frac{3}{7}$ ct.; A pays $23.50; B $59.09; C $50.49

REVIEW.—374. What is the rule for making out rate bills for schools? What is convenient in applying the rule?

2. The salary is \$180; other expenses, \$18.75; the public money, \$30; the whole number of days of attendauce, 2562: what is the rate per day? and A's bill, who sends for 87 da.?

Ans. $6\frac{501}{854}$ ct. rate; A pays \$5.73

3. The expenses are \$312; the whole number of days of attendance, 4160: what is the rate per day? and D's bill, who sends 1 pupil 116 days, and another 98 days?

Ans. Rate $7\frac{1}{2}$ ct.; D's bill \$16.05

4. The salary is \$35 per month; the other expenses for 1 quarter, \$6.80; the public money, \$10.20; the whole number of days of attendance, 1120: what is the rate per day? and E's bill, who sends 48 days?

Ans. Rate $9\frac{1}{14}$ ct.; E's bill \$4.35

XXXII. ALLIGATION.

ART. **375**. *Alligation* is a method of finding the value per lb., bu., &c., of a mixture, when the quantity and price of its several ingredients are known.

It is also used in getting the mean or average result of several quantities of the same kind, but differing in degree, such as observations by instruments, measurements of irregular figures, &c.

The price or other result obtained by Alligation is called the *mean or average.*

CASE I.—To find the average price of a mixture, when the quantity and cost of each ingredient are known.

RULE.—*Find the value of each ingredient at its price; add these values for the value of the mixture; divide the sum by the sum of the quantities of the ingredients.*

NOTE.—Express each ingredient in the same denomination.

If 3 lb. of sugar at 5 ct. a lb. and 2 lb. at $4\frac{1}{2}$ ct. a lb. be mixed with 9 lb. at 6 ct. a lb., what per lb. is the mixture worth?

SOLUTION.—The 3 lb. at 5 ct. per lb. = 15 ct.; the 2 lb. at $4\frac{1}{2}$ ct. per lb. = 9 ct.; the 9 lb. at 6 ct. per lb. = 54 ct.: therefore, the whole 14 lb. are worth 78 ct., = 78 ÷ 14 = $5\frac{4}{7}$ ct. per lb. *Ans.*

1. Find the average price of 6 lb. tea at 80 ct., 15 lb. at 50 ct., 5 lb. at 60 ct., 9 lb. at 40 ct. *Ans.* 54 ct. per lb.

REVIEW.—375. What is Alligation? What is the result called? What is Case 1? The rule? How must the quantity of each ingredient be expressed? Solve the example.

2. The average price of 40 hogs at $8 each, 30 at $10 each, 16 at $12.50 each, 54 at $11.75 each. *Ans.* $10.39 each.

3. How fine is a mixture of 5 pwt. of gold, 16 carats fine; 2 pwt. 18 carats fine; 6 pwt., 20 carats fine, and 1 pwt. pure gold? *Ans.* $18\frac{4}{7}$ carats fine.

4. Find the specific gravity of a compound of 15 lb. of copper, specific gravity, $7\frac{3}{4}$; 8 lb. of zinc, specific gravity, $6\frac{7}{8}$; and $\frac{1}{4}$ lb. of silver, specific gravity, $10\frac{1}{2}$. *Ans.* 7.445—

EXPLANATION.—The *Specific Gravity* of a body is its own weight divided by the weight of an equal bulk of water.

5. What per cent. of alcohol in a mixture of 9 gal., 86 % strong; 12 gal., 92 % strong; 10 gal., 95 % strong; and 11 gal., 98 % strong? *Ans.* 93 %.

ART. 376^a. CASE II.

To find what proportions of several ingredients, whose prices are known, must be used, to make a mixture of a given price.

1. What relative quantities of sugar at 9 ct. a lb. and 5 ct. a lb. must be used for a compound at 6 ct. a lb.?

OPERATION.

$$6\left|\begin{matrix}5\\9\end{matrix}\right) \quad \begin{matrix}\ldots\ 3 \text{ lb. at } 5 \text{ ct.} = 15 \text{ ct.}\\ \ldots\ 1 \text{ lb. at } 9 \text{ ct.} = \ \ 9 \text{ ct.}\\ \overline{4 \text{ lb.}} \quad \text{worth} \quad \overline{24 \text{ ct.}}\end{matrix}$$

which is $\frac{24}{4} = 6$ ct. a lb.

ANALYSIS.—If you put 1 lb. at 9 ct. in the mixture to be sold for 6 ct., you lose 3 ct.; if you put 1 lb. at 5 ct. in the mixture to be sold at 6 ct., you gain 1 ct.; 3 such lb. gain 3 ct.: the gain and loss would then be equal if 3 lb. at 5 ct. are mixed with 1 lb. at 9 ct.

RULE.—*Place the prices of the ingredients in a vertical row in regular order, having the smallest at the top, and the largest at the bottom. To the left of this row draw a vertical line, and on the other side of it set the price of the mixture opposite the place it would occupy, if it stood in the right-hand column of prices.*

Then connect by a curved line any two numbers in the right-hand column, one of which is greater and one less than the mean price; when each of the right-hand numbers has been thus connected with another, take the difference between each of them and the mean price on the left, and set this difference opposite the number with which it is connected.

After all the differences have been taken, the proportional quantity at each price will be the number standing opposite that price, unless there be several such numbers, in which case, their sum will be the proportional quantity at that price.

NOTE.—One number may be connected with two or more others; if so, several numbers will stand opposite it, and their sum must be taken for its proportional.

PROOF.—With the proportional quantities thus found, determine by Case I the mean price of the mixture; if it agrees with the mean price given, the work is right.

2. What relative quantities of tea worth 25, 27, 30, 32, and 45 ct. per lb. must be taken for a mixture worth 28 ct. per lb.?
Ans. 19, 4, 3, 1, 3 lb. respectively.

REM.—It is evident that other results may be obtained by making the connections differently; as 6, 17, 3, 3, 1 lb., or 17, 6, 1, 1, 3 lb.

3. What of sugar at 5, $5\frac{1}{2}$, 6, 7, and 8 ct. per lb. must be taken for a mixture worth $6\frac{3}{4}$ ct. per lb.?
Ans. 1, 5, 5, 7, 8 lb. respectively; or, 5, 1, 1, 8, 7 lb., &c.

SUGGESTION.—When the proportional quantities determined by the rule, contain fractions, multiply them all by the least common multiple of the denominators, which converts them into whole numbers having the same relative values.

4. What relative quantities of alcohol, 84, 86, 88, 94 and 96 % strong, must be taken for a mixture 87 % strong?
Ans. 10, 7, 3, 1, 3 gal.; or, 7, 10, 1, 3, 1 gal., &c.

5. What of gold and silver, whose specific gravities are $19\frac{1}{4}$ and $10\frac{1}{2}$, will make a compound whose specific gravity shall be 16.84? *Ans.* 723 lb. silver to 3487 lb. gold.

6. What of silver $\frac{3}{4}$ pure, and $\frac{9}{10}$ pure, will make a mixture $\frac{7}{8}$ pure? *Ans.* 1 lb., $\frac{3}{4}$ pure; 5 lb., $\frac{9}{10}$ pure.

7. What of pure gold (24 carats), and 18 carats, and 20 carats fine, must be taken to make 22 carat gold?
Ans. 1 part 18 carats, 1 part 20 carats, 3 pure.

SUG.—Any common factor may be omitted from the proportionals.

CASE III.

ART. 376^b. Given, the prices of the ingredients, and the quantity and price of the mixture, to find the quantity of each ingredient.

RULE.—*Find the relative quantities of the ingredients according to the last rule; divide the given quantity of the mixture into parts proportional to these numbers, by rule in Art.* 369: *these results will be the quantities of the ingredients required.*

PROOF.—Add the quantities of the ingredients; also add the cost of the ingredients at the given prices: these sums must agree with the quantity and cost of the mixture.

How many bushels of peaches at 20, 30, 37, 40, and 50 ct. a bu., will make a lot of 58 bu. 3 pk. at 35 ct. a bushel?

REVIEW.—376^a. What is Case 2? Solve the example. What is the rule? The proof? When will it be necessary to connect one price with two of the others? What is Case 3?

SOLUTION. -By last case the proportional quantities of the ingredients are 2, 20, 15, 5 and 5. Dividing 58 bu. 3 pk. or 235 pk. in proportion to these numbers, by rule in Art. 369, gives 2bu. 2pk.; 25 bu., 18 bu. 3 pk.; 6 bu. 1 pk.; 6 bu. 1 pk., respectively, at the prices mentioned.

1. How much copper, specific gravity $7\frac{3}{4}$, with silver, specific gravity $10\frac{1}{2}$, will make 1lb. Tr., of spe. grav. $8\frac{3}{5}$?

Ans. $7\frac{223}{473}$ oz. copper, $4\frac{250}{473}$ oz. silver.

2. How much gold 15 carats fine, 20 carats fine, and pure, will make a ring 18 carats fine, weighing 4 pwt. 16 gr.?

Ans. 2 pwt. 16 gr.; 1 pwt.; 1 pwt.

3. Hiero, king of Syracuse, gave his goldsmith 14 lb. of gold and $3\frac{1}{2}$ lb. of silver to make a crown: suspecting that the gold had not been all used, he requested Archimedes to find how much had been abstracted, the specific gravity of gold being $19\frac{1}{4}$; of silver $10\frac{1}{2}$; and of the crown, $14\frac{5}{8}$.

Ans. It contained $10\frac{67}{78}$ lb. of gold and $6\frac{25}{39}$ lb. of silver; $3\frac{11}{78}$ lb of gold had been replaced by silver.

CASE IV.

ART. **377.** Given, the price of the mixture, the prices of the ingredients, and the quantity of one ingredient, to find the quantities of the other ingredients, and of the mixture.

RULE.—*Find the relative quantities of the ingredients, by Case I; the quantity of each ingredient will be such a part of the given quantity, as its proportional is of the proportional belonging to the given quantity.*

PROOF.—By Case I.

NOTE.—After the quantities of all the ingredients have been found, their sum will be the quantity of the mixture.

How many bushels of hops, worth respectively 50, 60 and 75 ct. per bu., with 100 bu. at 40 ct. per bu., will make a mixture worth 65 ct. a bu.?

SOLUTION.—By Case II the proportionals are 10, 10 and 45 of the first three sorts to 10 of the last; hence, the quantities of the two first sorts must be $\frac{10}{10}$ of 100 bu. = 100 bu.; the quantity of the third sort is $\frac{45}{10}$ of 100 bu. = 450 bu.

1. How many railroad shares at 50% must A buy, who has 80 shares that cost him 72%, in order to reduce his average to 60%?

Ans. 96 shares.

2. I bought 2000 cwt. of pork at $5.80 a cwt.: how much must I buy at $4.75 a cwt., so as to average $5.25 per cwt.?

Ans. 2200 cwt.

REVIEW.—376[b]. What is the rule? The proof? 377. What is Case 4? The rule? The proof?

3. A jeweller has 3 pwt. 9 gr. of old gold, 16 carats fine; how much U. S. gold, 21$\frac{3}{5}$ carats fine, must he mix with it, to make it 18 carats fine? *Ans.* 1 pwt. 21 gr.

4. How much water (0 per cent.) will dilute 3 gal. 2 qt. 1 pt. of acid 91 % strong, to 56 %? *Ans.* 2 gal. 1 qt. $\frac{1}{8}$ pt.

5. I mixed 1 gal. 2 qt. $\frac{1}{2}$ pt. of water with 3 qt. 1$\frac{1}{2}$ pt. of pure acid: the mixture has 15 % more acid than desired: how much water will reduce it to the required strength?
Ans. 1 gal. 2 qt. 1$\frac{1}{3}$ pt.

CASE V.

Art. 378. Given, the price of the mixture, the price of each ingredient, and the quantities of two or more ingredients, to find the quantities of the remaining ingredients and of the mixture.

Rule.—*Find by Case I, the quantity and average price of a mixture composed of those ingredients whose quantities and prices are known; consider these as the quantity and price of a single ingredient, and find the quantities of the remaining ingredients, and of the whole mixture, by Case IV.*

Proof.—By Case I.

A buys 400 bbl. of flour at $7.50 each, 640 bbl. at $7.25, and 960 bbl. at $6.75: how many must he buy at $5.50, to reduce his average to $6.50 per bbl.?

Solution.—The mean price of the first three items, by Case I, is $7.06 a piece for 2000 bbl. Then with $7.06, and $5.50 for the prices of the ingredients, and $6.50 for the price of the mixture, and 2000 bbl. for the quantity of the first ingredient, find, by Case IV, the quantity of the other ingredient, = 1120 bbl.

1. How much lead, specific gravity 11, with $\frac{1}{2}$ oz. copper, sp. gr. 9, can be put on 12 oz. of cork, sp. gr. $\frac{1}{4}$, so that the 3 will just float, that is, have a sp. gr. (1) the same as water?
Ans. 2 lb. 7$\frac{1}{9}$ oz.

2. How much water, with 3 pt. of alcohol, 96 % strong, and 8 pt., 78 %, will make a mixture 60 % strong? *Ans.* 4$\frac{1}{5}$ pt.

3. How many shares of stock at 40 % must A buy, who has bought 120 shares at 74 %, 150 shares at 68 %, and 130 shares at 54 % so that he may sell the whole at 60 %, and gain 20 %?
Ans. 610 shares.

CASE VI.

Art. 379. Given, the quantity and price of the mixture, the quantities and prices of one or more ingredients, and the prices of the remaining ingredients, to find the quantities of the remaining ingredients.

Review.—378. What is Case 5? The rule? The proof?

RULE.—*From the quantity and cost of the whole mixture, deduct the quantities and costs of the given ingredients: the remainders will be the quantity and cost of a mixture composed of the remaining ingredients: from which, the quantities of those ingredients can be found by Case III.*

PROOF.—By Case I.

What quantities of sugar at 3 ct. per lb. and 7 ct. per lb.; with 2 lb. at 8 ct., and 5 lb. at 4 ct. per lb., will make 16 lb. worth 6 ct. per lb.?

SOL.—The 2 lb. at 8 ct. and the 5 lb. at 4 ct., make a mixture of 7 lb. worth 36 ct., which, deducted from the 16 lb. worth 96 ct., leaves 9 lb. worth 60 ct., or 9 lb. at $6\frac{2}{3}$ ct. a lb. Taking 9 lb. and $6\frac{2}{3}$ ct. as the quantity and price of the mixture whose ingredients are worth 3 ct. and 7 ct. a lb., find, by Case III, the quantities of the ingredients, $\frac{3}{4}$ lb. at 3 ct., and $8\frac{1}{4}$ lb. at 7 ct. *Ans.*

REMARK.—There must be at least *two* ingredients, whose quantities are required.

1. How many bbl. flour at $8, and $8.50; with 300 bbl. at $7.50, and 800 at $7.80, and 400 at $7.65, will make 2000 bbl. at $7.85 a bbl.? *Ans.* 200 bbl. at $8; 300 bbl. at $8.50

2. What quantities of tea at 25 ct., and 35 ct. a lb.; with 14 lb. at 30 ct., and 20 lb. at 50 ct., and 6 lb. at 60 ct., will make 56 lb. at 40 ct. a lb.? *Ans.* 10 lb. at 25 ct., and 6 lb. at 35 ct

XXXIII. INVOLUTION.

ART. **380**. INVOLUTION is the process of finding a *power*.

A *power*, is the product of a number by itself one or more times.

The number from which the powers arise, is called the *root* of those powers, and sometimes the *first power*.

Powers are of different *degrees*, 2d, 3d, 4th, &c., according to the number of times the root is used in their formation.

The degree of a power is indicated by an *exponent*, which is a small figure, placed to the right of the root; thus, 7^2 signifies the 2d power of 7; 5^3, the *third* power of 5.

The 2d power of a number is called its *square*, because the area of a square is obtained by forming a 2d power; viz., the product of the number of linear units in one side by itself.

The 3d power of a number, is called its *cube*, because the solidity

REVIEW.—379. What is Case 6? The rule? The proof? How many ingredients must there be, whose quantities are required?

of a cube is obtained by forming a 3d power; viz., the product of the number of linear units in one side by itself twice.

TO FIND ANY POWER OF A NUMBER,

ART. **381.** RULE.—*Multiply the number continually by itself, until it has been used as often, as the degree of the power indicates: the last product will be the power required.*

NOTES.—1. The number of multiplications will be *one less* than the exponent, because the root is used *twice* in the first multiplication, once as multiplicand and once as multiplier.

2. When the power to be obtained is of a high degree, multiply by some of the powers instead of by the root continually; thus, to get the 9th power of 2, multiply its 6th power (64) by its 3d power (8); or, its 5th power (32) by its 4th power (16): the rule being, that *the product of any two powers of a number is that power whose degree is equal to the sum of their degrees.*

3. Any power of 1 is 1; any power of a number *greater* than 1 is *greater* than the number itself: any power of a number *less* than 1, is *less* than the number itself.

ART. **382.** From Note 2, last Art., $4^3 \times 4^3 \times 4^3 \times 4^3 \times 4^3 = 4^{15}$, but the expression on the left is the 5th power of 4^3: hence, $(4^3)^5 = 4^{15}$; that is, *when the exponent of the power required is a composite number* (15), *raise the root to a power whose exponent is one of its factors* (3), *and this result to a power whose exponent is the other factor* (5).

ART. **383.** *Any power of a fraction is equal to that power of the numerator divided by that power of the denominator.*

To raise a mixed number to any power, reduce it to a fractional form, and then proceed as just directed.

ART. **384.** The *square* of a decimal must contain *twice*, its *cube, three times* as many decimal places as the root, &c.: *hence, to obtain any power of a decimal, proceed as if it were a whole number, and point off in the result a number of decimal places equal to the number in the root multiplied by the exponent of the power.*

EXAMPLES FOR PRACTICE.

		ANS.			ANS.
1.	$(5)^2$	$= 25.$	8.	$(\frac{7}{8})^5$	$= \frac{16807}{32768}$
2.	14^3	$= 2744.$	9.	$(.02)^3$	$= .000008$
3.	6^5	$= 7776.$	10.	$(5^4)^2$	$= 5^8$, or 390625.
4.	192^2	$= 36864.$	11.	$(.046)^3$	$= .000097336$
5.	1^{10}	$= 1.$	12.	$(\frac{1}{9})^7$	$= \frac{1}{4782969}$
6.	$(\frac{3}{5})^4$	$= \frac{81}{625}$	13.	2056^2	$= 4227136.$
7.	$(2\frac{1}{4})^3$	$= 11\frac{25}{64}$	14.	$(7.62\frac{1}{2})^2$	$= 58.1406\frac{1}{4}$

XXXIV. EVOLUTION.

ART. **385**. Evolution is the process of finding a *root.*

A *root* of a number is another number, of which the given number is some power.

Evolution is the reverse of Involution, and is sometimes called the *Extraction of roots.*

ART. **386**. *Roots*, like *powers*, are divided into degrees, 2d, 3d, 4th, &c.: the degree of a root is always the same as the degree of the power, to which that root must be raised, to produce the given number.

Thus, the 3d root of 343 is 7, since 7 must be raised to the 3d power, to produce 343: the 5th root of 1024 is 4, since 4 must be raised to the 5th power, to produce 1024.

Since the 2d and 3d powers are called the *square* and *cube*, so the 2d and 3d roots are called the *square* root and *cube* root.

ART. **387**. To indicate the root of a number, use the *radical sign* ($\sqrt{}$), or *fractional exponent.*

The radical sign is placed before the number; the degree of the root is shown by the small figure between the branches of the radical sign, called the *index* of the root.

Thus, $\sqrt[3]{18}$ signifies the cube root of 18; $\sqrt[5]{9}$ signifies the 5th root of 9. The square root is usually indicated without the index 2; thus, $\sqrt{10}$ is the same as $\sqrt[2]{10}$.

The root of a number may be expressed by a *fractional exponent whose numerator is* 1, *and denominator is the index of the root to be expressed.*

Thus, $\sqrt{7} = 7^{\frac{1}{2}}$, and $\sqrt[3]{5} = 5^{\frac{1}{3}}$, and $\sqrt[4]{64} = 64^{\frac{1}{4}}$, and so on.

NOTE.—Any root of 1 is 1; any root of a number *greater* than 1 is *less* than the number itself: any root of a number *less* than 1 is *greater* than the number itself.

THE SQUARE ROOT.

ART. **388**. The *square root* of a number is another number which, multiplied by itself, will produce the given number; thus, 7 is the square root of 49, because $7 \times 7 = 49$.

The square root of a number of two figures is found by trial and proof: thus $\sqrt{64} = 8$, because $8 \times 8 = 64$. If the root can not be exactly obtained, the number is called an *imperfect square*, and its root is obtained by trial *to within unity:* thus, 75 is an imperfect square, because, as it lies between 64 and 81, whose exact square roots are 8 and 9, its square root must be greater than 8 and less than 9, either of which is its square root *to within unity*, but neither its square root exactly.

ART. 389. A number of more than two figures has two figures in its square root; for, being equal to or greater than 100, its square root must be equal to or greater than 10. To find the square root of such, first prove this

PROPOSITION.

If a number is composed of tens and units, its square will consist of the square of those tens, with twice the product of the tens by the units, and the square of the units.

DEM.—Take any number composed of tens and units, as 47, and square it, first separating it into its tens and units. The work shows that the square of 47 consists of 1600, (*the square of the tens* 40), of 560, (*twice the product of the tens by the units*, 2 × 40 × 7), and of 49, (*the square of the units*): the same is true of any other number containing tens and units.

$$\begin{array}{rr} 40+\ 7 = & 47 \\ 40+\ 7 = & 47 \\ \hline 280+49 & 329 \\ 1600+280\quad & 188\ \\ \hline \end{array}$$
$$1600+560+49=2209$$

Extract the Square Root of 7396.

ANALYSIS.—Since the number 7396 has more than 2 figures, its root will be composed of tens and units, and the square will be made up of the *square of these tens, with twice the product of the tens and units, and the square of the units.* The square of tens is always hundreds, and since 73, the hundreds of the number, lies between 64 and 81, its square root must lie between theirs, being greater than 8 and less than 9; hence, 8 is the tens' figure of the root, 9 being too large. As the two right-hand figures (96) are not used in finding the tens' figure of the root, point them off so as to show distinctly the 73, the only figures to be examined for that purpose.

```
      73'96(86
      64
  166)996
      996
```

Subtract the *square of the tens*, 64 (hundreds), from the given number 7396, which is *the square of the tens, with twice the product of the tens by the units, and the square of the units:* the remainder 996, must be *twice the product of the tens by the units, and the square of the units.*

The square of the units being comparatively small, may be neglected for the present, and the 996 regarded as *twice the product of the tens by the units:* divide it by *twice the tens* (16 tens), to get the units. Dividing 996, by 160 (16 tens), or, what is more convenient, dividing 99 by 16, the units' figure is found to be 6.

To prove this figure correct, form with it not only *twice the product of the tens by the units*, but also *the square of the units*, since the 996 consists of both these parts. To do this conveniently, set the units' figure (6) on the right of the *trial divisor* (16), and multiply the complete divisor (166) by the units' figure (6), thereby making *twice the product of the tens by the units* (160 $\times$ 6), and the *square of the units* (6 $\times$ 6); since this produces 996 exactly, 86 must be the exact square root of 7936: the same process and explanation apply in obtaining the square root of any number, however large. Hence,

ART. **390**. TO EXTRACT THE SQUARE ROOT OF A NUMBER,

RULE.—1. *Separate the given number into periods of two figures, commencing at units; the left hand period may have* 1 *or* 2 *figures.*

2. *Take the square root of the nearest square below the left hand period: this will be the first figure of the root.*

3. *Subtract the square of this figure from the left hand period, bring down the next period; divide the result exclusive of the right hand figure, by twice the part of the root already found. the quotient will be the* 2d *figure of the root.*

4. *Set this figure of the root on the right of the divisor; multiply the divisor thus completed, by the* 2d *figure of the root; subtract the product from the last dividend, and bring down another period.*

5. *Double the root already found for a trial divisor, find another figure of the root, and proceed as before, until all the periods have been brought down.*

NOTES.—1. If any product is larger than the dividend from which it is to be taken, the last figure of the root is too large.

2. If any dividend, exclusive of its right-hand figure, is not large enough to contain its trial divisor, place a cipher in the root, and at the right of the divisor; bring down another period and continue as before.

3. When a remainder is greater than the previous divisor, it does not follow that the last figure of the root is too small, *unless that remainder is large enough to contain twice the part of the root already found, and* 1 *more;* for this would be the proper divisor to go into the remainder, if the root were increased by 1.

EXAMPLES FOR PRACTICE.

		ANS.			ANS.
1.	$\sqrt{2809}$ . . .	= 53.	4.	$\sqrt{57600}$ . .	= 240.
2.	$\sqrt{1444}$ . . .	= 38.	5.	$\sqrt{16499844}$.	= 4062.
3.	$\sqrt{11881}$ . . .	= 109.	6.	$\sqrt{49098049}$.	= 7007.

7. $\sqrt{185640625}$. = 13625.	11. $\sqrt{73005}$. = 270+
8. $\sqrt{80012304}$. = 8945—	12. $\sqrt{386^3}$ . . = 7584—
9. $\sqrt{6203794}$. = 2491—	13. $\sqrt{1245\times(252)^2}$ = 8892—
10. $\sqrt{3444736}$. = 1856.	14. $\sqrt{(96059601)^{\frac{1}{2}}}$ = 99.

15. $\sqrt{(126)^2 \times (58)^2 \times (604)^2}$ . . = 4414032.

ART. **391.** *The square root of a common fraction is the sq. oot of its numerator, divided by the sq. root of its denominator.*

NOTES.—1. Reduce the fraction to its lowest terms before commencing the operation, and if the denominator be not a square, make it so, by multiplying both terms by the denominator, or some smaller number that will answer the purpose.

2. The square root of a mixed number may be found by first converting it into a common fraction, and then proceeding by this rule.

Extract the Square Root of $\frac{540}{864}$.

SOL.—Reduce $\frac{540}{864}$ to $\frac{5}{8}$; multiplying both terms by 2, gives $\frac{10}{16}$, the square root of which is $\frac{3}{4}$ nearly.

1. $\sqrt{\frac{4}{7}}$ . . . = $\frac{5}{7}$ nearly.	4. $\sqrt{272\frac{3}{12}}$ . . . = $16\frac{1}{2}$.
2. $\sqrt{5\frac{2}{3}}$. . . = $2\frac{1}{3}$ nearly.	5. $\sqrt{\frac{770}{960}}$ . . = $\frac{7}{8}$ nearly.
3. $\sqrt{\frac{176}{275}}$ . . . . . = $\frac{4}{5}$.	6. $\sqrt{90\frac{5}{8}}$ . . = $9\frac{1}{2}$ nearly.

ART. **392.** Since the *square* of a decimal has just *twice* as many decimal places, as the root (Art. 384), the *square root* of a decimal must have exactly *half* as many decimal places, as the number itself; and as the mode of operation is the same as in whole numbers, it follows that,

To get the square root of a decimal, annex a cipher, to make the number of its decimal places even, (if it be not so already); then proceed as with a whole number, pointing off from the root half as many decimal places, as are in the given number.

NOTES.—1. The number of decimal places *must be even* before commencing the operation; otherwise it will be impossible to point off in the root *exactly half* as many decimal places as the number contains

2. The last rule applies also to mixed decimals; also to common fractions or mixed numbers, after changing them to decimals.

3. Since decimal ciphers may be annexed to any whole number or decimal, if the square root is not exact, the process may be continued by bringing down *two decimal ciphers* at every step, and pointing off one decimal place in the root for every pair of ciphers annexed; and the further the process is carried, the nearer the result is to the exact square root. The limit of the error is always *one of the lowest order in the root.*

1. Extract the square root of .07625. *Ans.* .276 +
2. Of $\frac{7}{8}$ in decimal hundredths. *Ans.* .93 +
3. Of 2.135 to thousandths. *Ans.* 1.461 +
4. Of 3 to six decimal places. *Ans.* 1.732051 —

OPERATION.		CONTRACTED METHOD.	
3	1.732051 nearly.	3	1.732051 nearly.
	1		1
27	200	27	200
	189		189
343	1100	343	1100
	1029		1029
3462	7100	34~~62~~	7100
	6924		6924
346405	1760000		176
	1732025		173
346410	27975		3

RULE FOR CONTRACTED METHOD.

Extract the square root as usual, until one more than half of the figures required in the root have been determined; to obtain the remaining figures, divide the last remainder by the last divisor, using the contracted method in Art. 158.

	ANS.		ANS.
5. $\sqrt{.0081}$	$= .09$	15. $\sqrt{\frac{6}{7}}$ . . .	$= .92582+$
6. $\sqrt{.451584}$ . . .	$= .672$	16. $\sqrt{34\frac{5}{8}}$ . .	$= 5.8843+$
7. $\sqrt{.08894}$ . .	$= .298+$	17. $\sqrt{.9}$ . . .	$= .94868+$
8. $\sqrt{17.7241}$ . . .	$= 4.21$	18. $\sqrt{6.4}$ . . .	$= 2.5298+$
9. $\sqrt{139.31655}$	$= 11.803+$	19. $\sqrt{1089^3}$ . . .	$= 35937$
10. $\sqrt{\frac{2}{3}}$ . .	$= .8165$ nearly.	20. $\sqrt{8^5}$ . . .	$= 181.02-$
11. $\sqrt{.030625}$ . . .	$= .175$	21. $\sqrt{6102815944}$	$= 78120.52+$
12. $\sqrt{13}$ . . .	$= 3.60555+$	22. $\sqrt{504125310742198}$	$= 22452735+$
13. $\sqrt{3.1415926}$	$= 1.7724+$		
14. $\sqrt{.00000625}$.	$= .0025$		

APPLICATIONS OF SQUARE ROOT.

ART. 393. To find the side of a square figure of given area.

RULE.—*Reduce the area, if necessary, to square units; the square root of the number thus obtained will be the side of the square in linear units of the same name.*

What is the side of a square field of 10 acres?

SOL.—10 A. = 1600 sq. rd.; then $\sqrt{1600}$ sq. rd. = 40 rd. *Ans.*

1. Find the side of a square field of 122 A. 2 R. *Ans.* 140 rd.
2. The fencing for a square field of 8 A. 2 R. 9 P. *Ans.* 148 rd.
3. A has 5 A. 1 R. 7 P.; 7 A. 2 R.; and 2 A. 3 R. 13 P. of land, in a square: how much fencing will enclose them? *Ans.* 200 rd.

ART. **394**. A triangle is the space bounded by three straight lines, called its *sides.*

A *right-angled* triangle has two sides forming a right angle; that is, perpendicular to each other.

The longest side of a right-angled triangle is called the *hypotenuse;* the other two, the *perpendicular* sides.

ART. **395**. To find the hypotenuse, when the perpendicular sides are known, use this

RULE.—*Square each of the given sides; add the results, and extract the square root of the sum.*

If the perpendicular sides of a right-angled triangle are **3** in. and **4** in., what is the hypotenuse?

SOL.—The square of 3 is 9; the square of 4 is 16: their sum is 25, the square root of which is 5 in., the hypotenuse.

ART. **396**. To find either of the perpendicular sides, when the other, and the hypotenuse, are known, use this

RULE.—*Square each of the given sides; take the difference of the results, and extract its square root.*

If the hypotenuse is 13, and one of the perpendicular sides 5, what is the other side?

SOL.—The square of 13 is 169; the square of 5 is 25: the difference is 144, whose square root is 12, the other side.

1. Find the length of a ladder reaching 12 ft. into the street, from a window 30 ft. high. *Ans.* 32.31+ft.
2. What is the *diagonal,* or line joining the opposite corners, of a square whose side is 10 ft.? *Ans.* 14.142+ft.
3. What is saved by following the diagonal instead of the sides (69 and 92 rd.) of a rectangle? *Ans.* 46 rd.
4. A boat in crossing a river 500 yd. wide, drifted with the current 360 yd.; how far did it go? *Ans.* 616+yd.

ART. **397**. It is known, that similar figures are to each other as the squares of their like dimensions; hence,

1st. *The ratio of the areas of two similar figures, is equal to the square of the ratio of any two like dimensions of them.*

2d. *The ratio of any two like dimensions of two similar figures, is equal to the square root of the ratio of their areas.*

One square has a side $3\frac{2}{5}$ times as large as another: how many times does it contain the smaller?

SOL.—Since $3\frac{2}{5}$ or $\frac{17}{5}$ is the ratio of the sides, $\frac{289}{25}$, is the ratio of their areas, and the larger contains the smaller $\frac{289}{25} = 11\frac{14}{25}$ times.

1. One square is $12\frac{1}{4}$ times another: how many times does the side of the 1st contain the side of the 2d? *Ans.* $3\frac{1}{2}$.

2. The diagonals of two similar rectangles are as 5 to 12: how many times does the larger contain the smaller? *Ans.* $5\frac{19}{25}$.

CUBE ROOT.

ART. **398**. The *cube root* of a number is another number, which, being cubed, will produce the given number.

The cube root of any number containing three figures or less, is found by trial and proof; thus, $\sqrt[3]{512}$ is 8, because $8^3 = 512$.

If the number does not have an exact cube root, as 247, it is called an *imperfect cube;* and its cube-root is obtained by trial, *to within unity*; thus, as 247 lies between 216 and 343, its cube root must lie between theirs, 6 and 7.

ART. **399**. If a number has more than three figures, its cube root will have two figures, tens and units; in order to extract its cube root, first prove this

PROPOSITION. — *The cube of any number containing tens and units, will consist of the cube of the tens, three times the square of the tens multiplied by the units, three times the tens multiplied by the square of the units, and the cube of the units.*

$$
\begin{array}{rrrrr}
(47)^2 = & & 1600 + & 560 + & 49 \\
47 = & & & 40 + & 7 \\
\hline
& & 11200 + & 3920 + & 343 \\
& 64000 + & 22400 + & 1960 & \\
\hline
& 64000 + & 33600 + & 5880 + & 343
\end{array}
$$

DEM.—Multiply the square of 47 (Art. 389) by $47 = 40 + 7$. The result is the cube of 47, which consists of 64000, (*the cube of the tens* 40), of 33600, (3 *times the square of the tens multiplied by the units* $3 \times 1600 \times 7$), of 5880, (3 *times the tens multiplied by the square of the units* $3 \times 40 \times 49$) and 343, (*the cube of the units*): and the same may be shown of any number containing tens and units.

Find the Cube Root of 238328.

$$
\begin{array}{rr|l}
& & 238'328(62 \\
& & 216 \\
\hline
36 \times 300 = & 10800 & 22328 \\
6 \times 2 \times 30 = & 360 & \\
2 \times 2 = & 4 & \\
\hline
& 11164 & 22328 \\
\hline
\end{array}
$$

ANAL.—Since 238328 has more than 3 figures, its cube root must consist of tens and units; and 238328 must contain the 4 parts mentioned in the last proposition, the first of which is *the cube of the tens' figure* of the root. The cube of *tens* being *thousands*, point off the three right hand figures, and regard only the 238, which are thou sands. Since this lies between 216 and 343, its cube root must lie between theirs, being greater than the former (6) and less than the latter (7); hence, 6 is the tens' figure of the root, 7 being too large.

Cubing the tens' figure, 6 (tens), gives 216 (thousands); subtract this from the given number; the remainder, 22328, must be the other 3 parts mentioned in the proposition; as the 3d and 4th parts are small compared with the 2d, neglect them for the present, and consider 22328 as the 2d part alone; that is, 3 *times the square of the tens multiplied by the units.*

Divide 22328 by 3 times the square of the tens (3 × 3600), to get the units' figure 2, which must not only be placed in the root, but also used to complete the divisor, so that when it is multiplied by the units' figure of the quotient, the product shall contain the 3d and 4th parts mentioned in the proposition, as well as the 2d part.

To do this conveniently, increase the trial divisor (10800) by 3 *times the tens by the units* (6 × 2 × 30) and also *by the square of the units'* (2 × 2); for these, when multiplied by the units' figure (2), produce the 3d and 4th parts that must be taken account of.

ART. **400.** The same process and analysis apply to any number; hence,

TO EXTRACT THE CUBE ROOT OF A WHOLE NUMBER,

RULE.—1. *Separate the number into periods of* 3 *figures, commencing at units; the left hand period may contain* 1, 2, *or* 3 *figures.*

2. *Find the cube root of the nearest perfect cube below the left hand period; this will be the* 1*st figure of the root.*

3. *Cube the* 1*st figure of the root, subtract it from the left hand period, and bring down the next period; this will be the first dividend.*

4. *Take* 3 *times the square of the* 1*st figure of the root, annexing* 2 *ciphers, for a trial divisor; see how often it is contained in the first dividend; the quotient will be the* 2*d figure of the root.*

5. *Under the trial divisor, write* 3 *times the* 2*d figure of the root multiplied by the figure before it, annexing one cipher, and also the square of the second figure of the root, and add these to the trial divisor for a complete divisor.*

6. *Multiply the complete divisor by the* 2*d figure of the root, and subtract the product from the first dividend, bringing down the next period; this will be the* 2*d dividend.*

7. *Take* 3 *times the square of that part of the root already found for a trial divisor, find another figure of the root, complete the divisor, and so on, until all the periods have been brought down:*

if there is no remainder, the exact cube root is obtained, but if there is, it is the cube root to within unity.

NOTES.—1. If any product is greater than the dividend above it, the last figure of the root is too large.

2. If any trial divisor is not contained in its dividend, put a cipher in the root, two ciphers at the right of the trial divisor, bring down another period, and proceed as before.

3. If any remainder is larger than the previous divisor, it does not follow that the last quotient figure is too small, *unless the remainder is large enough to contain* 3 *times the square of that part of the root already found, with* 3 *times that part of the root, and* 1 *more*; for this is the proper divisor to go into the remainder, if the root is increased by 1.

	ANS.		ANS.
1. $\sqrt[3]{512}$	$= 8.$	6. $\sqrt[3]{6506321}$.	$= 187-$
2. $\sqrt[3]{19683}$ . .	$= 27.$	7. $\sqrt[3]{98765^2}$.	$= 2137-$
3. $\sqrt[3]{7301384}$ . .	$= 194.$	8. $\sqrt[3]{782^5}$.	$= 66376+$
4. $\sqrt[3]{94818816}$.	$= 456.$	9. $\sqrt[3]{(10604499373)^{\frac{1}{3}}}$	$= 13$
5. $\sqrt[3]{1067462648}$.	$= 1022.$	10. $\sqrt[3]{(30840979456)^{\frac{1}{2}}}$	$= 56$

TO FIND THE CUBE ROOT OF A COMMON FRACTION,

ART. **401.** RULE.—*Reduce the fraction to its lowest terms; make the denominator a perfect cube, if it be not so, by multiplying both terms by the square of the denominator or some smaller number that will answer the purpose: extract the cube root of the numerator for a numerator, and the cube root of the denominator for a denominator.*

	ANS.		ANS.
1. $\sqrt[3]{\frac{4}{9}} =$	$\frac{2}{3}+$	4. $\sqrt[3]{\frac{2744}{35937}} =$ . . .	$\frac{14}{33}$
2. $\sqrt[3]{\frac{297}{528}} =$ . . .	$\frac{3}{4}+$	5. $\sqrt[3]{\frac{2808}{43875}} =$ . . .	$\frac{2}{5}$
3. $\sqrt[3]{\frac{45}{75}} =$ . . .	$\frac{4}{5}+$	6. $\sqrt[3]{\frac{672}{5670}} =$ . .	$\frac{7}{15}+$

ART. **402**. The cube of a decimal has just *three times* as many decimal places as the root; hence, the cube root of a decimal must contain just *one-third* as many decimal places as the number itself; hence,

TO FIND THE CUBE ROOT OF A DECIMAL,

RULE.—*Make the number of decimal places divisible by* 3, *if they be not so, by annexing ciphers; then proceed as in whole numbers, and point off from the root, one-third as many decimal places as are in the given number.*

NOTES.—1. The number of decimal places must be exactly divisible by 3, otherwise, it would be impossible to point off in the root, *exactly* $\frac{1}{3}$ as many decimal places as are in the number.

2. This rule applies also to mixed decimals, and to common fractions after changing them to a decimal form ; to extend the process, if the cube root is not exact, bring down 3 decimal ciphers at every step and make a decimal place in the root, for every 3 ciphers annexed.

What is $\sqrt[3]{189.3754}$ to 4 decimal places?

```
            189.375400(5.7426
            125
      7500| 64375
      1050|
        49| 60193
      8599|  4182400
    974700|
      6840|
        16| 3926224
    981556|  256176000
  98842800|
     34440|
         4| 197754488
  98877244|  58421512000
9891169200|
   1033560|
        36|
9892202796| 59353216776
```

CONTRACTED METHOD.

```
        189.375400(5.7426
        125
  7500| 64375
  1050|
    49| 60193
  8599|  4182400
974700|
  6840|
    16| 3926224
981556|  256176
          19631
           5987
            589
```

ART. 403. RULE FOR CONTRACTED METHOD.

Extract the cube root as usual, until one more than half of the figures required in the root have been ascertained, and then, with the last divisor and the last remainder, proceed as in contracted division to determine the other figures of the root, except that two figures are dropped instead of one from the divisor at every step and one from every remainder.

		ANS.			ANS.
1	$\sqrt[3]{5.088448}$ . . .	= 1.72	7.	$\sqrt[3]{\frac{23}{729}}$. :	= .315985
2.	$\sqrt[3]{22188.041}$ . .	= 28.1	8.	$\sqrt[3]{25}$ . .	= 2.924018
3.	$\sqrt[3]{32.65}$ . .	= 3.196154+	9.	$\sqrt[3]{11}$ . .	= 2.22398
4.	$\sqrt[3]{.0079}$ . .	= .1991632+	10.	$\sqrt[3]{\frac{2}{3}}$	= .87358
5.	$\sqrt[3]{3.0092}$.	= 1.443724	11.	$\sqrt[3]{\frac{4}{15}}$ . .	= .64366
6.	$\sqrt[3]{2315.68}$ . .	= 13.23+	12.	$\sqrt[3]{38\frac{22}{31}}$ . .	= 3.38277

APPLICATIONS OF CUBE ROOT.

ART. **404.** To find the side of a cube containing a given solidity, use the following

RULE.—*Reduce the solidity, if necessary, to a denomination of cubic units; extract the cube root of the result, and it will be the side of the cube in linear units of the same name as the cubic units in which the solidity is expressed.*

Find the side of a cubical box containing 25 bu.

$$25 \text{ bu.} = 25 \times 2150.42 \text{ cu. in.} = 53760.5 \text{ cu. in.};$$

$$\sqrt[3]{53760.5} = 37.75 \text{ in.} = 3 \text{ ft. } 1\tfrac{3}{4} \text{ in.}$$

1. Of a cube containing 76 cu. ft., 1323 cu. in. *Ans.* 4 ft. 3 in.

2. Of a cube equal to a rectangular solid 14 ft. 5 in. long, 6 ft. 8 in. wide, 3 ft. 2 in. high. *Ans.* 6 ft. 8.7 in.

3. Of a cubical tank, to hold 150 bbl. *Ans.* 8 ft. 7 in.

4. Find the sq. ft. in 1 face of a cube containing 39304 cu. ft. *Ans.* 1156.

5. The sq. in. in all the faces of a cube containing 8365427 cu. in. *Ans.* 247254.

ART. **405.** Any two similar solids are to each other as the cubes of their like dimensions; hence,

1st. *The ratio of two similar solids is equal to the cube of the ratio of any two like dimensions.*

2d. *The ratio of any two like dimensions of similar solids is equal to the cube root of the ratio of the solids.*

1. The lengths of two similar solids are 4 in. and 50 in.; the 1st contains 16 cu. in.: what does the 2d contain? *Ans.* 31250 cu. in.

2. The solidities of two balls are 189 cu. in. and 875 cu. in.; the diameter of the 2d is $17\frac{1}{2}$ in.; find the diameter of the 1st. *Ans.* $10\frac{1}{2}$ in.

EXTRACTION OF ANY ROOT.

ART. **406.** There is a general method of extracting roots, called *Horner's* method, from its inventor, which has great advantages. It is comprised in this

RULE FOR EXTRACTING ANY ROOT.

1. *Make as many columns as there are units in the index of the root to be extracted; place the given number at the head of the right hand column, and ciphers at the head of the others.*

2. *Commencing at the right, separate the given number into periods of as many figures as there are columns; extract the required root to within unity, of the left-hand period, for the* 1st *figure of the root.*

3. *Add this figure into the* 1st *column, multiply it then by itself, and set it in the* 2d *column; multiply this again by the same figure, and set it in the* 3d *column, and so on, placing the last pro-*

duct in the right hand column, under that part of the given number from which the figure was derived, and subtracting it from the figures above it.

4. *Add the same figure into the* 1*st column again, multiply the result by the figure again, adding the product to the* 2*d column, and so on, stopping at the next to the last column.*

5. *Repeat this process, leaving off one column at the right every time, until all the columns have been thus dropped; then annex one cipher to the number in the* 1*st column, two to the number in the* 2*d column, and so on, to the number in the last column, to which the next period of figures from the given number must be brought down.*

6. *Divide the number in the last column by the number in the previous column as a trial divisor (making allowance for completing the divisor); this will give the* 2*d figure of the root, which must be used precisely as the* 1*st figure of the root has been: and so on till all the periods have been brought down.*

ART. **407**. The process may often be shortened by this

RULE FOR CONTRACTED METHOD.—*Obtain one less than half of the figures required in the root as the rule directs; then, instead of annexing ciphers and bringing down a period to the last numbers in the columns, leave the remainder in the right hand column for a dividend; cut off the right hand figure from the last number of the previous column, two right hand figures from the last number in the column before that, and so on, always cutting off one more figure for every column to the left.*

With the number in the right hand column and the one in the previous column, determine the next figure of the root, and use it as directed in the rule, recollecting that the figures cut off are not used except in carrying the tens they produce. This process is continued until the required number of figures are obtained, observing that when all the figures in the last number of any column are cut off, that column will be no longer used.

REM.—Add to the 1st column mentally; multiply and add to the next column in one operation: multiply and subtract from the right hand column in like manner.

Extract the cube root of **44.6** to six decimals.

0	0	44.600(3.546323
3	9	17600
6	2700	1725000
90	3175	238136
95	367500	12182
100	371716	865
1050	375948	**111**
1054	37659	
1058	37723	
1062		

EXP.—The trial divisors may be known by ending in two ciphers; the complete divisors stand just beneath them. After getting 3 figures of the root, contract the operation by last rule.

Extract the 5th root of 1256, to 4 places of decimals.

0	0	0	0	1256 (4.1668
4	16	64	256	23200000
8	48	256	12800000	9743799
12	96	640000	13456201	1014765
16	16000	656201	14128805	113733
200	16201	672604	1454839	
201	16403	689210	1497403	
202	16606	6993	150172	
203	16810	7094	150604	
204		7195		
205				

Rem.—We have seen (Art. 382) that $(4^3)^5 = 4^{15}$, that is, if 4 be raised to the 3d power, and this product raised to the 5th power, the result will be the 15th power of 4; hence, conversely if the 5th root of 4^{15} be taken, and the 3d root of this result, it will give 4, which is the 15th root of 4^{15}, that is, the 3d root of the 5th root of a number is equal to the 15th root of that number. Hence,

Rule.—*Whenever the index of the root to be extracted can be separated into factors, extract successively the roots indicated by those factors, and the final result thus obtained will be the root required.*

In using this rule, begin with the smallest roots.

Thus, to get the 4th root, extract the square root of the square root.

And to get the 6th root, extract the cube root of the square root.

And to get the 8th root, extract the sq. root 3 times in succession.

And to get the 9th root, extract the cube root of the cube root, and so on.

		Ans.			Ans.
1.	$\sqrt[4]{97.41}$	$= 3.1416$	4.	$\sqrt[3]{.0004856}$	$= .0786007$
2.	$\sqrt[5]{3}$	$= 1.2457$	5.	$\sqrt[5]{\frac{5}{12}}$	$= .83938$
3.	$\sqrt[6]{21035.8}$	$= 5.254$	6.	$\sqrt[6]{43\frac{3}{8}}$	$= 1.87445$

XXXV. SERIES.

Art. **408.** A Series is a succession of numbers, each derived from the preceding, according to a fixed law.

The numbers which form a series are called its *terms*; if they increase toward the right, it is an *Ascending* series; if they decrease toward the right, it is a *Descending* series: the first and last terms are the *extremes*. Series are Arithmetical or Geometric.

ARITHMETICAL SERIES.

ART. **409**. An Arithmetical series is one whose terms increase or decrease by the addition or subtraction of a fixed number, called the *common difference* ; as, 2, 5, 8, &c.; or, 10, 7, 4 ; in which the common difference is 3.

ART. **410**. To find the last (or any) term of an arithmetical series, when the 1st term, common difference, and number of terms, are known.

RULE.—*Multiply the common difference by* 1 *less than the number of terms, and add the product to, or subtract it from, the* 1*st term, according as the series is ascending or descending.*

1. Find the 12th term of the series 3, 7, 11, &c. *Ans.* 47.
2. The 18th term of the series 100, 96, &c. *Ans.* 32.
3. The 64th term of the series $3\frac{1}{2}$, $5\frac{3}{4}$, &c. *Ans.* $145\frac{1}{4}$
4. The 10th term of the series .025, .037, &c. *Ans.* .133

ART. **411**. To find the 1st term, when the last term, common difference, and number of terms, are known.

RULE.—*Transpose the series, and then find its last term ; this will be the* 1*st term of the given series.*

NOTE.—If the given series is ascending, the transposed one will be descending, and *vice versa.*

1. Find the 1st term of .....68, 71, having 19 terms. *Ans.* 17.
2. Of ....117, $123\frac{1}{2}$, 130, having 6 terms. *Ans.* $97\frac{1}{2}$
3. Of ....$18\frac{3}{4}$, $12\frac{1}{2}$, $6\frac{1}{4}$, having 365 terms. *Ans.* $2281\frac{1}{4}$

ART. **412**. To find the common difference, when the extremes and number of terms are known.

RULE.—*Divide the difference of the extremes by the number of terms less one.*

1. Find the common difference of a series whose extremes are 8 and 28, and number of terms, 6. *Ans.* 4.
2. Extremes are $4\frac{1}{2}$ and $20\frac{3}{4}$, and number of terms, 14. *Ans.* $1\frac{1}{4}$
3. 44th term is 150, and 19th term, 30. *Ans.* $4\frac{4}{5}$
4. 4th term is 7, and 14th term, 39. *Ans.* $3\frac{1}{5}$

ART. **413**. To find the number of terms, when the extremes and common difference are known.

RULE.—*Divide the difference of the extremes by the common difference, and add* 1 *to the quotient.*

1. What is the number of terms in a series whose extremes are 9 and 42, and common difference, 3 ? *Ans.* 12.
2. Whose extremes are 3 and $10\frac{1}{2}$, and com. dif. $\frac{3}{8}$? *Ans.* 21.
3. In the series 10, 15 . . . 500 ? *Ans.* 99.

ART. **414**. To insert a given number of arithmetical means between two given numbers.

RULE.—*Take the given numbers as the extremes, and* 2 *more than the number of means as the number of terms, of an arithmetical series, and find the common difference by rule in Art.* 412. *Add this common difference to the smaller number, to get the* 1*st mean; add it to the* 1*st mean, to get the* 2*d mean, and so on.*

1. Insert 1 arithmetical mean between 8 and 54. *Ans.* 31.
2. 5 arith. means between 6 and 30. *Ans.* 10, 14, 18, 22, 26.
3. 2 arith. means between 4 and 40. *Ans.* 16, 28.
4. 4 arith. means between 2 and 3. *Ans.* $2\frac{1}{5}$, $2\frac{2}{5}$, $2\frac{3}{5}$, $2\frac{4}{5}$.

ART. **415**. To find the sum of the terms in an arithmetical series, when the extremes and number of terms are known.

RULE.—*Multiply half the sum of the extremes by the number of terms: the product will be the sum of the series.*

1. Find the sum of the arithmetical series whose extremes are 850 and 0, and number of terms, 57. *Ans.* 24225
2. Extremes, 100 and .0001: No. of terms, 12345. *Ans.* 617250.61725

NOTE.—It may be necessary to find the extremes or number of terms by one of the previous rules, before applying this rule.

3. What is the sum of the arithmetical series 1, 2, 3, &c., having 10000 terms? *Ans.* 50005000.
4. Of 1, 3, 5, &c., having 1000 terms? *Ans.* 1000000.
5. Of 999, 888, 777, &c., having 9 terms? *Ans.* 4995.
6. Of 4.12, 17.25, 30.38, &c., having 250 terms? *Ans.* 409701.25
7. Whose 5th term is 21; 20th term, 60; number of terms, 46? *Ans.* $3178\frac{3}{5}$.

GEOMETRIC SERIES.

ART. **416**. A GEOMETRIC SERIES is one, in which each term is formed by multiplying the previous one by a fixed number called the *common ratio.*

ART. **417**. To find the last (or any) term of a geometric series, when the 1st term, the common ratio, and the number of terms are known.

RULE.—*Raise the common ratio to a power whose degree is one less than the number of terms, and multiply the* 1*st term by it.*

NOTE.—The common ratio in a given geometric series may be found by dividing any term by the preceding.

1. Find the last term in the series 64, 32, &c., of 12 terms. *Ans.* $\frac{1}{32}$
2. In 2, 5, $12\frac{1}{2}$, &c., having 6 terms. *Ans.* $195\frac{5}{16}$
3. In 100, 20, 4, &c., having 9 terms. *Ans.* $\frac{4}{15625}$
4. 1st term, 4; com. ratio, 3: find the 10th term. *Ans.* 78732.

5. 3d term, 16; com. ratio, 6: find the 9th term. *Ans.* 746496.
6. 33d term, 1024; com. ratio $\frac{3}{4}$: find the 40th term. *Ans.* $136\frac{11}{16}$.

ART. **418**. To find the 1st term, when the last term, number of terms, and common ratio are known.

RULE.—*Transpose the series; find the last term by the previous rule: this will be the 1st term of the given series.*

NOTE.—The common ratio of the transposed series will be the common ratio of the given series inverted.

1. Find the 1st term of the series ...90, 180, of 6 terms. *Ans.* $5\frac{5}{8}$.
2. Of ...$\frac{729}{2048}$, $\frac{2187}{4096}$, having 11 terms. *Ans.* $\frac{1}{108}$
3. 9th term, 576; com. ratio, 3: find the 1st term. *Ans.* $\frac{64}{729}$
4. 18th term, 15625; com. ratio, $2\frac{1}{2}$: find the 10th term. *Ans.* $10\frac{6}{25}$
5. 56th term, 1440; com. ratio, 6: find the 49th term. *Ans.* $\frac{5}{972}$

ART. **419**. To find the common ratio, when the extremes and number of terms are known.

RULE.—*Divide the last term by the first, and extract that root of the quotient whose index is one less than the number of terms: this will be the common ratio.*

1. Find the common ratio: 1st term, 8; 4th term, 512. *Ans.* 4.
2. 1st term, $4\frac{15}{16}$; 11th term, 49375000000. *Ans.* 10.
3. 16th term, 729; 22d term, 1000000. *Ans.* $3\frac{1}{3}$.

ART. **420**. To insert a given number of geometric means between two given numbers.

RULE.—*Take the given numbers as the extremes of a geometrical series, and 2 more than the number of means for the number of terms, and find the common ratio by the last rule; multiply the first term by the common ratio, to get the 1st mean; multiply this by the common ratio to get the 2d mean, and so on.*

1. Insert 1 geometric mean between 63 and 112. *Ans.* 84.
2. 4 geometric means between 6 and 192. *Ans.* 12, 24, 48, 96.
3. 3 geometric means between $\frac{1}{36864}$ and $\frac{1}{9}$. *Ans.* $\frac{1}{4608}$, $\frac{1}{576}$, $\frac{1}{72}$
4. 2 geometric means between 14.08 and 3041.28. *Ans.* 84.48 and 506.88

ART. **421**. To find the sum of all the terms in a geometric series, when the extremes and common ratio are known.

RULE.—*Multiply the last term by the common ratio; find the difference between this product and the first term, and divide it by the difference between the common ratio and 1.*

NOTE.—It may be necessary sometimes to find one of the extremes by one of the previous rules.

1. Find the sum of 6, 12, 24, &c., to 10 terms. *Ans.* 6138.
2. Of 16384, 8192, &c., to 20 terms. *Ans.* $32767\frac{31}{32}$
3. Of $\frac{2}{3}$, $\frac{4}{9}$, $\frac{8}{27}$, &c., to 7 terms. *Ans.* $1\frac{1931}{2187}$

4. Extremes, 25 and 102400; No. of terms, 13. *Sum*, 204775.
5. 1st term, 7; 6th term, 1701: No. of terms, 9. *Sum*, 68887.
6. 3d term, 40; last (11th) term, 2621440. *Sum*, $3495252\frac{1}{2}$.
7. 4th term, 216; 8th term, $42\frac{2}{3}$; No. of terms, 10. *Sum*, $2149\frac{2}{27}$.

In an infinite decreasing geometrical series the last term is 0.

Find the sum of the following infinite geometrical series:

8. Of 1, $\frac{1}{2}$, $\frac{1}{4}$, &c. *Ans.* 2.
9. Of $\frac{3}{5}$, $\frac{9}{25}$, $\frac{27}{125}$, &c. *Ans.* $1\frac{1}{2}$.
10. Of $\frac{1}{2}$, $\frac{3}{8}$, $\frac{9}{32}$, &c. *Ans.* 2.
11. Of $\frac{7}{6}$, 1, $\frac{6}{7}$, &c. *Ans.* $8\frac{1}{6}$.
12. Of $.\dot{3}\dot{6} = .3636$, &c. $= \frac{36}{100} + \frac{36}{10000}$, &c. *Ans.* $\frac{4}{11}$.
13. Of $.\dot{3}4920\dot{6}$, of $\dot{4}8\dot{0}$, of $\dot{6}$. *Ans.* $\frac{22}{63}$ and $\frac{160}{333}$ and $\frac{2}{3}$

XXXVI. PERMUTATIONS.

ART. **422**. PERMUTATIONS are the changes of order which a number of things may undergo.

To find the number of permutations possible with a given number of objects, using them all each time.

RULE.—*Multiply together the series of natural numbers from 1 to the given number of objects inclusive.*

1. How long can 7 persons sit in different order at table, allowing $365\frac{1}{4}$ da. to a year, and 3 meals a day? *Ans.* 4 yr. 219 da.
2. How many changes of order are possible with the letters that compose the word *anthem?* *Ans.* 720.
3. How many different numbers can be expressed by the same figures as the number 1234567890? *Ans.* 3628800.
4. In how many different ways can the 8 notes of an octave be written? *Ans.* 40320.

ART. **423**. To find the number of permutations of a given number of objects, using a given number less than all, each time.

RULE.—*Multiply together the series of natural numbers, commencing with 1 more than the number of objects omitted in each permutation, and ending with the whole number of objects.*

1. How many permutations can be made of 7 letters, using 3 each time? *Ans.* 210.
2. Of 5 letters, using 2 each time? *Ans.* 20.
3. Of 6 letters, using 1 each time? *Ans.* 6.

XXXVII. COMBINATIONS.

ART. **424**. COMBINATIONS are the different sets containing the same number of objects, which may be selected from a given larger number of objects.

To find the number of combinations possible with a given number of objects, using a given number each time.

RULE.—*Find by the last rule, the number of permutations possible with the whole number of objects, using as many each time as are to be in each combination; then find by the rule in Art. 422, the number of permutations possible with the number of objects in a combination, using all each time: the former result divided by the latter will give the number of combinations required.*

1. How many combinations of 10 letters, taken 7 in a set? *Ans.* 120.
2. Of 9 letters, taken 4 in each set? *Ans.* 126.
3. Of 8 letters, taken 3 in a set; also, 5 in a set? *Ans.* Each 56.

XXXVIII. SYSTEMS OF NOTATION.

ART. **425**. The *radix* of a system of Notation, is the number of units of each order which make *one* of the next higher order.

The radix of the ordinary or decimal system is 10. Other systems, *binary*, *ternary*, &c., arise from other radices; as 2, 3, &c.

ART. **426**. To change a number in the decimal system, to any other system whose radix is known.

RULE.—*Divide the given number by the radix of the system to which it is to be reduced; divide this quotient by the radix again, and so on, until a quotient is obtained smaller than the radix: the last quotient, with the several remainders annexed, will be the number in the required system.*

NOTE.—When there is no remainder, place a cipher instead.

1. Change 8764 in the decimal system, to the binary (2), quinary (5), and nonary (9) systems. *Ans.* 10001000111100; 240024; 13017.

ART. **427**. To reduce a number in any system whose radix is known, to the decimal system.

RULE.—*Multiply the left hand figure of the given number by the radix of the given system, and to the product add the next figure; multiply this sum by the radix again, and add in the next figure, and so on, until all the figures have been added in: the last result will be the number in the decimal system.*

1. Reduce 7056341 in the octary (8) system, and 201221 in the ternary (3) system, to the decimal system. *Ans.* 1858785; 538.

ART. **428**. To change a number in any system whose radix is known, to any other system whose radix is known.

RULE.—*First reduce the given number to the decimal system by the last rule; then reduce this result to the required system by the rule in Art.* 426.

1. Change 4210532 from the senary (6) system to the quaternary (4) system. *Ans.* 301232120.

REMARK—Numbers expressed in any other than the decimal system, may be added, subtracted, multiplied, divided, &c., as in the decimal system, except that *in carrying, borrowing, and reducing, take* 1 *of any order, not equal to* 10 *of the next lower, but equal to as many of the next lower, as there are units in the radix of the system.*

XXXIX. DUODECIMALS.

ART. **429**. The *duodecimal* system is the one whose radix is 12. It is applied in practice to the measurement of surfaces and solids; the foot, square foot, or cubic foot, is the unit; the duodecimal divisions are called *primes, seconds, thirds,* &c.

ART. **430**. For all operations in duodecimals use this

RULE.—*Change the number of whole feet from the decimal to the duodecimal system, if necessary, after which set down the duodecimal divisions, in order, using* Ⓘ *for* 10, *and* ⊕ *for* 11, *and separating the units from the duodecimal orders by* (:) *instead of* (.).

Then proceed with the operation as with ordinary numbers, except that carrying, borrowing, reducing, &c., are performed on the basis of 12 *instead of* 10; *the result thus obtained will be in the duodecimal system, and the number of whole feet may then be changed to the decimal system if necessary.*

REMARK.—In adding and subtracting duodecimals, the number of whole feet need not be expressed in the duodecimal system. Primes are marked ('), seconds ("), thirds ('''), and so on.

1. Add 3 ft. 2' 6", 1 ft. 8" 10''', and 4 ft. 7' 9'''. *Ans.* 8 ft. 10' 3" 7'''.
2. Subtract 14 ft. 9' 7" 8''' from 20 ft. 10". *Ans.* 5 ft. 3' 2" 4'''.
3. Multiply 3 ft. 8' 4" by 7 ft. 2' 4". *Ans.* 26 sq. ft. 6' 11" 5''' 4''''.
4. What is the surface of a board 13 ft. 8' long by 1 ft. 9' 3" wide? *Ans.* 24 sq. ft. 2' 5".
5. Of a floor 32 ft. 8' 4" long by 21 ft. 6' 8" wide? *Ans.* 704 sq. ft. 8' 11" 6''' 8''''.
6. What is the solidity of a log 16 ft. 2' 4" long by 1 ft. 9' wide and 10' 6" thick? *Ans.* 24 cu. ft. 9' 6" 10''' 6''''.
7. Divide 14 ft. 3' 11" 4''' by 8. *Ans.* 1 ft. 9' 5" 11'''.
8. Divide 22 ft. 1' 2" 9''' by 3 ft. 7". *Ans.* 7.25 = 7¼.
9. Divide 53 sq. ft. 9' 2" 2''' by 6 ft. 3' 2". *Ans.* 8 ft. 7'.
10. Divide 99 cu. ft. 9' 11" by 17 sq. ft. 4' 4". *Ans.* 5 ft. 9'.
11. Divide 64 cu. ft. 10" 10''' by 26 ft. 6' 2". *Ans.* 2 sq. ft. 5'.
12. How long must a board be, that is 1 ft. 5' wide, to contain 19 sq. ft. 1' 2''' 7''''? *Ans.* 13 ft. 5' 7" 11'''.
13. What is the hight of a rectangular log which contains 165 cu. ft. 2' 2" 8''', and is 21 ft. 4' long by 3 ft. 1' 2" wide? *Ans.* 2 ft. 6'.

REMARK.—If the pupil prefers, he may solve these examples according to the method explained in Ray's Arithmetic, Third Book, Art. 276, similar to the operations in compound numbers.

XL. MENSURATION OF SURFACES.

ART. **431**. A *Parallelogram* is a figure bounded by four straight lines, and whose opposite sides are equal and parallel.

If the adjacent sides are perpendicular to each other, the figure is *rectangle;* if they are also equal, it is a square.

To find the area of any parallelogram, rectangle, or square,

RULE.—*Multiply the base and altitude together, after expressing them in the same denomination: the product will be the area in square units of the same name.*

NOTE.—The *base* is any side; the *altitude* is the perpendicular or shortest distance from the opposite side to the base. In a rectangle and square, two adjacent sides are the base and altitude.

1. Find the area of a parallelogram whose base is 9 ft. 4 in. and altitude 2 ft. 5 in. *Ans.* 22 sq. ft. 80 sq. in.
2. Of an oil cloth 42 ft. by 5 ft. 8 in. *Ans.* $26\frac{4}{9}$ sq. yd.
3. How many tiles 8 in. square in a floor 48 ft. by 10 ft.? *Ans.* 1080.

ART. **432**. A *Triangle* is a figure of 3 sides; any side is the *base;* the *altitude* is the perpendicular or shortest distance to the base from the opposite vertex, or corner.

To find the area of a triangle with the base and altitude.

RULE.—*Take half the product of the base and altitude, after expressing them in the same denominations; this will be the area in square units of the same name.*

To find the area of a triangle with the three sides.

RULE.—*Take half the sum of the sides; subtract each side from it; multiply together the three remainders and the half sum; extract the square root of the product; this will be the area in square units.*

1. Find the area of a triangle whose base is 72 rd. and altitude 16 rd. *Ans.* 3 A. 2 R. 16 P.
2. Base 13 ft. 3 in.; altitude 9 ft. 6 in. *Ans.* 62 sq. ft. 135 sq. in.
3. Sides 1 ft. 10 in.; 2 ft.; 3 ft. 2 in. *Ans.* 1 sq. ft. 102—sq. in.
4. Sides 15 rd.; 18 rd.; 25 rd. *Ans.* 3 R. 13.66—P.

ART. **433**. A *Trapezoid* is a figure of 4 sides, two of which are parallel but not equal, and are called the bases.

To find the area of a trapezoid.

RULE.—*Take half the sum of the bases and multiply it by the altitude, after expressing them in the same denomination: the product will be the area in square units of the same name.*

NOTE.—The altitude is the perpendicular between the bases.

1. What is the area of a trapezoid whose bases are 9 ft. and 21 ft and altitude 16 ft.? *Ans.* 240 sq. ft.
2. Bases 43 rd. and 65 rd.; altitude 27 rd.? *Ans.* 9 A. 18 P.

ART. **434**. To find the area of any irregular figure bounded by four or more straight lines.

RULE.—*Separate the figure into triangles by joining its angular points; find the area of each triangle, and take their sum for the area of the figure.*

1. What is the area of a figure made up of 3 triangles whose bases are 10, 12, 16 rd. and altitudes 9, 15, $10\frac{1}{2}$ rd.? *Ans.* 1 A. 1 R. 19 P.

2. Whose sides are 10, 12, 14, 16 rd. in order, and distance from the starting point to the opposite corner, 18 rd.? *Ans.* 1 A. 3.9—P.

PLASTERERS', PAINTERS', AND PAVERS' WORK,

ART. **435**. Is computed by the sq. yd.; glaziers' work by the sq. ft. or pane; carpenters' and joiners' work by the sq. yd., and sometimes by the *square*, which is 10 ft. square, and contains 100 sq. ft.

1. Find the cost of roofing a house 60 ft. long and 22 ft. 9 in. from the ridge to the eaves, at 36 ct. a sq. yd. *Ans.* $109.20

2. How much wainscoting in a room 25 ft. long, 18 ft. wide, and 14 ft. 3 in. high, allowing a door 7 ft. 2 in. by 3 ft. 4 in., and two windows, each 5 ft. 8 in. by 3 ft. 6 in., and a chimney 6 ft. 4 in. by 5 ft. 6 in.; charging the door and windows half-work? *Ans.* $128\frac{62}{81}$ sq. yd.

3. How many squares in a partition 156 ft. 9 in. long by 10 ft. 4 in. high? *Ans.* 16.1975

4. Find the cost of flooring and joisting a house of 3 floors, each 48 ft. by 27 ft. deducting from each floor for a stairway 12 ft. by 8 ft. 3 in., allowing 9 in. rests for the joists; estimating the flooring and joisting *between* the walls at $1.46 a sq. yd., and the joisting *in* the walls at 76 ct. a sq. yd.; each row of rests being measured 48 ft. long by 9 in. wide. *Ans.* $600.78

5. What is the cost of plastering a partition 7 ft. 8 in. long and 10 ft. 3 in. high, at 45 ct. a sq. yd., deducting a door 6 ft. 3 in. by 2 ft. 10 in.? *Ans.* $3.04

6. How many sq. yd. of plastering in a room 30 ft. long, 25 ft. wide, and 12 ft. high, deducting 3 windows, each 8 ft. 2 in. by 5 ft., 2 doors each 7 ft. by 3 ft. 6 in., and a fire-place 4 ft. 6 in. by 4 ft. 10 in.; the sides of the windows being plastered 15 in. deep? And what will it cost, at 25 ct. a sq. yd.? *Ans.* $215\frac{1}{3}$ sq. yd.; cost $53.83

7. Find the cost of painting a wall 14 ft. by $9\frac{1}{2}$ ft., at 18 ct. a sq. yd., except a chimney 4 ft. 6 in. by 3 ft. 10 in. *Ans.* $\$2.31\frac{1}{2}$

8. How much painting on the sides of a room 20 ft. long, 14 ft. 6 in. wide, and 10 ft. 4 in. high, deducting a fire-place 4 ft. 4 in. by 4 ft., and 2 windows each 6 ft. by 3 ft. 2 in.? *Ans.* $73\frac{2}{27}$ sq. yd.

9. Find the cost of glazing the windows of a house of 3 stories, at 20 ct. a sq. ft. Each story has 4 windows, 3 ft. 10 in. wide; those in the 1st story are 7 ft. 8 in. high; those in the 2d, 6 ft. 10 in. high; in the 3d, 5 ft. 3 in. high. *Ans.* $\$60.56\frac{2}{3}$

10. The cost of lining a tank 2 ft. 10 in. long, 2 ft. 6 in. deep, 2 ft. wide, with zinc, at 10 lb. to the sq. ft., and 6 ct. a lb. *Ans.* $17.90

11. Find the cost of paving a court 50 ft. by 20 ft. 6 in., at 75 ct. a sq. yd. *Ans.* $85.42

12. Of paving a court 150 ft. square; a walk 10 ft. wide around the whole being paved with flag-stones at 54 ct. a sq. yd., and the rest at $81\frac{1}{2}$ ct. a sq. yd. *Ans.* $927.50

ART. **436**. A *Circle* is a figure bounded by a *circumference*, which is every-where equally distant from a *center* within.

The *diameter* is any line passing through the center and terminated on each side by the circumference; the *radius* is half the diameter, being the distance from the center to the circumference.

ART. **437**. To find the circumference from the diameter.

RULE.—*Multiply the diameter by* 3.14159265, *or* $\frac{355}{113}$.

NOTE.—$\frac{355}{113}$ may be used for 3.14159265, being nearly the same.

1. What are the circumferences whose diameters are 16; $22\frac{1}{4}$; 72.16; and 452 yd.? *Ans.* 50.265482; 69.900436; 226.6973; 1420 yd.

ART. **438**. To find the diameter from the circumference.

RULE.—*Divide the circumference by* 3.14159265, *or* $\frac{355}{113}$.

1. What are the diameters whose circumferences are 56, $182\frac{1}{2}$, 316.24 and 639 ft.? *Ans.* 17.82539, 58.09, 100.66232 and 203.4 ft.

ART. **439**. To find the area of a circle from the diameter.

RULE.—*Square half of the diameter and multiply it by* 3.14159265 (*or* $\frac{355}{113}$).

To find the area of a circle from the circumference.

RULE.—*Square half the circumference and divide it by* 3.14159265 (*or* $\frac{355}{113}$).

To find the area of a circle from the circum. and diameter.

RULE.—*Multiply the circumference by* $\frac{1}{4}$ *of the diameter.*

NOTE.—The last rule can be shown to be identical with the others by Art. 437; to prove it, consider a circle made up of small triangles, having their bases on the circumference and their vertices at the center. The area of each triangle is equal to the base multiplied by half its altitude (the radius). (Art. 432.) All of them, or the circle, will be equal to the sum of their bases, that is, the circumference, multiplied by half the radius, or, which is the same, by one-fourth of the diameter.

1. Find the areas of the circles with diameters 10 ft.; 2 ft. 5 in.; 13 yd 1 ft. *Ans.* 78.54 sq. ft.; 660.52 sq. in.; 139 sq. yd. 5.637 sq. ft.

2. Whose circumferences are 46 ft.; 7 ft. 3 in.; 6 yd. 1 ft. 4 in. *Ans.* 168.386 sq. ft.; 4 sq. ft. 26.322 sq. in.; 29.7443 sq. ft.

3 Circum., 47.124 ft., diameter 15 ft. *Ans.* 176.715 sq. ft.

4. Find the area of a ring whose breadth is 2 in., and diameter inside, 9 in. *Ans.* 69.115+ sq. in.

ART. **440**. To find the diameter or circumference from the area.

RULE.—*Divide the area, expressed in square units, by* 3.14159265 (*or* $\frac{355}{113}$); *the square root of the quotient will be the radius; twice the radius will be the diameter; and* 3.14159265 (*or* $\frac{355}{113}$) *times the diameter will be the circumference.*

1. Find the diameter and circumference of a circular field containing 10 A. *Ans.* D. 45.14 rd., circ. 141.8 rd

2. Of a circle containing 8 sq. ft. 116 sq. in. *Ans.* D. 40.18 in., circ. 126.23 in.

MENSURATION OF SOLIDS.

ART. **441**. A *Prism* is a solid having two equal and parallel bases, and whose longitudinal faces are parallelograms.

The bases may be triangles, quadrilaterals, or figures with any number of sides, and the prism is called *triangular*, *quadrangular*, &c., accordingly.

If the bases are parallelograms, the solid is called a *parallelopiped;* if they are circles, it is called a *cylinder*.

The altitude of any prism or cylinder is the perpendicular between the levels of the bases.

ART. **442**. To find the solidity of any prism or cylinder.

RULE.—*Find the area of one base in square units, and multiply it by the altitude expressed in linear units of the same name; the product will be the solidity in cubic units of that name.*

NOTE.—The area of the base, whether it be a triangle, parallelogram, or circle, is found by one of the previous rules.

1. What is the solidity of a prism whose bases are squares 9 in. on a side, and whose altitude is 1 ft. 7 in.? *Ans.* 1539 cu. in.

2. Whose altitude is $6\frac{1}{2}$ ft., and whose bases are parallelograms 2 ft. 10 in. long by 1 ft. 8 in. wide? *Ans.* 30 cu. ft. 1200 cu. in.

3. Whose altitude is 7 in. and whose base is a triangle with a base of 8 in. and an altitude of 1 ft.? *Ans.* 336 cu. in.

4. Whose altitude is 4 ft. 4 in. and whose base is a triangle with sides of 2, $2\frac{1}{2}$ and 3 ft.? *Ans.* 10.75 cu. ft.

5. What is the solidity of a cylinder whose altitude is $10\frac{1}{2}$ in. and the diameter of whose base is 5 in.? *Ans.* 206.167 cu. in.

ART. **443**. A *Pyramid* is a solid with a single base, and tapering regularly to a point called the *vertex*.

The base may be a triangle, quadrilateral, &c., and the pyramid is called *triangular*, *quadrangular*, &c., accordingly.

If the base is a circle, the solid is called a *cone*.

The *altitude* of a pyramid or cone is the perpendicular from the vertex to the base or the level of the base.

ART. **444**. To find the solidity of any pyramid or cone.

RULE.—*Find the area of the base in square units; multiply it by one-third of the altitude expressed in linear units of the same name; the product will be the solidity in cubic units of that name.*

1. Find the solidity of a pyramid whose altitude is 1 ft. 2 in., and whose base is a square $4\frac{1}{2}$ in. to a side. *Ans.* $94\frac{1}{2}$ cu. in.

2. Whose altitude is 15.24 in. and whose base is a triangle having each side 1 ft. *Ans.* 316.76 cu. in.

3. Whose altitude is 69 ft. and whose base is a parallelogram 34 ft. long by 26 ft. wide. *Ans.* 20332 cu. ft.

4. The solidity of a cone whose altitude is 5 ft. 3 in. and the diameter of whose base is 2 ft. 1 in. *Ans* 5 cu. ft. 1668.35 cu. in.

ART. **445**. A *right* prism or *right* cylinder stands perpendicular to its bases, the altitude being equal to the length.

A *right* pyramid is one whose vertex is equally distant from the angular points of the base, and the sides of whose base are all equal.

A *right* cone is one whose vertex is equally distant from all points in the circumference of the base.

In a right pyramid and right cone the altitude falls at the center of the base.

ART. **446**. The surface of a solid is its outside or visible part.

The *convex* surface is all the surface but the base or bases.

The *slant hight* of a right pyramid or right cone is the shortest distance from the vertex to the boundary of the base.

ART. **447**. To find the convex surface of a right prism or right cylinder.

RULE.—*Multiply the boundary of the base by the altitude, after expressing them both in the same denomination; the product will be the convex surface in square units of the same name.*

NOTES.—1. To get the whole surface, add in the areas of the two bases.

2. To prove the rule, consider that the convex surface of a right prism or right cylinder when rolled out on a plane, becomes a rectangle whose adjacent sides are the altitude of the solid and the boundary of its base.

1. Find the convex surface of a right prism with altitude $11\frac{1}{4}$ in. and sides of base, $5\frac{1}{4}$, $6\frac{1}{2}$, $8\frac{3}{4}$, $10\frac{1}{2}$, 9 in. *Ans.* 450 sq. in.

2. Of a right cylinder whose altitude is $1\frac{3}{4}$ ft. and the diameter of whose base is 1 ft. $2\frac{1}{2}$ in. *Ans.* 6 sq. ft. 92.6 sq. in.

3. Find the whole surface of a right triangular prism, the sides of the base 60, 80, and 100 ft.; altitude 90 ft. *Ans.* 26400 sq. ft.

4. The whole surface of a cylinder; altitude 28 ft.; circumference of the base 19 ft. *Ans.* 589.455 sq. ft.

ART. **448**. To find the convex surface of a right pyramid or right cone.

RULE.—*Multiply the boundary of the base by half the slant hight, after expressing them in the same denomination; the product will be the convex surface in square units of the same name.*

NOTES.—1. To get the whole surface, add in the area of the base.

2. To prove the rule, consider the convex surface as made up of triangles reaching from the vertex to the boundary of the base, their altitude being the slant hight of the solid, and their bases making the boundary of its base.

1. Find the convex surface and whole surface of a right pyramid whose slant hight is 225 ft.; the base 640 ft. square.
Ans. Conv. surf. 288000 sq. ft.; whole surf. 697600 sq. ft.

2. Of a right cone whose slant hight is 66 ft. 8 in.; radius of the base 4 ft. 2 in. *Ans.* 125663.706 sq. in.; 133517.6876 sq. in.

ART. **449.** A *Sphere* or *Globe* is a solid bounded by a curved surface, which is at all points equally distant from its *center*.

The *diameter* of a sphere is any line passing through the center and terminated both ways by the surface.

The *radius* is half the diameter, being the distance from the center to the surface.

ART. **450.** To find the surface of a sphere from its diameter.

RULE.—*Multiply the square of the diameter by* 3.14159265 (*or* $\frac{355}{113}$); *the product will be the surface in square units.*

NOTE.—The surface of any sphere is just equal to the area of four circles having the same diameter as the sphere.

1. What are the surfaces of two spheres whose diameters are 27 ft. and 10 in.? *Ans.* 2290.221+ sq. ft., and 314.16 sq. in.

ART. **451.** To find the solidity of a sphere from its diameter.

RULE.—*Multiply the cube of the diameter by* 3.14159265 (*or* $\frac{355}{113}$); *take* $\frac{1}{6}$ *of the product: this will be the solidity in cubic units.*

To find the solidity of a sphere from its surface and diameter,

RULE.—*Multiply the surface expressed in square units by* $\frac{1}{6}$ *of the diameter expressed in linear units of the same name; the product will be the solidity in cubic units of that name.*

NOTES.—1. If the surface only is given, divide it by 3.14159265 (or $\frac{355}{113}$); the square root of the quotient will be the diameter: then apply the rule.

2. To prove the second rule, which is identical with the first, consider the sphere to consist of small pyramids having their vertices at the center and their bases on the surface, the altitude of each being the radius of the sphere.

1. Find the solidity of a sphere whose diameter is 6 mi., and surface, 113.097335 sq. mi. *Ans.* 113.097335 cu. mi.

2. Of a sphere whose diameter is 4 ft. *Ans.* 33.5103 cu. ft.

3. Of a sphere whose surface is 40115 sq. mi. *Ans.* 755499$\frac{1}{6}$ cu. mi.

Art. 452. A *Frustum* of a pyramid or cone is what remains after cutting off its top, making the upper base parallel to the lower

The *altitude* of a frustum is the perpendicular or shortest distance between the levels of the bases.

The *slant hight* of a frustum of a right pyramid or of a right cone, is the shortest distance between the boundaries of its bases, measured on its convex surface.

Art. 453. To find the convex surface of a frustum of a right pyramid or of a right cone.

Rule.—*Take half the sum of the boundaries of its two bases, and multiply it by the slant hight, after expressing them in the same denomination: the product will be the convex surface in square units of the same name.*

Notes.—1. To get the whole surface, add in the areas of the bases.

2. The reason of this rule is seen from that of the trapezoid.

1. Find the convex surface of a frustum of a pyramid with slant hight, $3\frac{1}{3}$ in.; lower base, 4 in. square; upper base, $2\frac{1}{2}$ in. square.
Ans. $43\frac{1}{3}$ sq. in.

2. The convex surface and whole surface of the frustum of a cone; the diameters of the bases, 7 in. and 3 in.; the slant hight, 5 in.
Ans. Conv. surf. 78.5398 sq. in., whole surf. 124.0929 sq. in.

Art. 454. To find the solidity of a frustum of a right pyramid or frustum of a right cone.

Rule.—*Find the areas of the two bases, also of a mean base equal to the square root of the product of the other two; take the sum of the three, and multiply it by $\frac{1}{3}$ of the altitude; the product will be the solidity in cubic units.*

1. Find the solidity of a frustum of a pyramid whose altitude is 1 ft. $4\frac{1}{2}$ in.; lower base, $10\frac{2}{3}$ in. square; upper, $4\frac{1}{4}$ in square.
Ans. $974\frac{131}{288}$ cu. in.

2. Of a frustum of a cone; the diameters of the bases being 18 in. and 10 in.; the altitude 16 in. *Ans.* 2530.03 cu. in.

MASONS' AND BRICKLAYERS' WORK.

Art. 455. Masons' work is sometimes measured by the cubic foot, and sometimes by the perch, which is $16\frac{1}{2}$ ft. long, $1\frac{1}{2}$ ft. wide, and 1 ft. deep, and contains $16\frac{1}{2} \times 1\frac{1}{2} \times 1 = 24\frac{3}{4}$ cu. ft., or 25 cu. ft., nearly.

Art. 456. To find the No. of perches in a piece of masonry.

Rule.—*Find the solidity of the wall in cubic feet by the rules given for mensuration of solids, and divide it by $24\frac{3}{4}$.*

Note.—Brick work is generally estimated by the thousand bricks; the usual size being 8 in. long, 4 in. wide, and 2 in. thick. When bricks are laid in mortar, an allowance of $\frac{1}{10}$ is made for the mortar.

1. How many perches of 25 cu. ft., in a pile of building stone 18 ft. long, $8\frac{1}{2}$ ft. wide, and 6 ft. 2 in. high? *Ans.* 37.74 = $37\frac{3}{4}$ nearly.

2. Find the cost of laying a wall 20 ft. long, 7 ft. 9 in. high, and with a mean breadth of 2 ft., at 75 ct. a perch. *Ans.* $9.39

3. The cost of a foundation wall 1 ft. 10 in. thick, and 9 ft. 4 in. high, for a building 36 ft. long, 22 ft. 5 in. wide outside, allowing for 2 doors 4 ft. wide, at $2.75 a perch. *Ans.* $192.98

4. The cost of a brick wall 150 ft. long, 8 ft. 6 in. high, 1 ft. 4 in. thick, allowing $\frac{1}{10}$ for mortar, at $7 a thousand. *Ans.* $289.17

5. How many bricks of ordinary size will build a square chimney 86 ft. high, 10 ft. wide at the bottom and 4 ft. at the top, outside, and 3 ft. wide inside all the way up? *Ans.* 89861+

SUGGESTION.—Find the solidity of the whole chimney, then of the hollow part; the difference will be the solid part of the chimney.

GUAGING.

ART. **457.** *Guaging* is finding the contents of vessels, in bushels, gallons, or barrels.

To guage any vessel in the form of a rectangular solid, cylinder, cone, frustum of a cone, &c.

RULE.—*Find the solidity of the vessel in cubic inches by the rules already given; this divided by* 2150.42 *will give the contents in bu.; by* 282 *will give it in beer gal.; by* 231 *will give it in wine gal., which may be reduced to bbl. by dividing them by* $31\frac{1}{2}$.

NOTE.—In applying the rule to cylinders, cones, and frustums of cones, instead of multiplying the square of half the diameter by 3.14159265, and dividing it by 231, *multiply the square of the diameter by* .0034, which amounts to the same, and is shorter.

1. How many bushels in a bin 8 ft. 3 in. long, 3 ft. 5 in. high, and 2 ft. 10 in. wide? *Ans.* 64.18 bu.

2. How many wine gallons in a bucket in the form of a frustum of a cone, the diameters at the top and bottom being 13 in. and 10 in., and depth 12 in.? *Ans.* 5.4264 gal.

3. How many barrels in a cylindrical cistern 11 ft. 6 in. deep, and 7 ft. 8 in. wide? *Ans.* 126.0733

4. In a vat in the form of a frustum of a pyramid, 5 ft. deep, 10 ft. square at top, 9 ft. square at bottom? *Ans.* 107.26 bbl.

5. In a well 3 ft. wide, and 12 ft. deep? *Ans.* 20.1435 bbl.

ART. **458.** To find the contents in gallons of a cask or barrel.

When the staves are straight from the bung to each end, consider the cask as two frustums of a cone, and calculate its contents by last rule; but when the staves are curved, use this

RULE.—*Add to the head diameter (inside) two-thirds of the difference between the head and bung diameters; but if the staves are only slightly curved, add six-tenths of this difference; this gives the mean diameter; express it in inches, square it, multiply it by the length in inches and this product by* .0034: *the product will be the contents in wine gallons.*

NOTE.—After finding the mean diameter, the contents are found as if the cask were a cylinder.

1. Find the number of gallons in a cask of beer whose staves are straight from bung to head, the length being 26 in., the bung diameter 16 in., and head diameter 13 in. *Ans.* 15.28 gal.

2. In a barrel of whisky, with staves slightly curved, length 2 ft. 10 in., bung diameter 1 ft. 9 in., head 1 ft. 6 in. *Ans.* 45.32 gal.

3. In a cask of wine, with curved staves, length 5 ft. 4 in., bung diameter 3 ft. 6 in., head diameter 3 ft. *Ans.* 348.16 gal.

TUNNAGE OF VESSELS.

CARPENTERS' RULE.

ART. **459**. *For single-decked vessels, multiply together the length of the keel, the breadth at the main beams, and the depth of the hold, all in feet, and divide the product by* 95.

If the vessel is double-decked, proceed as for single-decked vessels, except that the depth of the hold is not measured, but assumed to be half the breadth at the main beam.

GOVERNMENT RULE.

If the vessel is double-decked, measure its length from the fore part of the main stem, to the after part of the stern-post, above the upper deck; measure the breadth at the broadest part above the main wales, half of which breadth shall be accounted the depth; then deduct from the length three-fifths of the breadth, multiply the remainder by the breadth, and the product by the depth, and divide the result by 95 *for the tunnage.*

If the vessel is single-decked, proceed as with double-decked vessels, except that the depth, instead of being assumed half of the breadth, should be measured from the under side of the deck-plank to the ceiling of the hold.

ART. **460**. To find the tunnage more accurately, use this

RULE.—*Find the plumb or perpendicular hight, from the lowest point of the deck to the water-line; take a mean breadth between the breadth on deck and the breadth at the water line; also a mean length of hull between the length on deck and the length at the water-line; express these dimensions in feet; multiply them together, and divide the product by* 35.84 *for the tunnage.*

NOTES.—1. The water line is the line made on the outside of the hull, by the surface of the water, when the vessel floats without any load.

2. The product of the mean length, breadth, and depth to the water-line, gives the number of cubic feet displaced, by putting a full load in the vessel; this, multiplied by $62\frac{1}{2}$, gives the weight of that water in lb., and this divided by 2240 becomes tuns; but, instead of multiplying by $62\frac{1}{2}$, and dividing by 2240, get the tuns at once by dividing by 35.84, since $\frac{62\frac{1}{2}}{2240}=\frac{1}{35.84}$. This water is of the same weight as the load which displaces it; that is, the tunnage of the vessel.

1. What are the carpenters' and government tunnage of a single-decked vessel, whose length of keel is 84 ft., breadth of beam 28 ft., and depth of hold 9 ft.? *Ans.* Carp. 223 T. nearly; Gov. $178\frac{1}{4}$ T.

2. Also a double-decked vessel, 150 ft. keel and $36\frac{1}{4}$ ft. breadth of beam? *Ans.* Carp. $1037\frac{1}{2}$ T. nearly; Gov. 887 T.

3. Also, by the 3d rule, of a vessel, mean length, 93 ft. 4 in., mean breadth, 26 ft. 7 in., depth to the water line, 4 ft. 3 in.? *Ans.* 294 T.

4. How many bushels of coal (80 lb.) can be shipped on a boat 78 ft. long, 22 ft. wide, and 5 ft. to the water line, leaving 1 ft. 8 in. above the water? *Ans.* $4468\frac{3}{4}$ bu.

XLI. MECHANICAL POWERS.

ART. **461.** The Mechanical Powers are six, viz.: the Lever, Wheel and Axle, Inclined Plane, Screw, Pulley, and Wedge.

All machines contain one or more of these powers.

The *power* in a machine is the force applied, as steam, horses, weights, hand-power, &c.

The *weight* is the obstacle overcome.

One of the chief obstacles to the working of machinery is *friction;* that is, the rubbing and interference of the parts of the machine itself: this is generally estimated at $\frac{1}{3}$ of the power applied.

ART. **462.** The power and weight in a machine are to each other inversely as their velocities; hence,

TO COMPARE THE POWER AND WEIGHT IN A MACHINE,

RULE.—*The power : the weight : : the distance the weight moves in the direction of resistance : the distance the power passes through in the same time.*

NOTE.—Any three terms being given in this proportion, the fourth may be found by Art. 249.

THE LEVER

ART. **463.** Is a bar movable about a fixed point called a *fulcrum;* there are three kinds:

1st. When the power and weight are on different sides of the fulcrum.

2d. When the power and weight are on the same side of the fulcrum, the power being nearer the fulcrum.

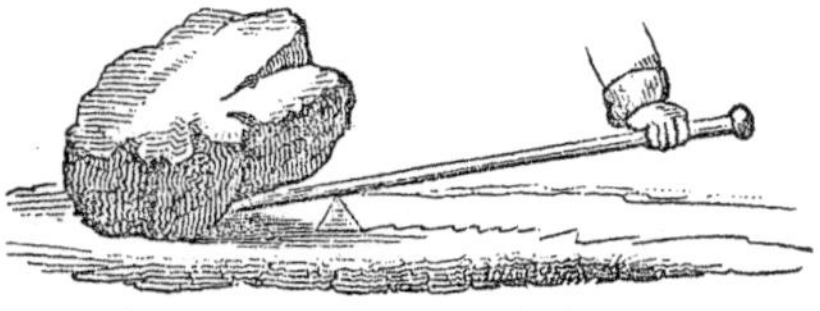

3d. When the power and the weight are on the same side of the fulcrum, the weight being nearer the fulcrum.

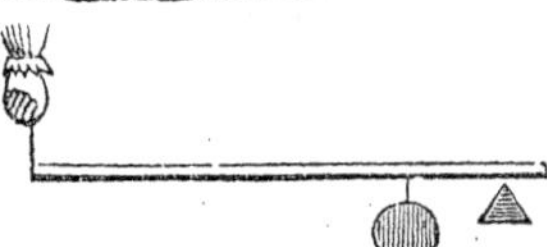

ART. **464.** To compare the power and weight in any lever.

RULE.—*The power : the weight : : the distance of the weight from the fulcrum : the distance of the power from the fulcrum.*

NOTE.—This rule is the same in fact as the general rule (Art. 462); for the distances passed over by the power and weight in the same time, depend on their distances from the fulcrum.

1. What weight 10 in. from the fulcrum, is balanced by a power of 40 lb., 4 ft. 8 in. from the fulcrum? *Ans.* 224 lb.

2. What power, 6 ft. from the fulcrum, will sustain a weight of 1080 lb., 1 ft. 4 in. from the fulcrum? *Ans.* 240 lb.

3. How far from the fulcrum must a person weighing 160 lb. stand, to balance a weight of 1200 lb., 2 ft. 8 in. from the fulcrum? *Ans.* 20 ft.

THE WHEEL AND AXLE

ART. **465.** Is a sort of lever. The radius of the wheel and the radius of the axle are parts of the lever, and the center of the axle is the fulcrum.

The power acts at the circumference of the wheel, and the weight at the circumference of the axle, by means of a rope. The distances passed over by the power and weight in the same time, will be the circumferences of the wheel and of the axle; hence,

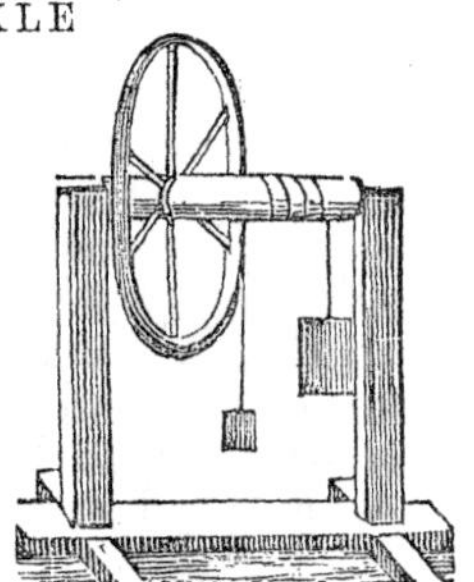

RULE.—*The power : the weight : : the circumference (or diameter) of the axle : the circumference (or diameter of the wheel).*

NOTE.—The diameters may be used instead of the circumferences, if desired, since they vary as the circumferences.

1. The diameter of the wheel is 6 ft., of the axle, 8 in.; the weight is 720 lb.: find the power, allowing $\frac{1}{4}$ for friction. *Ans.* $106\frac{2}{3}$ lb.

2. The diameter of the wheel is 9 ft. 4 in., of the axle 1 ft. 2 in.; the power is 185 lb.: find the weight, friction $\frac{1}{5}$. *Ans.* 1184 lb.

3. The diameter of the axle is 10 in., the power, 140 lb., the weight, 1750 lb.: find the diameter of the wheel. *Ans.* 10 ft. 5 in.

THE INCLINED PLANE

ART. **466**. Is used to raise or lower heavy bodies. The power must fall a distance equal to the length of the plane, in order to raise the weight a distance equal to its hight; hence,

RULE.—*The power : the weight : : the hight of the plane : the length of the plane.*

1. What power will roll a bbl. of flour (196 lb.) up an inclined plane 7 ft. 6 in. long, into a wagon 3 ft. 4 in. high? *Ans.* $87\frac{1}{9}$ lb.

2. What weight can a man capable of lifting 300 lb. roll up an inclined plane 8 ft. long, and 5 ft. 4 in. high? *Ans.* 450 lb.

THE SCREW

ART. **467**. May be considered an inclined plane, called the *thread*, wrapped round a cylinder.

The power is generally applied by a lever fixed firmly into the head of the cylinder. The power describes the circumference of a circle whose radius is the lever, while the weight is moved the distance between two contiguous threads; hence,

RULE.—*The power : the weight : : the distance between two contiguous threads : $\frac{355}{113}$ times twice the length of the lever.*

NOTE.—The last term is the circumference described by the power.

1. What weight will be moved by a power of 500 lb. acting at the end of a lever 15 ft. long, by means of a screw whose threads are $1\frac{1}{3}$ in. apart, allowing $\frac{1}{3}$ for friction? *Ans.* 251327 lb.

2. What power will lift 28400 lb. by a screw with threads $\frac{3}{4}$ of an inch apart, and a lever 6 ft. long? *Ans.* $47\frac{1}{12}$ lb.

THE PULLEY

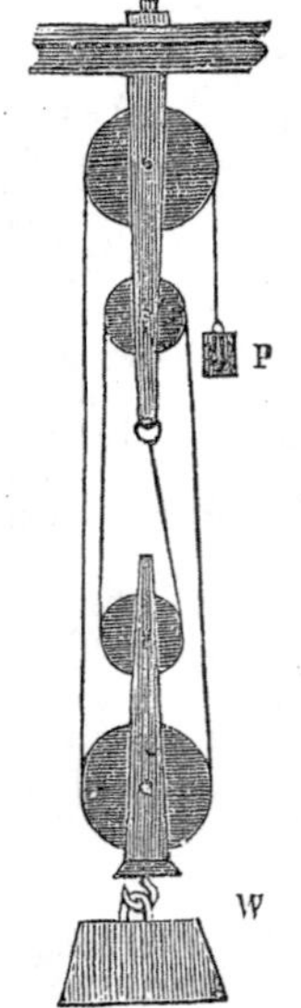

ART. **468**. Is a grooved wheel, movable about its center, the power and weight being connected by a cord running in the groove.

The single pulley affords no advantage except in the direction of applying the power.

In a system of pulleys, observe that for every movable pulley around which the cord runs, the power passes over twice the distance the weight is lifted; hence,

RULE.—*The power : the weight : :* **1** *: twice the number of movable pulleys.*

1. The power is 25 lb., the number of movable pulleys 8: what is the weight? *Ans.* 400 lb.

2. The weight is 1500 lb., the number of movable pulleys 6: what is the power? *Ans.* 125 lb.

NOTE.—The Wedge is a double inclined plane, and the power : the weight : : the head of the wedge : its length; but the friction is so great in this machine, and the power is so much influenced by other considerations, that it is very difficult to obtain satisfactory practical results by any method of calculation, and hence no examples are given.

ART. 469. ILLUSTRATIONS OF MECHANICAL POWERS.

PROBLEM 1.—To find the pressure of the steam in pounds, on each square inch of the surface of the boiler of a steam engine.

RULE.—*Multiply the weight at the end of the steam guage by its distance from the fulcrum, and divide the product by the distance of the valve from the fulcrum; this quotient divided by* $\frac{355}{113}$ *times the square of half the diameter of the valve, gives the pressure on each square inch of the boiler.*

NOTE.—The distances and diameter of the valve must be in inches.

1. The weight on the guage is 58 lb., and is 1 ft. 10 in. from the fulcrum; the valve is 4 in. from the fulcrum, and its diameter is 3 in.: what is the pressure? *Ans.* 45 lb. to the sq. in.

2. The weight is 36 lb., its distance from the fulcrum 2 ft. 4 in.; the valve is $4\frac{1}{2}$ in. in diameter, and is 3 in. from the fulcrum: what is the pressure? *Ans.* 21 lb. to the sq. in.

PROB. 2.—To find the horse power of a steam engine.

RULE.—*Multiply the pressure in lb. on the square inch of the boiler by* $\frac{355}{113}$ *times the square of half the diameter of the cylinder in inches, and this product by twice the length of the cylinder in feet, and this product by the number of double strokes made by the piston in a minute, and divide this by* 44000; *the quotient is the horse-power of the engine.*

NOTE.—This rule makes no allowance for the cooling of the steam in passing from the boiler to the cylinder, by which its elastic force i diminished. In high-pressure engines this is considerable, and in some more than others. An allowance may be made for this, based on experience, or the pressure of the steam in the cylinder may be determined by its temperature.

3. If the pressure is 34 lb. per sq. in., the diameter of the cylinder 2 ft. 6 in., its length 7 ft., and the piston makes 18 double-strokes per minute, what is the power of the engine? *Ans.* 138 horse-power.

4. If the pressure is 44 lb. per sq. in., the diameter of the cylinder 1 ft. 8 in., its length 6 ft., and the paddle-wheel makes 20 revolutions a minute, what is the power of the engine? *Ans.* 75 horse-power.

TO TEACHERS.—The Student is supposed to have learned the demonstrations of the Rules in series and subsequent subjects, from "Ray's Arithmetic, Third Book": hence they are omitted here.

XLII. PROMISCUOUS EXERCISES.

ART. 470. FIFTY EXAMPLES TO BE ANALYZED.

1. If I gain $\frac{1}{4}$ ct. apiece by selling eggs at 7 ct. a dozen, how much apiece will I gain by selling them at 9 ct. a dozen? *Ans.* $\frac{5}{12}$ ct.

2. If I gain $\frac{1}{2}$ ct. apiece by selling apples at 3 for a dime, how much apiece would I lose by selling them 4 for a dime? *Ans.* $\frac{1}{3}$ ct.

3. If I sell potatoes at $37\frac{1}{2}$ ct. per bu., my gain is only $\frac{3}{5}$ of what it would be, if I charged 45 ct. per. bu.: what did they cost me? *Ans.* $26\frac{1}{4}$ ct. per bu.

4. If I sell my oranges for 65 ct., I gain $\frac{3}{8}$ ct. apiece more than if I sold them for 50 ct.: how many oranges have I? *Ans.* 40.

5. If I sell my pears at 5 ct. a dozen, I lose 16 ct.; if I sell them at 8 ct. a dozen, I gain 11 ct.: how many have I, and what did they cost me? *Ans.* 9 dozen at $6\frac{7}{9}$ ct. per dozen.

6. If I sell eggs at 6 ct. per dozen, I lose $\frac{2}{3}$ ct. apiece; how much per dozen must I charge, to gain $\frac{2}{3}$ ct. apiece? *Ans.* 22 ct.

7. $\frac{1}{8}$ of a dime is what part of 3 ct.? *Ans.* $\frac{5}{12}$

8. If I lose $\frac{3}{8}$ of my money and spend $\frac{4}{7}$ of the remainder, what part have I left? *Ans.* $\frac{15}{56}$

9. A's land is $\frac{1}{7}$ less in quantity than B's, but $\frac{1}{20}$ better in quality: how do their farms compare in value? *Ans.* A's $=\frac{9}{10}$ of B's.

10. If $\frac{2}{3}$ of A's money equals $\frac{4}{5}$ of B's, what part of B's equals $\frac{5}{8}$ of A's? *Ans.* $\frac{3}{4}$

11. I gave A, $\frac{5}{14}$ of my money, and B, $\frac{7}{12}$ of the remainder: who got the most, and what part? *Ans.* B got $\frac{1}{56}$ of it more than A.

12. A is $\frac{2}{3}$ older than B, and B, $\frac{2}{3}$ older than C: how many times is A as old as C? *Ans.* $2\frac{7}{9}$

13. $\frac{2}{3}$ of my money equals $\frac{4}{5}$ of yours; if we put our money together, what part of the whole will I own? *Ans.* $\frac{6}{11}$

14. How many thirds in $\frac{1}{2}$? *Ans.* $1\frac{1}{2}$

15. Reduce $\frac{4}{5}$ to thirds; $\frac{5}{6}$ to ninths; and $\frac{3}{5}$ to a fraction, whose numerator shall be 8. *Ans.* $2\frac{2}{5}$ thirds; $7\frac{1}{2}$ ninths; $\frac{8}{13\frac{1}{3}}$

16. What fraction is as much larger than $\frac{4}{5}$, as $\frac{2}{3}$ is less than $\frac{4}{5}$? *Ans.* $\frac{14}{15}$

17. After paying out $\frac{1}{4}$ and $\frac{1}{5}$ of my money, I had left $8 more than I had spent: what had I at first? *Ans.* $80.

18. In 12 yr., I shall be $\frac{7}{5}$ of my present age; how long since was I $\frac{5}{7}$ of my present age? *Ans.* $8\frac{4}{7}$ yr.

19. Four times $\frac{2}{9}$ of a number, is 12 less than the number: what is the number? *Ans.* 108.

20. A man left $\frac{5}{11}$ of his property to his wife, $\frac{2}{3}$ of the remainder to his son, and the balance, $4000, to his daughter: what was the estate? *Ans.* $22000

21. I sold an article for $\frac{1}{4}$ more than it cost me, to A, who sold it for $6, which was $\frac{2}{5}$ less than it cost him: what did it cost me? *Ans.* $8

22. A is $\frac{2}{9}$ older than B; their father, who is as old as both of them, is 50 yr. of age: how old are A and B? *Ans.* A, $27\frac{1}{2}$ yr.; B, $22\frac{1}{2}$ yr.

23. A pole was $\frac{2}{7}$ under water; the water rose 8 ft., and then there was as much under water as had been above water before: how long is the pole? *Ans.* $18\frac{2}{3}$ ft.

24. A is $\frac{3}{4}$ as old as B; if he were 4 yr. older, he would be $\frac{9}{10}$ as old as B; how old is each? *Ans.* A, 20 yr.; B, $26\frac{2}{3}$ yr.

25. A's money is $4 more than $\frac{2}{3}$ of B's, and $5 less than $\frac{3}{4}$ of B's: how much has each? *Ans.* A, $76; B, $108.

26. $\frac{2}{3}$ of A's age is $\frac{3}{4}$ of B's, and A is $3\frac{1}{2}$ yr. the older: how old is each? *Ans.* A, $31\frac{1}{2}$ yr.; B, 28 yr.

27. If 3 boys do a work in 7 hr., how long will it take a man who works $4\frac{1}{2}$ times as fast as a boy? *Ans.* $4\frac{2}{3}$ hr.

28. If 6 men can do a work in $5\frac{1}{2}$ days, how much time would be saved, by employing 4 more men? *Ans.* $2\frac{1}{5}$ days.

29. A man and 2 boys do a work in 4 hr.; how long would it take the man alone, if he worked equal to 3 boys? *Ans.* $6\frac{2}{3}$ hr.

30. A man and a boy can mow a certain field in 8 hr.; if the boy rests $3\frac{3}{4}$ hr., it takes them $9\frac{1}{2}$ hr.: in what time can each do it? *Ans.* Man, $13\frac{1}{3}$ hr.; boy, 20 hr.

31. Five men were employed to do a work; two of them failed to come, by which the work was protracted $4\frac{1}{2}$ days: in what time could the 5 have done it? *Ans.* $6\frac{3}{4}$ days.

32. 3 men can do a work in 5 days; in what time can 2 men and 3 boys do it, allowing 4 men to work equal to 9 boys? *Ans.* $4\frac{1}{2}$ da.

33. A man and a boy mow a 10 acre field; how much more does the man mow than the boy, if 2 men work equal to 5 boys? *Ans.* $4\frac{2}{7}$ A.

34. 6 men can do a work in $4\frac{1}{3}$ days; after working 2 days, how many must join them so as to complete it in $3\frac{2}{5}$ da.? *Ans.* 4 men.

35. 8 men can do a work in $6\frac{3}{4}$ days; after beginning, how soon must they be joined by 2 more, so as to complete it in $5\frac{7}{8}$ days?. *Ans.* In $2\frac{3}{8}$ days.

36. 7 men can build a wall in $5\frac{1}{2}$ days; if 10 men are employed, what part of his time can each rest, and the work be done in the same time? *Ans.* $\frac{3}{10}$

37. 9 men can do a work in $8\frac{1}{3}$ days; how many days may 3 remain away, and yet finish the work in the same time by bringing 5 more with them? *Ans.* $5\frac{5}{24}$

38. 10 men can dig a trench in $7\frac{1}{2}$ days; if 4 of them are absent the first $2\frac{1}{2}$ days, how many others must they then bring with them, to complete the work in the same time? *Ans.* 2.

39. At what times between 6 and 7 o'clock are the hour hand and minute hand 20 min. apart? *Ans.* $10\frac{10}{11}$ min. after 6, and $54\frac{6}{11}$ min. after 6.

40. At what times between 4 and 5 o'clock is the minute hand as far from 8 as the hour hand is from 3? *Ans.* $32\frac{4}{13}$ after 4; and $49\frac{1}{11}$ min. after 4.

41. At what time between 5 and 6 o'clock, is the minute hand midway between 12 and the hour hand? when is the hour hand midway between 4 and the minute hand?
Ans. $13\frac{1}{23}$ min. after 5; and 36 min. after 5.

42. A, B and C dine on 8 loaves of bread; A furnishes 5 loaves; B, 3 loaves; C pays the others 8d. for his share: how must A and B divide the money? *Ans.* A takes 7d.; B, 1d.

43. A boat makes 15 mi. an hour down stream, and 10 mi. an hour up stream: how far can she go and return, in 9 hr.? *Ans.* 54 mi.

44. I can pasture 10 horses or 15 cows on my ground; if I have 9 cows, how many horses can I keep? *Ans.* 4.

45. A's money is 12 % of B's, and 16 % of C's; B has $100 more than C: how much has A? *Ans.* $48.

46. Eight men hire a coach; by getting 6 more passengers, the expense to each is diminished $\$1\frac{3}{4}$: what do they pay for the coach?
Ans. $\$32\frac{2}{3}$

47. A company engage a supper; being joined by $\frac{2}{5}$ as many more, the bill of each is 60 ct. less: what would each have paid, if none had joined them? *Ans.* $2.10

48. By mixing 5 lb. of good sugar with 3 lb. worth 4 ct. a lb. less, the mixture is worth $8\frac{1}{2}$ ct. a lb.: find the prices of the ingredients.
Ans. 10 ct. and 6 ct. a lb.

49. By mixing 10 lb. of good sugar with 6 lb. worth only $\frac{2}{3}$ as much, the mixture is worth 1 ct. a lb. less than the good sugar: find the prices of the ingredients and of the mixture.
Ans. Ingredients, 8 ct. and $5\frac{1}{3}$ ct.; Mixt., 7 ct. per lb.

50. A and B have the same money; if A had $20 more, and B, $10 less, A would have $2\frac{1}{3}$ times as much as B: what has each?
Ans. $\$32\frac{1}{2}$

Art. 471. PRACTICAL EXAMPLES.

1. What is the least sum that can be paid with quarters, dimes, or 3 ct. pieces? *Ans.* $1.50

2. A and B pay $1.75 for a quart of varnish, and 10 ct. for the bottle; A contributes $1, B, the rest: they divide the varnish equally, and A keeps the bottle: which owes the other, and how much? *Ans.* B owes A $2\frac{1}{2}$ ct.

3. Find the difference between $\frac{1}{3}$ sq. ft. and $\frac{1}{3}$ foot square; also between $\frac{1}{4}$ cu. ft. and $\frac{1}{4}$ ft. cubed. *Ans.* 32 sq. in.; 405 cu. in.

4. How far does a man walk, while planting a field of corn 285 ft square, the rows being 3 ft. apart and 3 ft. from the fences?
Ans. 5 mi. 6 rd. 6 ft.

5. How far apart should the knots of a log-line be, to indicate every half-minute a speed of 1 mi. per hour? *Ans.* 44 ft.

6. I bought 25 sheep for $56: if I had got 3 more for the same sum, how much less per head would I have paid? *Ans.* 24 ct.

7. Find the least number which, divided by 2, 3, 4, 5, and 6, leaves a remainder 1 each time. *Ans.* 61.

8. How many carats in gold consisting of 1 pwt. $1\frac{7}{8}$ gr. pure gold and 3.5 gr. alloy? *Ans.* $21\frac{33}{235}$.

9. Until recently the pay of a congressman was \$8 a day during the session, and \$8 for every 20 mi. he traveled in going to and coming from the capital: how far did one live from the seat of government, whose pay for a session of 158 days, was \$2104?
Ans. 1050 mi.

10. Land worth \$1000 an acre, is worth how much a foot front of 90 ft. depth; reserving $\frac{1}{10}$ for streets? *Ans.* \$2.295+.

11. Reduce $\frac{2}{3}$, $\frac{6}{7}$, $\frac{8}{9}$, $\frac{10}{13}$, $\frac{9}{16}$ to the least common numerator.
Ans. $\frac{360}{540}$, $\frac{360}{420}$, $\frac{360}{405}$, $\frac{360}{468}$, $\frac{360}{640}$.

12. I buy stocks at 20 % discount, and sell them at 10 % premium: what per cent. do I gain? *Ans.* $37\frac{1}{2}$ %.

13. I invest and sell at a loss of 15 %; I invest the proceeds again, and sell at a gain of 15 %: do I gain or lose on the two speculations, and how many per cent.? *Ans.* Lose $2\frac{1}{4}$ %.

14. I sell at 8 % gain, invest the proceeds and sell at an advance of $12\frac{1}{2}$ %; invest the proceeds again, and sell at 4 % loss, and quit with \$1166.40: what did I start with? *Ans.* \$1000.

15. I can insure my house for \$2500, at $\frac{8}{10}$ % premium annually, or permanently, by paying down 12 annual premiums: which should I prefer, and how much will I gain by it, if money is worth 6 % per annum to me? *Ans.* The latter; gain, \$113.33$\frac{1}{3}$

16. A owes B \$1500, due in 1 yr. 10 mon. He pays him \$300 cash, and a note of 6 mon. for the balance: what is the face of the note, allowing interest at 6 %? *Ans.* \$1080.56

17. If I charge 12 % per annum compound interest, payable quarterly, what rate per annum is that? *Ans.* $12\frac{550881}{1000000}$ %.

18. How many square inches in one face of a cube which contains 2571353 cu. in.? *Ans.* 18769.

19. What is the side of a cube which contains as many cubic inches, as there are square inches in its surface? *Ans.* 6 in.

20. If $12\frac{1}{2}$ % of what I receive for an article, is gain, what is my rate of gain? *Ans.* $14\frac{2}{7}$ %.

21. The boundaries of a square and circle are each 20 ft.; which contains the most, and how much? *Ans.* Circle; 6.831 sq. ft. nearly.

22. If I pay \$1000 for a 5 yr. lease and \$200 for repairs, how much rent payable quarterly is that equal to, allowing 10 % interest?
Ans. \$307.92 a year.

23. What is the value of a widow's dower in property worth \$3000, her age being 40, and interest 5 %? *Ans.* \$669.50

24. Bought eggs on credit; the first time one dozen, and every succeeding time 3 more. My last purchase was $7\frac{1}{2}$ dozen. The bill was presented for 120 doz.: how much too large was it? *Ans.* $5\frac{1}{4}$ doz.

25. What principal must be loaned Jan. 1, at 9 %, to be re-paid by 5 installments of \$200 each, payable on the first day of the five succeeding months? *Ans.* \$978.10

26. After spending 25 % of my money, and 25 % of the remainder, I had left \$675: what had I at first? *Ans.* \$1200.

27. I had a 60 day note discounted at 1 % a month and paid \$4.80 above true interest: what was the face of the note? *Ans.* \$11112.93

28. If the draft is 7 lb., the tare 18 %, and the net weight 14 cwt. 1 qr. 3 lb., what is the gross weight? *Ans.* 17 cwt. 1 qr. 25 lb.

29. Bought stock at 5 % discount and sold at 3 % premium, gaining $180: what did I invest? *Ans.* $2137.50

30. Bought stock at 20 % discount; sold out at 45 % gain, realizing $2204: what did I invest? *Ans.* $1520.

31. Invested $10000; sold out at a loss of 20 %: how much must I borrow at 4 %, so that by investing all I have at 18 %, I may retrieve my loss? *Ans.* $4000.

32. If $\frac{1}{4}$ of an inch of rain falls, how many bbl. will be caught by a cistern which drains a roof 52 ft. by 38 ft? *Ans.* 9.776+ bbl.

33. The tax on a lot, with 2 years' penalty, amounts to 275.92\frac{1}{2}$: if the rate of taxation for the 1st year was $1\frac{35}{100}$ %, and for the 2d year $1\frac{48}{100}$ %, and the penalty is 30 % of the tax; what is the lot assessed? *Ans.* $7500.

34. A father left $20000 to be divided among his 4 sons, aged 6 years, 8 years, 10 years, and 12 years, respectively, so that each share, placed at $4\frac{1}{2}$ % compound interest, should amount to the same when its possessor became of age (21 yr.): what were the shares?
Ans. $4360.34; $4761.59; $5199.78; $5678.29

35. $30000 of bonds bearing 7 % interest, payable semi-annually, and due in 20 yr. are bought so as to yield 8 % payable semi-annually: what is the price? *Ans.* $27031.08

36. I sell a lot for $4850; $250 payable at 6 mon., $1\frac{1}{2}$ yr., $2\frac{1}{2}$ yr., $3\frac{1}{2}$ yr., and $4\frac{1}{2}$ yr. each, and the balance in 5 yr.: if money is worth 10 % per annum to me, what is the cash value of the sale?
Ans. $3228.14

37. A, B, and C are partners; A puts in $240 for 6 mon., B a certain sum for 12 mon., C, $160 for a certain time; when they close up, A has $300, B, $600; C, $260: find B's stock and C's time.
Ans. $400 and 15 mon.

38. The stocks of 3 partners, A, B and C, are in trade 8, 10, and 7 mon. respectively; and their gains are $115.50, $204.75 and $183.75 respectively: find their stocks, the difference between B's and C's being $220. *Ans.* A's, $550; B's, $780; C's, $1000.

39. The stocks of 3 partners, A, B, and C, are $350, $220 and $250, and their gains $112, $88 and $120 respectively: find the time each stock was in trade, B's time being 2 mon. longer than A's.
Ans. A's, 8 mon.; B's, 10 mon.; C's, 12 mon.

40. By discounting a note at 20 % per annum, I get $22\frac{1}{2}$ % per annum interest: how long does the note run? *Ans.* 200 days.

41. A debt is to be paid in 4 equal installments at 4, 9, 12, 20 months respectively; its cash value is $750, allowing 5 % simple interest: what is the debt? *Ans.* $784.74

42. I bought a horse for $156 due in 8 months, and sold him at once for $180: find the gain per cent.; int. $4\frac{1}{2}$ %. *Ans.* $18\frac{1}{13}$ %.

43. A receives $57.90, and B, $29.70, from a joint speculation: if A invested 7.83\frac{1}{3}$ more than B, what did each invest?
Ans. A, 16.08\frac{1}{3}$; B, $8.25

44. A borrows a sum of money at 6 %, payable semi-annually, and lends it at 12 %, payable quarterly, and clears $2450.85 a year what is the sum? *Ans.* $38485.87

45. Find the sum, whose true discount by simple interest for 4 yr., is $25 more at 6 %, than at 4 % per annum. *Ans.* $449.50

46. I invested $2700 in stock at 25 % discount, which pays 8 % annual dividends: how much must I invest in stock at 4 % discoun and paying 10 % annual dividends, to secure an equal income? *Ans.* $2764.80

47. Exchanged $5200 of stock bearing 5 % interest at 69 %, fo stock bearing 7 % interest at 92 %, the interest on each stock having been just paid: what is my cash gain, if money is worth 6 % to me? *Ans.* 216.66\frac{2}{3}$

48. Bought goods on 4 mon. credit; after 7 mon. I sell them for $1500, 2$\frac{1}{2}$ % off for cash; my gain is 15 %, money being worth 6 %: what did I pay for the goods? *Ans.* $1252.94

49. The 8th term of an Arithmetical series is 76, and the 34th term, 206: what is the 19th term? *Ans.* 131.

50. The 9th term of a geometric series is 137781, and the 13th term, 11160261: what is the 4th term? *Ans.* 567.

51. My capital increases every year by the same per cent.; at the end of the 3d year, it was $13310; at the end of the 7th year, it was $19487.171: what was my original capital, and the rate of gain? *Ans.* $10000 and 10 %.

52. Find the length of a minute-hand, whose extreme point moves 4 inches in 3 min. 28 sec. *Ans.* 11.02— in.

53. Three men own a grindstone, 2 ft. 8 in. in diameter: how many inches must each grind off, to get an equal share, allowing 6 in. waste for the aperture?

Ans. 1st, 2.822— in., 2d, 3.621+ in.; 3d, 6.557— in.

54. A, B and C were partners: A put in $600 for 3 mon.; B, $400 for 4 mon.; C, $700: the gain was $120, of which C got $35: how long was C's money in trade? *Ans.* 2 mon.

55. I sold an article at 20 % gain; had it cost me $300 more, I would have lost 20 %: find the cost. *Ans.* $600.

56. A boat goes 16$\frac{1}{4}$ miles an hour down stream, and 10 mi. an hour up stream: if it is 22$\frac{1}{2}$ hr. longer in coming up than in going down, how far down did it go? *Ans.* 585 mi.

57. If an article had cost me 10 % less, my rate of gain would have been 15 % more: what was my rate of gain? *Ans.* 35 %.

58. Bought a check on a suspended bank at 55 %; exchanged it for railroad bonds at 60 %, which bear 7 % interest; what rate of interest do I receive on the amount of money invested? *Ans.* 21$\frac{7}{33}$ %.

59. Bought sugar for refinery; 6 % is wasted in the process; 30 % becomes molasses, which is sold at 40 % less than the same weight of sugar cost; at what % advance on the first cost, must the clarified sugar be sold, so as to yield a profit of 14 % on the investment? *Ans.* 50 %.

☞ Many questions usually solved by Algebra, are susceptible of easy and elegant Arithmetical solutions. The following are from 'Ray's Algebra.'

60. A servant was hired for 1 yr. at £8 and his livery; in 7 mon. he left, and received £2 13 s. 4 d. and his livery; what was the value of the livery? *Ans.* £4 16 s.

61. A smuggler had brandy which he held at $198; after selling 10 gal., a revenue officer seized $\frac{1}{3}$ of the remainder, by which his receipts on the whole were cut down to $162; how many gallons had he, and at what price? *Ans.* 22 gal. at $9 a gal.

62. A workman was engaged for 28 da.; every day he worked he received 75 ct., and every day he was absent he forfeited 25 ct. At the end of 28 days he received $12: how many days did he work? *Ans.* 19.

63. From a bag of money which contained a certain sum, there was taken $20 more than its half; from the remainder, $30 more than its third; and from the remainder, $40 more than its fourth, and then nothing was left. What sum did it contain? *Ans.* $290.

64. A bought eggs at 18 ct. a dozen; had he bought 5 more for the same sum, they would have cost him $2\frac{1}{2}$ ct. a dozen, less. How many eggs did he buy? *Ans.* 31.

65. A person bought sheep for $94; having lost 7, he sold $\frac{1}{4}$ the remainder at prime cost, for $20. How many sheep had he at first? *Ans.* 47.

66. For every 8 sheep a farmer keeps, he plows 1 A. of land, and keeps 1 A. of pasture, for every 5 sheep; how many sheep does he keep on 325 A.? *Ans.* 1000.

67. A person has just 2 hr. spare time; how far may he ride in a stage which travels 12 mi. an hour, so as to return home in time, walking back at the rate of 4 mi. an hour? *Ans.* 6 mi.

68. If 65 lb. of sea-water contain 2 lb. of salt, how much fresh water must be added to these 65 lb., so that 25 lb. of the new mixture shall contain but 4 oz. salt? *Ans.* 135 lb.

69. A grocer knows neither the weight nor first cost of a box of tea. He only recollects that if he had sold the whole at 30 ct. a lb., he would have gained $1, but if he had sold it at 22 ct. a lb., he would have lost $3. Find the number of lb. in the box, and the first cost per lb. *Ans.* 50 lb. at 28 ct.

70. A, B and C are paid $16.74 for doing a piece of work; A and B together could do $\frac{4}{5}$ of it, in the same time that A and C do $\frac{2}{3}$ of it, or B and C do $\frac{3}{5}$ of it; how must they divide the money?
Ans. A gets $7.02; B, $5.94; C, $3.78

THE END.